Landscape Planning

ENVIRONMENTAL APPLICATIONS

William M. Marsh
University of Michigan, Flint

ADDISON-WESLEY PUBLISHING COMPANY
Reading, Massachusetts • Menlo Park, California
London • Amsterdam • Don Mills, Ontario • Sydney

Sponsoring Editor: Tom Robbins
Production Editor: Lynda Griffiths

Art Coordinator: Robert Forget
Illustrators: Anthony Tellez, James Marsh, Brian Larson
Cover Designer: Brian Larson

Production Manager: Robert Duchacek

This text was composed in Zapf International
by *TKM Productions.*

Library of Congress Cataloging in Publication Data

Marsh, William M.
Landscape planning: environmental applications

Includes index.
1. Land use—Planning. 2. Landscape architecture.
3. Landscape protection. 4. Physical geography.
I. Title.
HD108.6.M37 1983 712 82-13889
ISBN 0-201-04102-2

ISBN 0-201-04102-2
ABCDEFGHIJ-HA-89876543

CONTENTS

PREFACE

The idea for this book grew from three observations: the need expressed by colleagues for a set of applied problems in physical geography, planning, and landscape architecture for the student without advanced training; the need in many departments for an applications book that encourages the student to go beyond soil maps, topographic maps, and climatic data; and my own experiences as a consultant on planning, architecture, and engineering projects that have led to the realization that there is much progress to be made toward integrating the "technical" fields such as geography and geomorphology with the "professional" fields such as urban planning and landscape architecture.

The book is aimed at the student who has either completed or is taking a course in physical geography, natural resources, or environmental science. The chapters are not intended to be comprehensive treatments of the topics addressed; for that the student is directed to a standard text in physical geography or a related field. This book can be used as a supplement in various courses or as a text in applications courses. For the problems at the end of the chapters, the necessary data and maps are either printed herein or made available in the Instructor's Guidebook.

Many colleagues contributed to the development of this book and I extend my sincere thanks to each of them. The authors of the case studies graciously donated their time and effort when called on for contributions. The manuscript was reviewed by John M. Grossa (Central Michigan University), Walter A. Schroeder (University of Missouri), Donald Tilton (Smith, Hinchman, Grylls Associates), and Tony Davis and Gillian Watkins (University of Toronto). Denise Flynn, James G. Marsh, Antonio Tellez, Brian Larson, Joseph Smigiel, Mark L. Hassett, and Josephine Kearns managed the production and graphics. Despite delays by the author, Addison-Wesley provided patient support; in particular, Thomas Robbins, senior editor, and Richard Kitowski, who provided valuable input from Canadian sources. As always, thanks to Susan A., Katherine W., and Christopher R. Marsh for their help and patience in preparing this book.

Flint, Michigan
January 1983

W. M. M.

INTRODUCTION

BACKGROUND CONCEPT

Virtually every modern field of science makes contributions toward the resolution of societal problems. In some cases the contributions are not very apparent, even to the practitioners in that field, because they are made via second and third parties. These parties are usually specialists in applied professions, such as urban planning, landscape architecture, architecture, and engineering, which synthesize, reformat, and adapt knowledge generated by scientific investigations.

Geography holds such a relationship with the various fields of landscape planning. For more than a century, geographers have studied the world's physical features, learning about their makeup, how to measure them, the forces that change them, and how we humans use them. Planning is concerned with the use of resources, especially those of the landscape, and how to allocate them in a manner consistent with people's goals. Thus planning and geography are linked together because of a mutual interest in resources, land use, and the nature and dynamics of the landscape.

CONTENT AND ORGANIZATION

This book addresses selected topics and problems of concern to physical geographers, planners, and landscape designers. The coverage is not intended to represent the full range of existing applications; such a book would be too large and cumbersome for most college courses. The choice of topics was guided by three considerations. First, the traditional areas of physical geography are represented; namely, topography, climate, soils, hydrology, and vegetation. Second, the topics are pertinent to modern planning as articulated by urban planners, landscape architects, and related professionals. Third, the topics do not demand advanced training in analytical techniques, data collection, mapping, and field techniques.

The book is organized into three parts. The first part, comprised essentially of Chapter 1, describes the nature of the problem and the types of activities associated with planning in the United States and Canada. The second part, Chapters 2 through 14, deals with the topics and problems in landscape planning. Each chapter introduces a problem area, describes its application to planning problems and issues, and ends with a problem set. The third part, Chapters 15, 16, and 17, is devoted to techniques: remote sensing, map reading, and interpreting graphs.

Where one should begin this book depends on the reader's familiarity with the techniques and problems of the landscape sciences and professions. For those in need of a review of topographic maps, aerial photographs, and the interpretation of graphs, it may be advantageous to read the last three chapters first; otherwise, these chapters would serve as references in reading and dealing with the problems in the earlier chapters. All readers are encouraged to examine Chapter 1, not only to acquire some orientation on the various types of planning activities in North America, but also to gain some insight on how different scientific and professional fields contribute to these activities.

PROBLEMS OF GEOGRAPHIC SCALE

Because of the need to address topics and problems of the landscape that are pertinent to modern planning as articulated by the practicing professional, most material is presented at the site or community scale. Sites are local parcels, usually ranging in size from less than an acre to tens of acres, with a simple ownership or stewardship arrangement (individuals, families, or organizations). They are the spatial units of land use planning, the building blocks of communities.

Regions, in the vocabulary of land use planning and landscape design, are variously defined as the geographic settings housing communities, either a single community and its rural hinterland, several communities and the systems connecting them (roads or streams, for example), or a metropolitan area with its inner city, industrial and suburban sectors. This differs from the geographer's notion of a region, which encompasses a much larger area; for example, the Midwest, the Great Plains, or the Hudson Bay region. Many environmental problems have a regional scope (geographer's version); for instance, acid rain in the eastern midsection of the continent and water supplies for irrigation in the Southwest, but for a variety of reasons, planning programs have generally not been very effective at this scale in North America. Most of our examples of effective or promising planning of the environment are of regional (planner's version), community, or site scales. For the present era at least these appear to be the operational scales of most landscape planning efforts.

LANDSCAPE PLANNING, ENVIRONMENTALISM, AND ENVIRONMENTAL PLANNING

The term *landscape planning* is used in this book to cover a variety of activities associated with development and land use as they pertain to landscape features, processes, and systems. Only two decades ago, the term *land use planning* was generally used for this sort of activity, but today, because of new knowledge, the recognition of new problems, the changing needs of society, and the modern propensity for the proliferation of specialty fields, several new and alternative titles have emerged. The Environmental Crisis in the United States in the 1960s and 1970s was brought on by a flurry of concern over the quality of the environment. Much of this took the form of a political movement to protect the "environment" from the onslaught of industry, government, and urbanization. Loosely translated, "environment" was taken by the movement to mean things of natural origins in the landscape, i.e., air, water, forests, animals, river valleys, mountains, canyons, and so on. From this emerged the *environmentalist,* a person who believes in or works for the protection and preservation of the environment. *Environmentalism,* it follows, is a philosophy, a political or social ideology, that implies nothing in particular about one's training, knowledge, or professional credentials in matters of the environment.

The environmental crisis also paved the way for stronger and broader environmental legislation at all levels of government. New types of professional skills were called for to provide various services in connection with environmental assessments, waste disposal planning, air and water quality management, and so on. In response, several new "environmental" fields emerged while many established fields, such as civil engineering and chemistry, developed "environmental" subfields. Taken as a whole, the resultant environmental fields fall roughly under three main headings: environmental science, environmental engineering and technology, and environmental planning.

Environmental planning is a "catch all" sort of title applied to planning and management activities in which environmental (as opposed to social, cultural, or political, for example) factors are central considerations. It is often confused with environmentalism and with the preparation of environmental impact studies, but in reality, environmental planning covers an enormous variety of topics, ranging, for instance, from health issues in work places to global population and food supply problems. *Landscape planning* represents one of the major areas of environmental planning that addresses both relatively new topics associated with development and land use, such as toxic waste disposal and urban microclimate, as well as traditional ones, such as watershed management and site planning. To some extent, landscape planning is also a term of convenience used to distinguish the activities of what we might call the landscape fields (geography, landscape architecture, geomorphology, and urban planning) from other areas of environmental planning.

UNITS OF MEASUREMENT

Finally, a note on units of measurement. Geography and related fields in the earth sciences have adopted the metric system whereas in planning and the design fields the United States Customary System continues to be widely used. Rather than employing one system or the other, we chose to mix the two, using the most appropriate units for the topic and both in many cases. Conversions may be necessary in some problems, but this is a relatively simple matter, and the reader is referred to Appendix B for equivalent units.

I

PLANNING AND PROBLEMS OF THE LANDSCAPE

1

LANDSCAPE PLANNING: ACTIVITIES AND FIELDS

THE PROBLEM: LAND USE AND LANDSCAPE CHANGE

The rate at which North Americans have developed this continent is unprecedented in the history of the world. Virtually every sort of landscape has been probed and settled in some fashion, and in the vast woodlands and grasslands of the continent's midsection, scarcely a wit of the original landscape remains. Wholesale transformation of natural landscapes represents only part of the story, however; the introduction of synthetic materials and forms represents the other part.

The age of materialism and economic expansionism has produced a colossal system of resource extraction which, for the United States, reaches over most of the world. At the output end of the system is the manufacture of new materials in the form of both saleable products and undesirable byproducts and residues of various compositions. Both end up in the landscape: steel, glass, concrete and plastics (in the form of buildings and cities), waste residues (in the form of chemical contaminants in air, water, soil, and biota), and solid and hazardous wastes (in landfills, waterbodies, and wetlands).

The landscapes that are ultimately created are essentially new to the earth. Cities, for example, are often built of materials that are thermally and hydrologically unusual to the land, and in structural forms that are geomorphically extreme in most landscapes. It is a landscape distinctly different from the ones it displaced and, in many respects, decidedly inferior as a human habitat. The modern metropolitan landscape tends to be less healthy, less safe, and less emotionally secure than people desire. Moreover, the very existence of such landscapes poses a serious uncertainty to future generations owing to the high cost of maintaining them and the quality of human life. Herein lies much of the basis for land use planning, landscape design, and urban and regional planning.

The planning problems we are facing today are many and complex, and not all are tied directly to the landscape. For those that are, most seem to result from some sort of a mismatch between land use and environment. The mismatches appear to be of mainly three origins: (1) those that stem from initially poor land use decisions because of ignorance or misconceptions about the environment, as exemplified by the person who unwittingly builds a house on an active fault or unstable slope; (2) those that stem from environmental change after a land use has been established, as illustrated by the property owner who comes to be plagued by flooding or polluted water because of new development upstream from his or her site; and (3) those that stem from social change, including technological change, after a land use has been established, represented, for example, by the resident living along a street initially designed for horse-drawn wagons but now used by automobiles and trucks and plagued by noise, air pollution, and safety problems.

THE PURPOSE OF PLANNING

In general, the primary objective of planning is to make decisions about the use of resources. Over the past twenty years, the need for land use and environmental planning has increased dramatically with rising competition for scarce land, water, and energy resources. The problems and issues are diverse in both type and scope, ranging from worldwide issues, such as the use of the continental shelves for oil or food production, to problems of land use on a two-acre parcel of ground in a modern city. In America, despite the political undertones traditionally associated with planning in general, land use planning has gained a toehold of legitimacy at the community and county levels in most states.

Who does planning? Actually, the majority of planning today is probably not done by professional planners. Most of it is done by corporation officers, government officials and their agents, and the military. Professional planners, that is, those with formal credentials in planning, usually function in a technical and advisory capacity to the decision-makers, providing data, forecasting futures, defining alternative courses of actions, and structuring strategies for implementation of formal plans. The overall direction of a plan, however, always represents some sort of policy decision—one based on a formal concept of what a company, city, or neighborhood intends for itself and thus will strive to become. These concepts about the future are called planning goals, and they are the driving force behind the planning process.

DECISION MAKING AND TECHNICAL PLANNING

Two broad categories of planning studies or activities are definable; one is that related to the decision-making process itself. This involves building the methods and means for making planning decisions, formulating plans, then providing the information necessary for carrying out decisions. Among the tasks commonly undertaken in *decision-making planning* are attitude surveys about present and future conditions, definition of relevant policies, articulation of goals, formulation of alternative courses of action, and selection of a preferred plan.

The other type of planning can be called *technical planning* and it involves the various types of analysis and related activities that are used in support of the decision-making process. This includes environmental inventories, such as soils and vegetation mapping, engineering analysis, such as soil suitability for construction, and assessment of the impacts that proposed land uses may have on the environment. Technical planning is usually carried out by a variety of specialists including cultural geographers, physical geographers, geologists, ecologists, hydrologists, wildlife biologists, archeologists, economists, sociologists, as well as professional planners from urban planning, landscape architecture, and architecture. The line separating decision making from technical planning activities is often indistinct, and in most projects the two types of planning merge into one another.

ENVIRONMENTAL IMPACT STUDIES

The use of the "technical" specialist in planning has increased rapidly in the past decade as a result of environmental impact legislation. Under the National Environmental Policy Act of 1970, it became necessary to forecast and evaluate the impact on the environment from proposed projects (actions) involving federal monies. This requires the services of the specialist in: (1) defining the makeup of a proposed project area in terms of vegetation, land use, water features, and so on; (2) forecasting the changes in these features as a result of the proposed action; and (3) finding alternatives that would reduce the impact on the environment where the impact is judged to be excessive, objectionable, unacceptable, or forbidden by law. The results of such studies are reported in a document called an *environmental impact statement.*

Today environmental impact analysis of some sort is also required by many states and communities in the United States for both privately and publicly funded projects. Typically, the process begins with an "environmental assessment," which is a brief description of the proposed action and the environment for which it is intended. Should the assessment show cause for further study, an environmental impact statement may be required.

TYPES OF LANDSCAPE PLANNING ACTIVITIES

For planning that deals with the environment, several types of projects and activities have become conventional in America. One of the best known of these is the so-called *environmental inventory,* an activity designed to provide a catalog and description of the features and resources of a project area. The basic idea behind the inventory is that one must know what exists in an area before one can formulate planning alternatives for it. Among the features consistently called for in environmental inventories are rare and endangered species, valued habitats, floodplains, archeological sites, soil types, vegetation, and land use. In the preparation of environmental impact statements, inventories also include an evaluation of the phenomena recorded. This is supposed to indicate the comparative importance or value of a feature or resource (Fig. 1.1).

Under the methodology for environmental impact statements, another planning activity has gained prominence: *forecasting impacts.* This involves identification of the changes called for or implied by a proposed action followed by an evaluation of the type and magnitude of the environmental impact. The process is a tough one because of the difficulty in deriving accurate forecasts by analytical means. As a result, forecasts of impacts are usually "best estimates" and the significance assigned to them seems to be as much a matter of perspective (for example, engineer versus environmentalist) as anything else (Fig. 1.2). Nevertheless, the *process* is an important one because it often leads to: (1) clarification of complex issues and their environmental implications; (2) modification of a proposed action to lessen its impact; or (3) abandonment of a proposed project.

A third type of planning activity is aimed at the *discovery of opportunities and constraints.* This is often undertaken after a use (such as recreation or residential housing) has been assigned to an area but the density, layout, and appropriate design of that use are undetermined. The study involves searching the environment both for features and situations that would

UNITED STATES DEPARTMENT OF THE INTERIOR
GEOLOGICAL SURVEY

INSTRUCTIONS

1- Identify all actions (located across the top of the matrix) that are part of the proposed project.
2- Under each of the proposed actions, place a slash at the intersection with each item on the side of the matrix if an impact is possible.
3- Having completed the matrix, in the upper left-hand corner of each box with a slash, place a number from 1 to 10 which indicates the MAGNITUDE of the possible impact; 10 represents the greatest magnitude of impact and 1, the least, (no zeroes). Before each number place + if the impact would be beneficial. In the lower right-hand corner of the box place a number from 1 to 10 which indicates the IMPORTANCE of the possible impact (e. g. regional vs. local); 10 represents the greatest importance and 1, the least (no zeroes).
4- The text which accompanies the matrix should be a discussion of the significant impacts, those columns and rows with large numbers of boxes marked and individual boxes with the larger numbers.

SAMPLE MATRIX

	a	b	c	d	e
a		2/1			8/5
b		7/2	5/8	3/1	9/7

II PROPOSED ACTIONS WHICH MAY CAUSE ENVIRONMENTAL IMPACT

A. MODIFICATION OF REGIME: a. Exotic flora or fauna introduction; b. Biological controls; c. Modification of habitat; d. Alteration of ground cover; e. Alteration of ground water hydrology; f. Alteration of drainage; g. River control and flow modification; h. Canalization; i. Irrigation; j. Weather modification; k. Burning; l. Surface or paving; m. Noise and vibration

B. LAND TRANSFORMATION AND CONSTRUCTION: a. Urbanization; b. Industrial sites and buildings; c. Airports; d. Highways and bridges; e. Roads and trails; f. Railroads; g. Cables and lifts; h. Transmission lines, pipelines and corridors; i. Barriers including fencing; j. Channel dredging and straightening; k. Channel revetments; l. Canals; m. Dams and impoundments; n. Piers, seawalls, marinas, and sea terminals; o. Offshore structures; p. Recreational structures; q. Blasting and drilling; r. Cut and fill; s. Tunnels and underground structures

C. RESOURCE EXTRACTION: a. Blasting and drilling; b. Surface excavation; c. Subsurface excavation and retorting; d. Well drilling and fluid removal; e. Dredging; f. Clear cutting and other lumbering; g. Commercial fishing and hunting

D. PROCESSING: a. Farming; b. Ranching and grazing; c. Feed lots; d. Dairying; e. Energy generation; f. Mineral processing; g. Metallurgical industry; h. Chemical industry; i. Textile industry; j. Automobile and aircraft; k. Oil refining; l. Food; m. Lumbering; n. Pulp and paper; o. Product storage

E. LAND ALTERATION: a. Erosion control and terracing; b. Mine sealing and waste control; c. Strip mining rehabilitation; d. Landscaping; e. Harbor dredging; f. Marsh fill and drainage

PROPOSED ACTIONS

I EXISTING CHARACTERISTICS AND CONDITIONS OF THE ENVIRONMENT

A. PHYSICAL AND CHEMICAL CHARACTERISTICS

1. EARTH: a. Mineral resources; b. Construction material; c. Soils; d. Land form; e. Force fields and background radiation; f. Unique physical features

2. WATER: a. Surface; b. Ocean; c. Underground; d. Quality; e. Temperature; f. Recharge; g. Snow, ice, and permafrost

3. ATMOSPHERE: a. Quality (gases, particulates); b. Climate (micro, macro); c. Temperature

4. PROCESSES: a. Floods; b. Erosion; c. Deposition (sedimentation, precipitation); d. Solution; e. Sorption (ion exchange, complexing); f. Compaction and settling; g. Stability (slides, slumps); h. Stress-strain (earthquake); i. Air movements

B. BIOLOGICAL CONDITIONS

1. FLORA: a. Trees; b. Shrubs; c. Grass; d. Crops; e. Microflora; f. Aquatic plants; g. Endangered species; h. Barriers; i. Corridors

2. FAUNA: a. Birds; b. Land animals including reptiles; c. Fish and shellfish; d. Benthic organisms; e. Insects; f. Microfauna; g. Endangered species; h. Barriers; i. Corridors

C. CULTURAL CONDITIONS

1. LAND USE: a. Wilderness and open spaces; b. Wetlands; c. Forestry; d. Grazing; e. Agriculture; f. Residential; g. Commercial; h. Industrial; i. Mining and quarrying

Fig. 1.1 An excerpt from a matrix designed for environmental inventory that provides for an evaluation of the magnitude of impact anticipated from a proposed action. (From Luna B. Leopold et al., "A Procedure for Evaluating Environmental Impact," *U.S. Geological Survey Circular 645,* 1971.)

facilitate the proposed land use and those that would represent drawbacks or threats (Fig. 1.3). Basically, the objective is to find the potential matches and mismatches between land use and environment and recommend the most appropriate relationship between the two. *Land capability* studies are a form of opportunity/constraint activity in which a development capacity, sometimes called carrying capacity, is assigned to different land types.

Hazard assessment is a specialized type of constraint study. The objective in hazard studies is to identify dangerous zones in the environment where land use is or would be in jeopardy of damage or destruction. Hazard research has been concerned with both the nature of threatening environmental phenomena, namely floods, earthquakes, and storms, and the nature of human responses to these phenomena. Zoning and disaster relief planning have benefitted from hazard assessment at the national, state, and local levels (Fig. 1.4).

Site selection is a traditional planning problem based on an opportunities and constraints approach. In this case, one would begin with a program for a facility or enterprise in hand and attempt to find an appropriate place to put it. Often this entails no more than an exercise in locational analysis based on economic factors, but in some instances, such as in highway location and residential housing projects, it may also involve environmental factors. On thc opposite hand are *feasibility studies,* which begin with a known site and, with the aid of various forecasting techniques, attempt to determine the most appropriate use for it.

Facility planning is usually carried out by engineers because it involves structural and mechanical systems of some sort. Sewage treatment plants are a prime example. Under the Water Pollution Control Act, planning of treatment facilities was enabled through a combination of federal and local funding. The typical problem involves selection of a site for a treatment

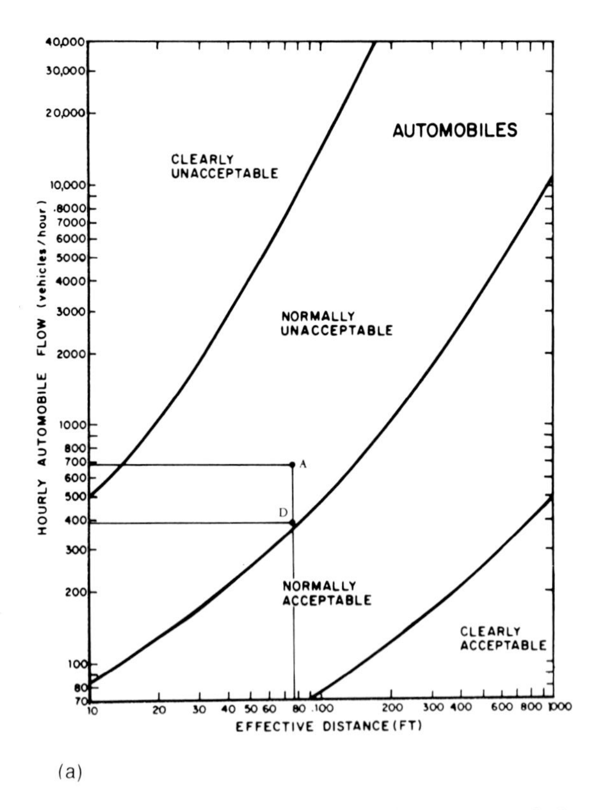

(a)

Fig. 1.2 Two examples of forecasting techniques: (a) quantitatively-based model relating automobile flow and distance (from the road) to noise levels; (b) gaming/simulation approach which asks "players" with different perspectives on the problem to estimate the relative impact of a proposed action on various aspects of their community. ((a) From

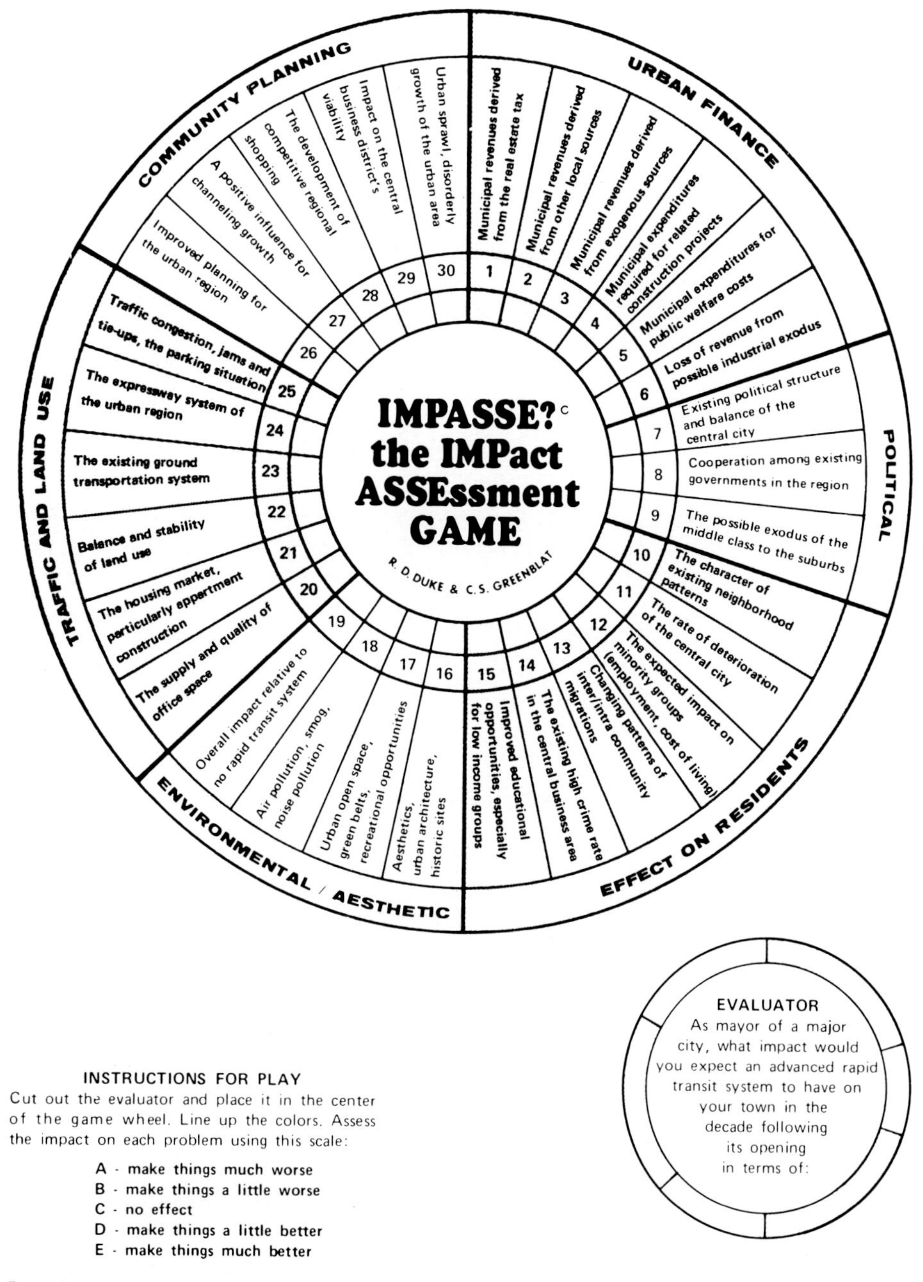

(b)

T. J. Schultz and N. M. McMahon, *Noise Assessment Guidelines*, Washington, D.C.: Department of Housing and Urban Development, 1971. (b) From C. S. Greenblat and R. D. Duke, *Gaming—Simulation: Rationale, Design and Applications*, New York: Halsted Press, 1975. Used by permission of R. D. Duke.)

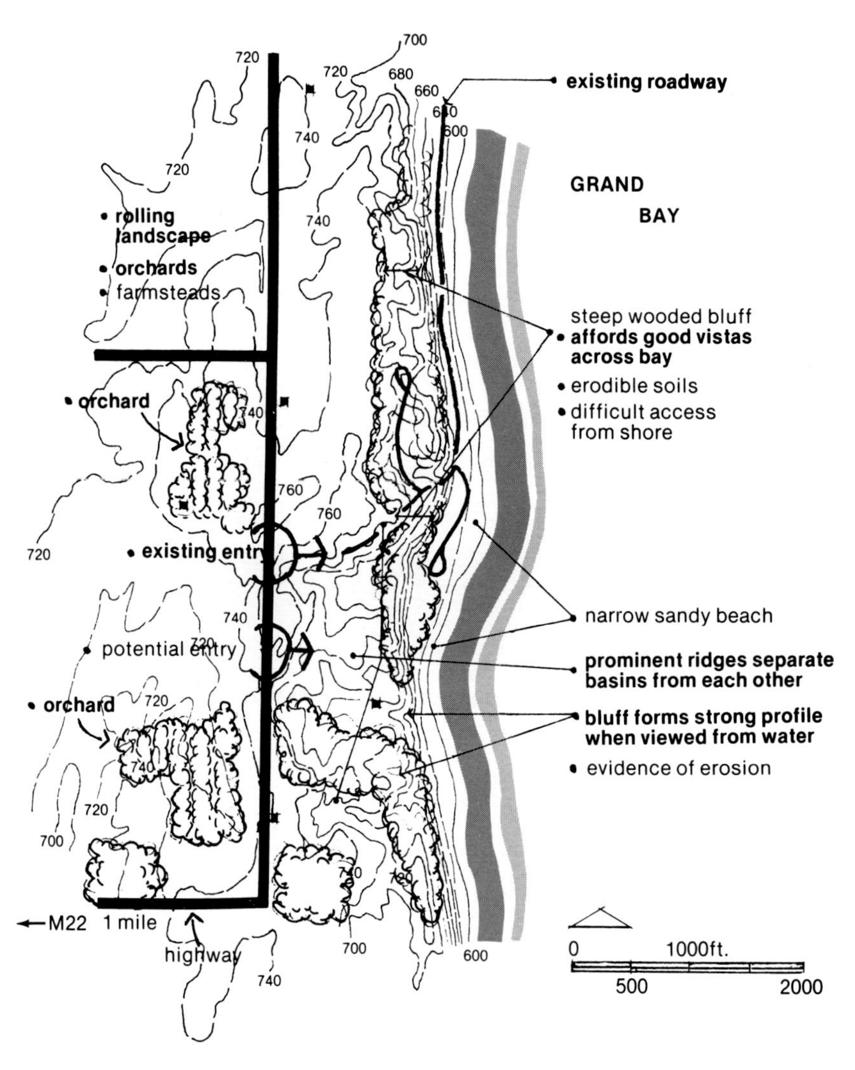

Fig. 1.3 An example of an opportunities and constraints map for a section of hilly coastal land that was under consideration for residential development. (Used by permission of Johnson, Johnson and Roy, Inc.)

plant, designing the plant, and providing plans for additional sewer lines to accommodate changing land use patterns.

Finally, we must consider *master planning*. This activity may include all of the planning activities previously mentioned, for its overriding aim is to present a comprehensive framework to guide land use changes. Early in the master planning process, goals are formulated relating to land use, economics, environment, demographics, and transportation. Existing conditions are analyzed and alternative plans are formulated; the alternatives are then tested against goals and existing conditions, and one is adopted (Fig. 1.5). The master plan is usually comprised of three parts: (1) a program proposal consisting of recommendations and guidelines; (2) a physical plan, showing the recommended locations, configurations, and interrelationships of the proposed land uses; and (3) a scheme for implementing the master plan which identifies funding sources, enabling legislation, and how the changes are to be phased over time (Fig. 1.6).

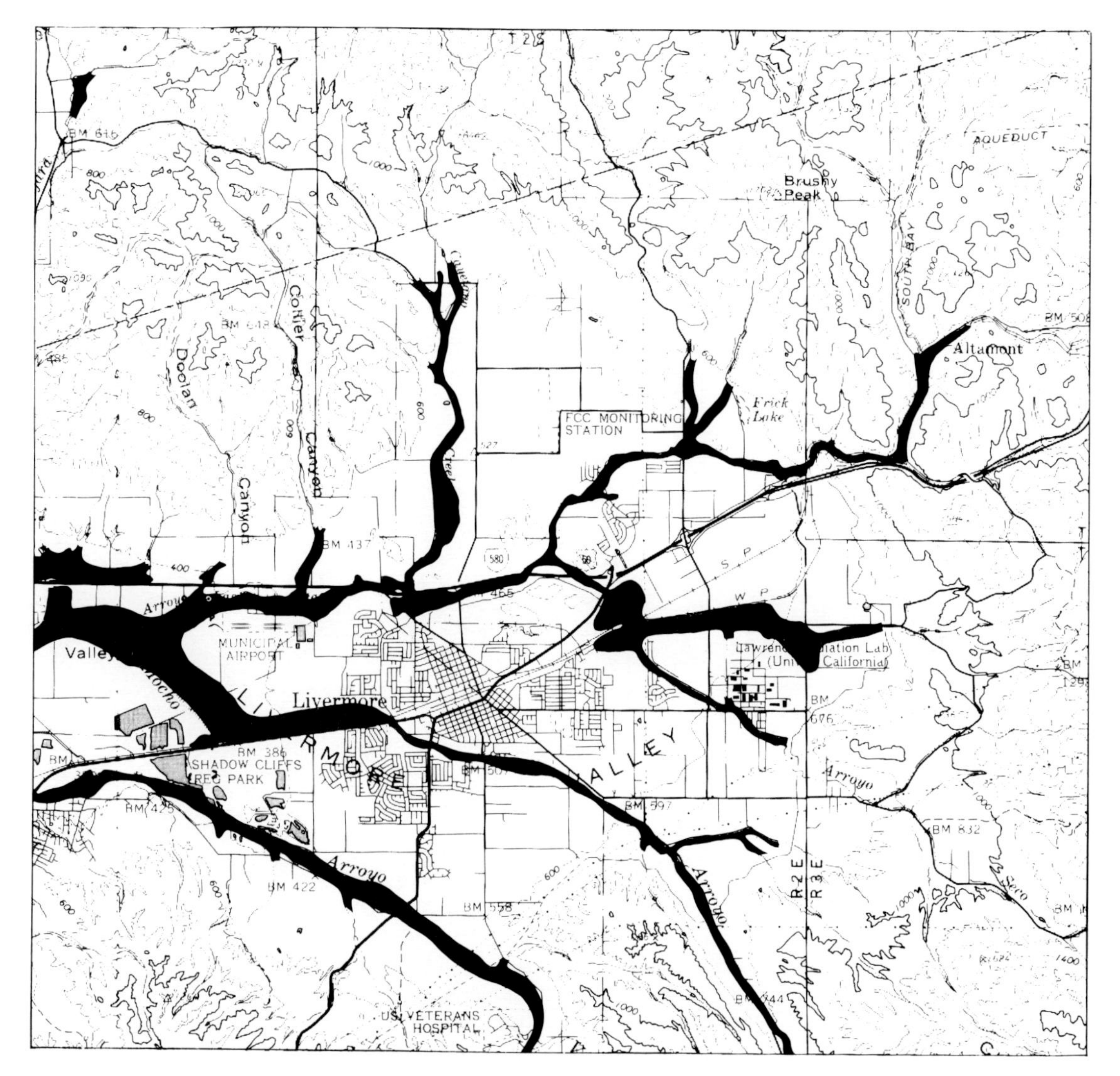

Fig. 1.4 Excerpt from a U.S. Geological Survey map showing the extent of flooding expected from the one-hundred-year flood flow in Livermore, California. Such maps have been prepared for many urbanized areas of the United States as part of a joint program between the Geological Survey and the Department of Housing and Urban Development. (From "Flood-Prone Areas in the San Francisco Bay Region, California," *U.S. Geological Survey Open-File Report*, 1973.)

THE PLANNING PROFESSIONS AND PARTICIPATING FIELDS

Each decade a growing number of academic fields participate in planning in North America. The 1970s saw increased participation from geography, geology, biology, and sociology in the formal arenas of planning. Still, only a handful of fields are actually involved in training the professionals who guide the decision-making processes, structure plans, and design buildings; among these, urban planning, architecture, and landscape architecture are the leading disciplines.

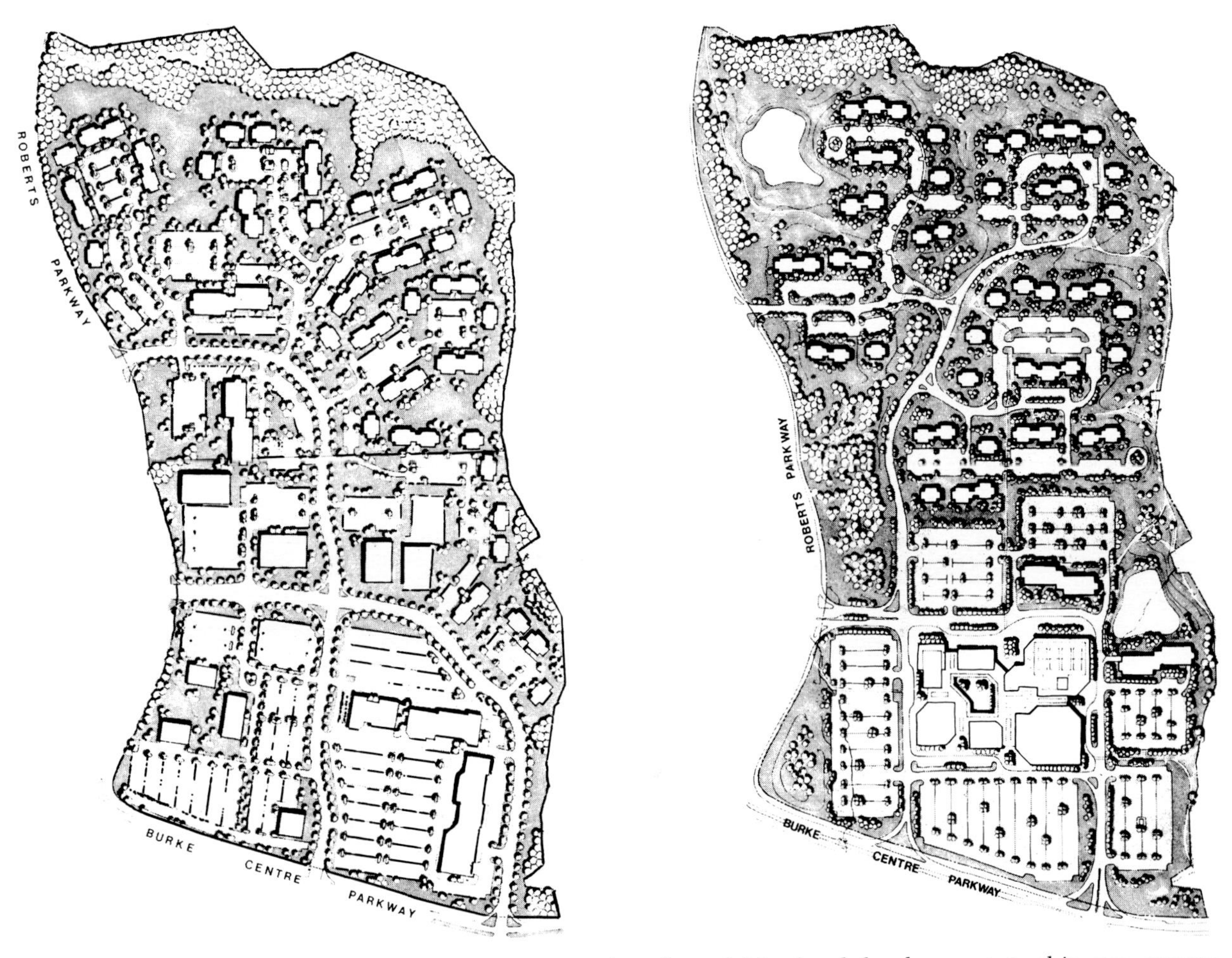

Fig. 1.5 Alternative master plans for neighborhood development. In this case, energy conservation was the major goal and thus became the primary criterion in evaluating the two alternatives. The one on the left is the energy conserving plan. (From U.S. Department of Energy, prepared by Land Design/Research, Inc., 1979.)

Let us briefly examine three fields that play important roles in planning today: geography, landscape architecture, and urban planning. Geography and landscape architecture are similar in that they focus on the phenomena of the landscape, including both the human and the natural, but are different in their perspectives, geography's being scientific (technical in planning) and landscape architecture's being largely professional (decision making in planning) with a strong bent toward landscape design. Urban planning also bridges the human and natural aspects of the landscape but, because of its focus on metropolitan areas, is more concerned with the built environment and its processes than geography and landscape architecture are.

Because of its long-time concern with landscape, land use, and human-land studies, geography is probably closer to the subject matter of planning than any of the traditional fields of natural and social science. This is underscored by the fact that in some countries university programs in planning evolved in geography departments. In the United States, universities were generally slow to develop planning programs, and when they did appear

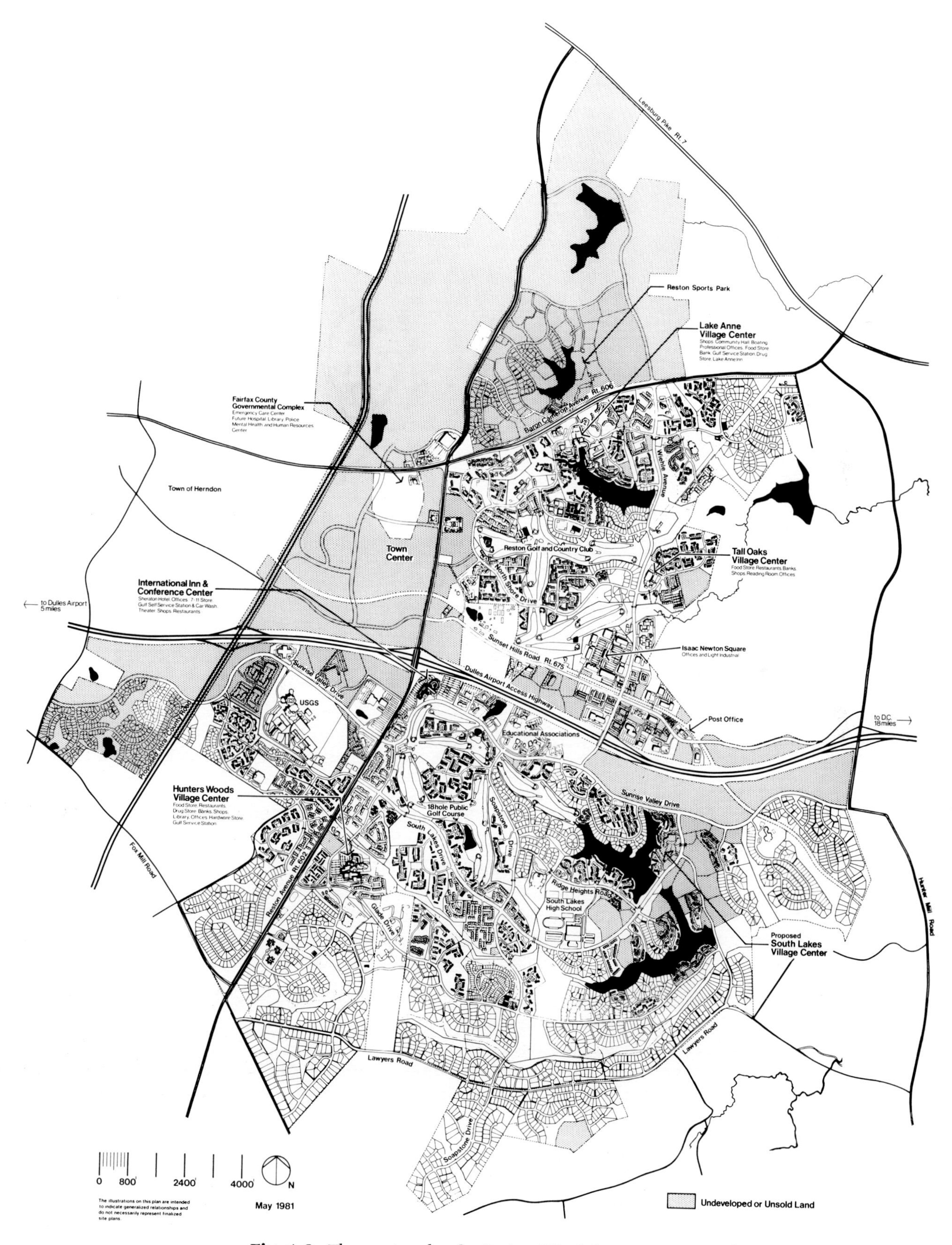

Fig. 1.6 The master plan for Reston, Virginia, a new town in the Washington, D.C., area. (Used by permission of Gulf Reston, Inc.)

in the 1950s and 1960s, they were often affiliated with schools of architecture or engineering. Geography chose to maintain its scientific orientation rather than adopt an applied field and the responsibility for training professional practitioners.

Today geography is one of many academic fields that participates in the various aspects of technical planning. Among the many areas of research in geography that are drawn on by the professional planning community, six may be singled out and briefly described:

Monitoring land use change: Land mapping is essential to many facets of land planning; in particular, identifying changes over time and measuring the effectiveness of planning decisions. Land use mapping today is being advanced rapidly with the use of remote sensing techniques that offer new types of data and faster data processing. (See Fig. 15.2 in Chapter 15.)

Site analysis: The geographer's eclectic perspective on the landscape makes his or her approach to site analysis highly valuable in planning problems. Site analysis requires not only an understanding of the internal makeup of a parcel of space, such as the distribution and spatial relations among bedrock, soils, and vegetation, but insight into the relationship of the parcel to the larger spatial framework, both natural and human, of which it is a part.

Analysis of relationships among physical features and processes in the landscape: The knowledge derived from analytical investigations of various aspects of soils, land forms, vegetation, climate, and so on lends itself to many practical applications in planning. For instance, based on the results of analysis defining the relationship between impervious surface area and runoff rates, it is possible to forecast changes in stream flow brought on by development, thereby enabling one to make forecasts about the impacts of development on the aquatic environment.

Analysis of human response to environmental conditions and events: Behavioral responses by humans to events and conditions of the environment, especially hazardous events, is an area in which geography has excelled in the past several decades. This knowledge has helped in shaping realistic alternatives for land use plans as well as providing perspectives and guidelines for more specialized efforts such as disaster relief planning.

Modeling spatial systems: Understanding the interconnectedness of process and features in the landscape is imperative to intelligent planning. Thus the efforts of geographers in describing and modeling economic, transportation, hydrologic, and geomorphic systems are important building blocks in impact studies, site analysis, opportunity and constraints studies, and master planning.

Graphic systems: Maps are the principal tools of both geographers and planners. Accordingly, advances in mapping techniques such as computer mapping systems are usually important to the planning community, not only as analytic devices but as communication tools as well (Fig. 1.7).

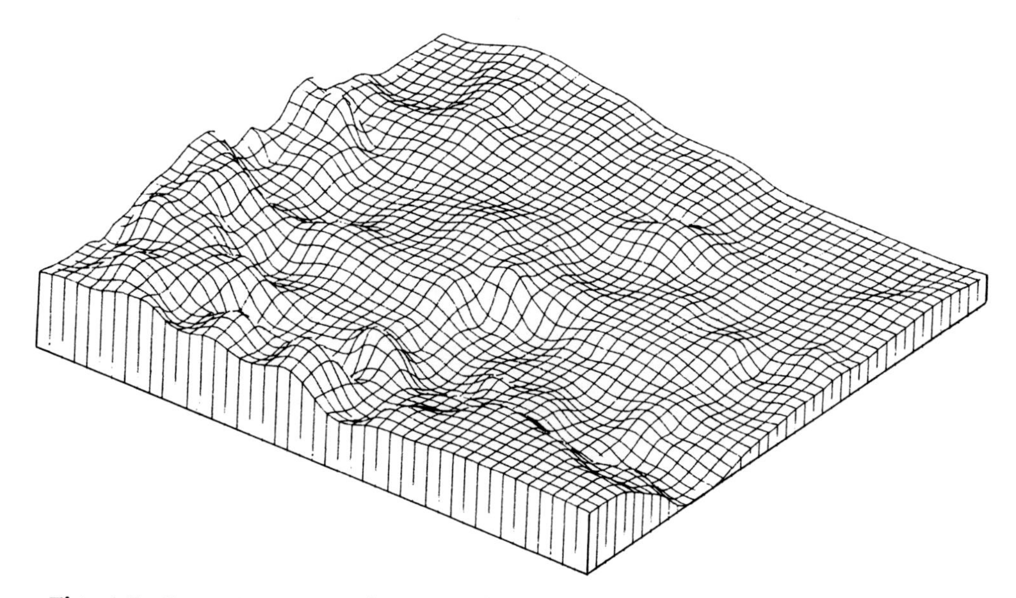

Fig. 1.7 Computer-generated topographic map in an isometric format. (From LeRoy Klopcic. Used by permission.)

Landscape architecture is described as both a planning and design field because it participates in the decision-making process as well as the formulation of physical plans and design schemes of various types. Like geography, landscape architecture deals with both natural and human phenomena of the landscape; but unlike geography, its principal goal is to exercise a guiding hand in changing the landscape to create pleasing and workable environments.

Landscape architecture in the United States emerged as a formal profession in the late nineteenth century as more or less the land-oriented counterpart to architecture. Today's areas of activity include the following:

Park and recreational planning: Among the field's early successes were the creation and design of urban parks such as Central Park in New York City. Park and recreational planning remains one of the primary areas of professional activity in landscape architecture today.

Site planning: This encompasses the formulation of plans for specific sites, usually small areas, for which a use or development program has been designated. The process is intended to yield a physical plan in which natural features, automobile and pedestrian circulation, building and facility locations, and landscaping are addressed in an integrated framework. For large sites and complex development programs, this often includes a master plan that addresses implementation strategies and project phasing.

Urban design: This area is the planning and designing of street, pedestrian, and exterior building spaces in urban settings. Today urban design is usually practiced in cities undergoing downtown redevelopment and neighborhood renovation, and often includes historic restoration.

Campus planning: The layout and design of college and university campuses takes into consideration pedestrian and automotive circulation, building sites, academic programs, and living and recreational spaces.

Residential landscaping: This includes the arrangement and design features such as walks, gardens, lawns, and woody plants in residential areas. This is one of the principal activities of the profession, and much of it centers around plant materials, their selection, placement, and maintenance.

Both landscape architecture and geography participate in urban planning—a broad field concerned with the analysis, planning, and management of metropolitan regions. Urban planning also deals with environmental and human phenomena, but the principal setting is the built environment, and the concern is as much with social, administrative, and fiscal matters as it is with land use and physical design. Because most of the people of the United States and Canada live in and around urban areas, most technical and decision-making activity in planning today is closely tied to urban planning.

Among the areas of research and professional activity in urban planning are *urban design,* which is shared with architecture and landscape architecture; *neighborhood development,* which draws heavily on sociology, political science, and anthropology; *economic development,* which is usually tied to economics and business administration; *historical preservation,* which is also part of architecture and landscape architecture; *transportation planning,* which is usually undertaken jointly with engineering; and *land use* and *environmental planning,* which are supported by geography and landscape architecture.

CASE STUDY

Use of Simulation/Games in Urban Planning

Allan G. Feldt

Because of the complexity of many planning problems, it is sometimes necessary to employ simulation and gaming techniques in decision making. Used first by the military for battle-training exercises, simulation/games are now widely used in business and urban planning to gain an overview of a problem and clarify and broaden the perspectives of individuals with different and often conflicting backgrounds. Games also collapse time, thereby enabling a player to see in the course of a few hours the effects of decisions representing many years. Most simulation/games consist of three principal elements plus a variety of artifacts in written, mathematic, graphic, or physical form whose purpose is to record or convey information.

The first principal element is the *scenario,* the data and description of the time and place being represented in the game. The second element is the *roles,* the primary actors in the game, including descriptions of who they are, what their relative interests and strategies may be, and what their status is at the beginning of the game. The third element is some form of *accounting system,* which records the decisions of players and determines what impact they have on the scenario and the roles. The account-

CASE STUDY (cont.)

ing system must also feed some or all of this information back to the players on a regular basis as a guide to future decision making.

Most games are played over a series of time periods called "rounds," approximating some amount of time in the "real world," ranging from hours to decades. Similar activities usually take place each round, with players making more rapid and sophisticated decisions in later rounds as they become more familiar with the game and its rules. The first round is often very confusing since players must learn to understand both the mechanics of the game and the factors influencing their decisions. In later rounds, confusion pertaining to rules and game mechanics dissipates while the complexity of decisions may increase as players grow more sophisticated and the influence of decisions accumulated from previous rounds grows.

Simulation/games can be reasonably accurate representations of the "real world" situations they depict, but their outcomes should not be interpreted as predictive of how real events are likely to turn out. Simulation/games are merely representations of how various components interact. The assumptions of the persons designing them, plus the decisions of the players themselves, strongly affect the outcome. Such an outcome represents, at best, only one of many possible worlds that could emerge in the "real world" system being modeled. Thus all simulation models, regardless of their complexity, can never be anything more than instructional devices allowing us to observe and experiment with possible futures.

A good example of a simulation/game is *CLUG,* the Community Land Use Game, which has been used in both teaching and planning applications for almost two decades. In CLUG, each player or team of players represents a private real estate developer and operator. They may buy land through competitive bidding, develop it into one of five basic land uses, and operate it at whatever profit or loss the sum of their business decisions yields within the structure of the game and the actions of other players. Major elements of an economic system are represented in the form of costs for land, labor, and capital, plus costs incurred for taxes and transportation of goods and labor to the marketplace. Shortage of land at prime locations soon forces players into reconsidering their positions and undertaking efforts to more rationally plan present and future development. The game has no clear-cut winners, and most players find that they are actually competing against the economic system and the environmental constraints represented in the game rather than against each other. Three to four hours of play gives participants a good overview of the major growth processes for a local region and some of the major alternatives confronting local decision makers in both the public and private sectors.

Allan G. Feldt, a human ecologist by training, is a Professor of Urban and Regional Planning at the University of Michigan.

SUMMARY

The overriding goal in landscape planning is to provide a rational basis for guiding land use change; more specifically, to create landscapes that are safer and healthier as human habitats, more resilient to deteriorating forces, and more consistent or harmonious with natural processes, features, and systems than would be possible in an unplanned world. To achieve these ends requires an understanding of the predevelopment landscape, both natural and human components, and the nature of the forces that are imposed on it when development takes place. It also requires understanding how land use decisions are made, by whom, and how the decisions are translated into the actions that become part of the landscape.

Modern planning is comprised of two types of activities: technical and decision making. The latter is largely a political process, though not necessarily a formal one, that is carried out by policy-makers and their agents with the assistance of professional planners. Planning is conceived as a means to attaining a desired future, but it is a hollow process unless it is founded in reality, and herein lies the role of the technical planner. We must know about past and current conditions in order to formulate realistic alternatives, make forecasts, and gauge impacts. In addition, technical understanding is necessary in order to formulate the designs called for by planning decisions.

Design may be thought of as the third type of planning activity. Design is the process of devising the physical solution to a planning decision. A decision that calls for a new highway linking one side of a community with the other for the purpose of relieving the commuter traffic through residential areas can usually be solved with several different design schemes. In order for a design scheme to be effective, however, it must respond to the original plan and be technically sound (Fig. 1.8). The designer must, therefore, understand the plan, its goals, and its limitations as well as the technical analysis upon which it is based. In addition, to formulate and test design schemes the designer must also draw directly on the technical specialist for data, design evaluations, and additional analysis.

SELECTED REFERENCES FOR FURTHER READING

Catanese, Anthony J., and Snyder, James C. *Introduction to Urban Planning.* New York: McGraw-Hill, 1979, 354 pp.

Feldt, Allan G. *The Community Land Use Game: Experiments in Urban Process.* New York: Macmillan, 1972.

Godschalk, David R. *Planning in America: Learning from Turbulence.* Chicago: APA Planners Press, 1974, 240 pp.

Goodman, Robert. *After the Planners.* New York: Simon and Schuster, 1971.

Holling, C. S. (ed.), *Adaptive Environmental Assessment and Management.* New York: Wiley, 1978, 377 pp.

Lang, Reg, and Armour, Audrey. *Environmental Planning Resourcebook.* Montreal: Environment Canada, 1980, 355 pp.

Lynch, Kevin. *Site Planning.* Boston: MIT Press, 1971, 384 pp.

Marsh, William M. *Environmental Analysis for Land Use and Site Planning.* New York: McGraw-Hill, 1978, 292 pp.

McHarg, Ian L. *Design With Nature.* New York: Doubleday, 1969.

Platt, Robert S. *Field Study in American Geography.* Chicago: University of Chicago, 1959, 405 pp.

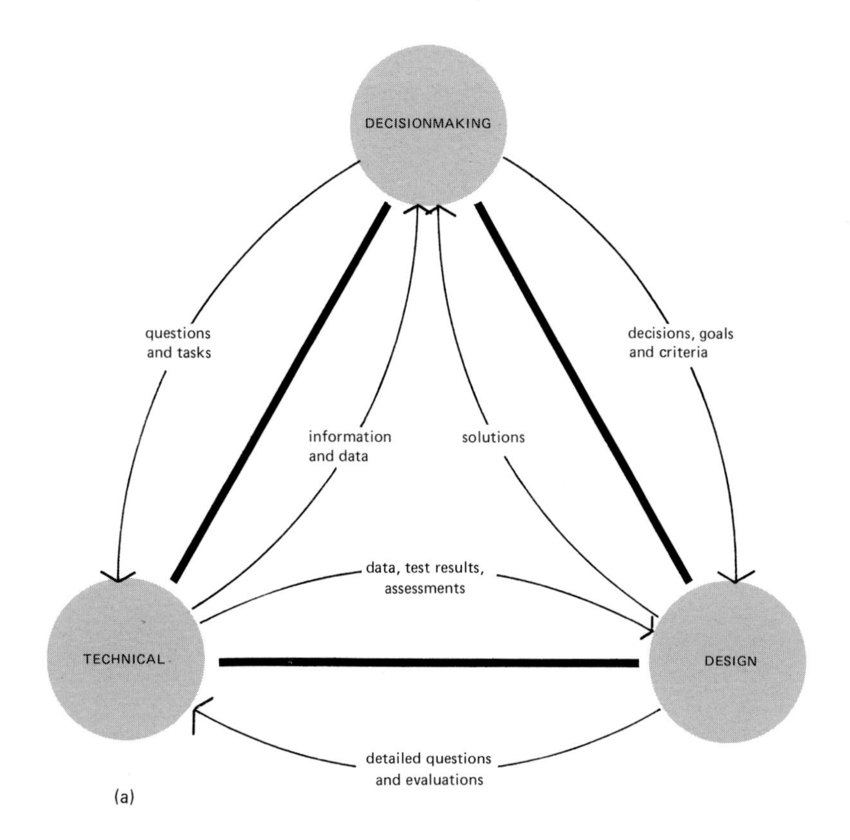

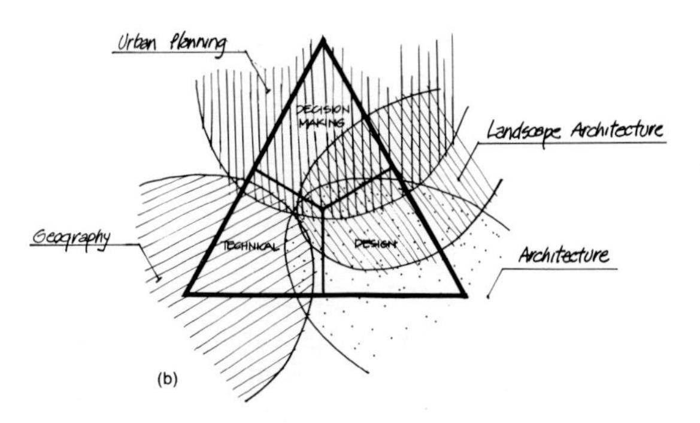

Fig. 1.8 (a) The relationship between the technical, decision-making, and design aspects of modern planning, and (b) the scope of responsibility for these aspects assumed by urban planning, geography, architecture, and landscape architecture.

II

TOPICS AND PROBLEMS IN LANDSCAPE PLANNING

2

SUN ANGLES, SOLAR HEATING, AND ENVIRONMENT

INTRODUCTION

Never in the twentieth century have we been more concerned about solar radiation than in the past several years. *Solar heating, solar energy,* and *solar collector* have become "buzz words" in virtually every segment of society as people debate the issues and alternatives of energy. Political, commercial, military, and educational institutions all have a stake in the debate, and each has its own perspectives on the problems. Which directions and alternatives are actually pursued by nations is, like it or not, largely a matter of political policy, especially at the national level. Whatever the politics of energy, solar energy is receiving serious attention in local planning and development and in turn requires data on solar radiation for different locations and geographic settings. This often brings with it an examination of sun angles, which, in addition to applications to energy problems *per se,* have an important bearing on microclimate, architecture, soils, and ecological conditions.

In building architecture, sun angles are a traditional consideration because of the designer's concern with both heating and lighting problems. In fact, great debates may be waged among architects over window sizes, building orientation, roof pitches, exterior skin materials, and whether or not to "go solar." In the past decade, sun angles and solar factors in general have also gained status among planners and landscape architects, and today it is not uncommon to find such variables incorporated into the data base for a project along with soils, drainage, and topography. The translation of solar variables into meaningful information for planning decisions can be a difficult exercise and, unfortunately, one that few practitioners are able to accomplish effectively. Approached properly, the problem often requires computation of the radiation balance and the heat balance, taking into consideration radiation gains and losses, surface reflection, ground materials and energy flows in the form of ground heat, sensible heat, and latent heat. Our discussion here begins with rudimentary sun angle concepts and then goes on to solar heating of the landscape and its implications for local environments.

SUN ANGLE AND INCIDENT RADIATION

The angle formed between sunlight approaching the earth's surface and the surface itself is called the *sun angle.* To envision this, think of straight rays of light striking a flat surface such as an airfield. A more direct angle, one that is closer to 90 degrees, causes a greater concentration of solar radiation on the surface. Conversely, smaller angles have weaker solar intensities.

Sun angle, understandably, is an important factor in the heating of the earth's surface. For example, visually compare the areas bombarded by the beam in diagrams (a) and (b) of Fig. 2.1. The beam in (a), which is the same strength as the beam in (b), spreads over more surface area because it

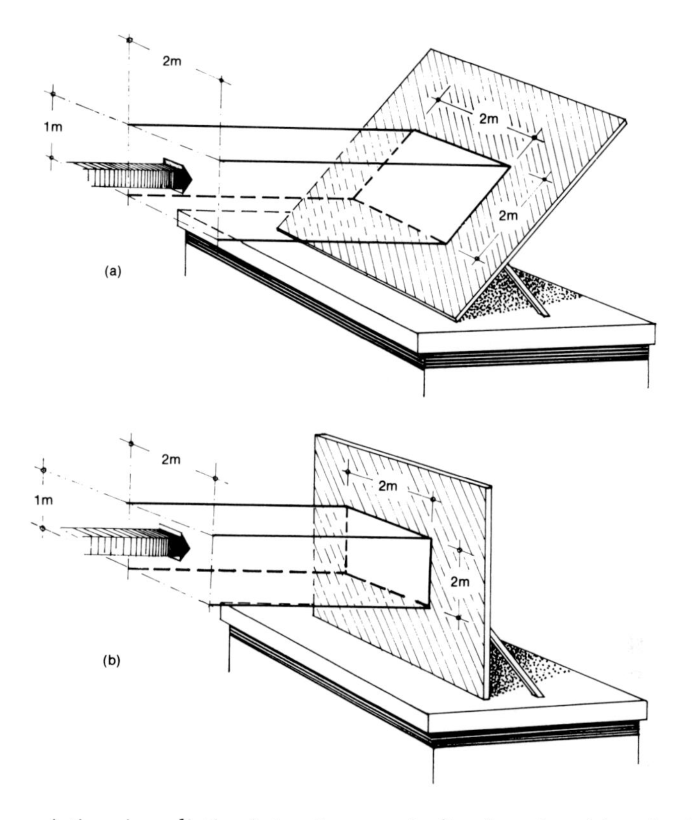

Fig. 2.1 The variations in radiation intensity on an inclined surface, (a), and a flat surface, (b).

strikes the surface at an angle. Since (a) spreads over more area, its density or incidence is lower. We can show this with a simple computation that involves dividing the quantity of energy in the beam (S_i) by the area (A) that it strikes:

$$S_I = \frac{S_i}{A}$$

where:

S_I = solar radiation incident on the surface, cal/cm^2 · min or J/m^2 · sec
S_i = quantity of energy in the beam, cal/min or J/sec
A = surface area intercepted by the beam, cm^2 or m^2

Because the earth is curved, most solar radiation enters the atmosphere and hits the surface at an angle. As one goes farther poleward, the angle becomes smaller and the beam of radiation becomes more diffuse. Figure 2.2 shows this by contrasting a beam at the equator (B) with one near the Arctic Circle (A). Beam *A* not only spreads over more surface area, but it also passes through a greater distance of atmosphere, thereby giving the atmosphere a greater opportunity to reflect and scatter radiation before it reaches the ground.

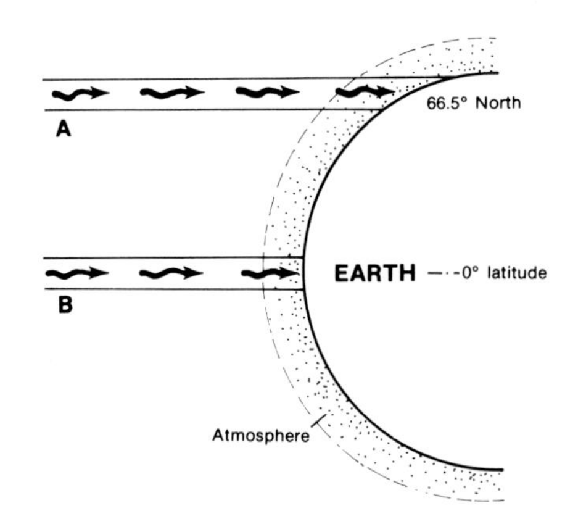

Fig. 2.2 The relationship between sun angle and the curvature of the earth.

VARIATIONS IN SUN ANGLE WITH SEASONS AND TOPOGRAPHY

To understand sun angles more completely, some additional factors must be taken into account: one is seasonal change in the earth's tilt with respect to the sun. Because of the inclination of the earth's axis (23.5 degrees off vertical), the earth appears to tip toward and away from the sun as it orbits around the sun. This produces seasonal changes in sun angle for all locations on the earth.

Four seasonal sun angles are important at any location on the planet, and they tend to correspond to the seasons in the mid-latitudes (Fig. 2.3).

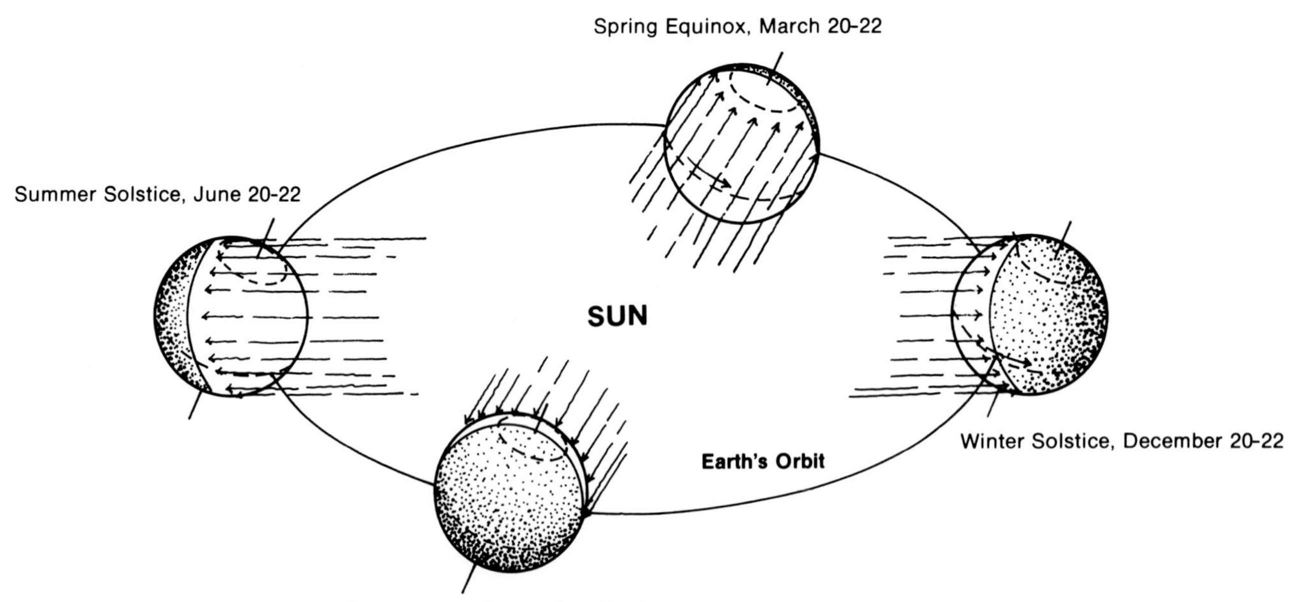

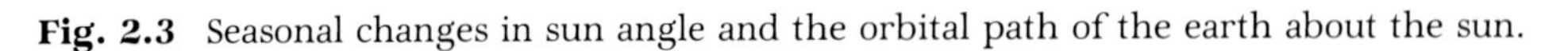
Fig. 2.3 Seasonal changes in sun angle and the orbital path of the earth about the sun.

For the northern hemisphere, the highest and lowest angles occur each year on June 20 to 22 and December 20 to 22, respectively. These dates are called *summer* and *winter solstices*. In fall and spring the sun angles are intermediate, and there are two dates on which they are exactly intermediate, March 20 to 22 and September 20 to 22, called the *equinoxes*.

From summer solstice to winter solstice, the sun angles for any location in the middle latitudes (defined, for convenience, as the zone between 23.5 degrees and 66.5 degrees latitude) vary by 47 degrees. Sun angle readings are normally given as the high noon position of the sun represented by the angle formed between one's outstretched arm pointed at the sun and the horizon on the landscape.[1] Figure 2.4 illustrates the principal sun angles in the year for 50 degrees north latitude.

Computing the sun angle for any latitude and date involves three basic steps. First, the declination of the sun must be known. This is the latitude on the earth where the sun angle is vertical (90 degrees) on a given date. For this information we consult Fig. 2.5. Next, the zenith angle must be determined, which can be done by counting the number of degrees that separate the latitude of the location in question from the declination (Fig. 2.6). *Zenith angle* is the angle formed between a vertical line (perpendicular to the ground) and the position of the sun in the sky. In Fig. 2.4, it is the angle between the sun and the broken line, for the summer solstice. The last step is to subtract the zenith angle from 90 degrees. This gives us the

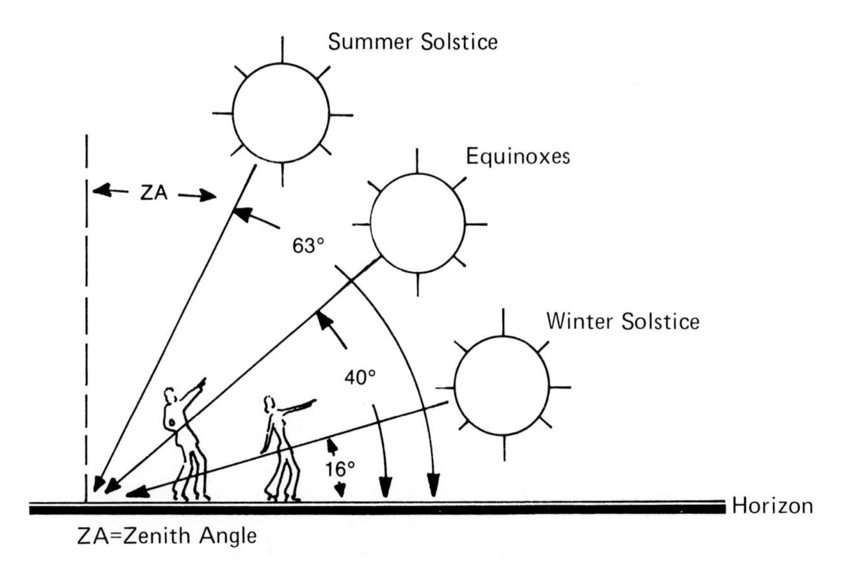

Fig. 2.4 The annual changes in sun angle for 50 degrees north latitude.

[1]This would actually be an approximation because solar radiation is refracted (bent) somewhat as it passes through the atmosphere.

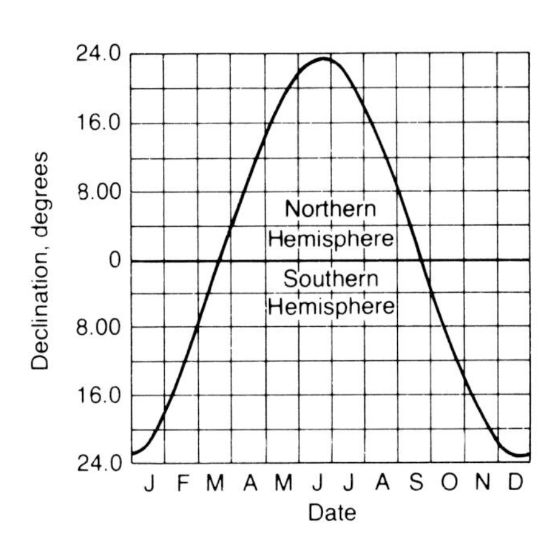

Fig. 2.5 Sun declination chart. Read the chart by finding the date in question on the bottom line, then follow the nearest vertical line upward to the curved graph line. At that point take the nearest horizontal line to the side of chart and read the appropriate number. Be sure to read northern or southern hemisphere. (Data from U.S. Nautical Almanac, 1976.)

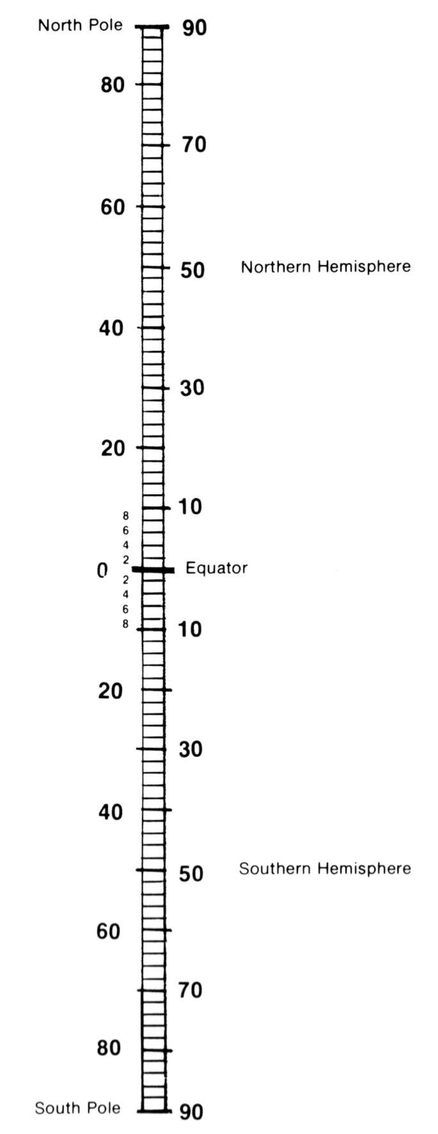

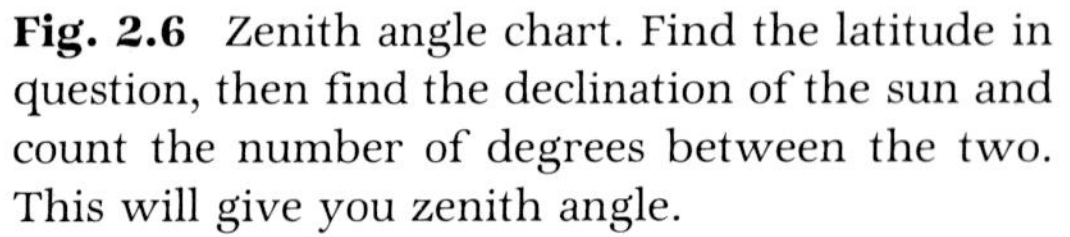

Fig. 2.6 Zenith angle chart. Find the latitude in question, then find the declination of the sun and count the number of degrees between the two. This will give you zenith angle.

sun angle. The following example shows the steps to be followed in computing a sun angle:

Location	= 50 degrees N. latitude (given)
Date	= June 15 (given)
Declination of sun	= 23 degrees (Fig. 2.5)
Zenith angle, *ZA*	= 27 degrees (Fig. 2.6)
Sun angle, *SA*	= 90 degrees minus *ZA*
	= 90 degrees minus 27 degrees
	= 63 degrees

Once the sun angle of a location is known, we can move to the local scale and examine the influence of the landscape; that is, how sun angle varies with hills, valleys, buildings, and the like. Hillslopes and roofs that face the sun are brighter and warmer than those that face away from the sun. In addition, the angle changes from dusk to dawn so that slopes with an eastward component to their orientations are favored by the morning sun and those with westward components are favored by the afternoon sun.

To compute the influence of an inclined surface on local sun angle, let us call it *ground sun angle*, we must first determine: (1) the sun angle on flat ground for that latitude; (2) the direction in which the slope faces; and (3) the angle of the slope (that is, its inclination in degrees). If the slope faces the noon sun, the sun angle on the slope is equal to the flat ground angle plus the angle of the slope. If the product is greater than 90 degrees, then subtract it from 180 degrees to get the appropriate angle. For slopes that face away from the sun, the sun angle is equal to the flat surface sun angle minus the angle of the slope. If the product is negative, the slope is in shadow.

$$\text{Ground sun angle} = \text{SA} \pm \alpha$$

where:

Ground sun angle = sun angle on slope face
SA = sun angle on flat ground
α = angle of slope in degrees

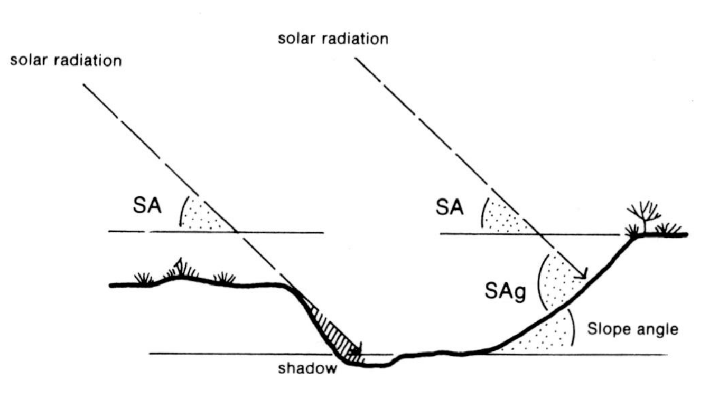

RADIATION BALANCE AND SOLAR HEATING

To determine the amount of solar heating on a surface, it is first necessary to understand that at ground level solar radiation can be disposed of in only two ways: reflection and absorption. The reflective capacity of a surface, called *albedo*, is expressed as the percentage of incoming solar (shortwave) radiation rejected by the surface:

$$A = \frac{S_o}{S_i} \times 100$$

where:

A = albedo
S_i = incoming shortwave (solar)
S_o = outgoing shortwave (solar)

All earth materials reflect a portion of the solar radiation that strikes them, but the values vary widely (Table 2.1).

Table 2.1 Albedos for Various Surfaces

Material	*Albedo, Percent*
Soil	
dune sand, dry	35–75
dune sand, wet	20–30
dark (e.g., topsoil)	5–15
gray, moist	10–20
clay, dry	20–35
sandy, dry	25–35
Vegetation	
broadleaf forest	10–20
coniferous forest	5–15
green meadow	10–20
tundra	15–20
chaparral	15–20
brown grassland	25–30
tundra	15–30
crops (e.g., corn, wheat)	15–25
Man-Made	
dry concrete	17–27
blacktop (asphalt)	5–10
Water	
fresh snow	75–95
old snow	40–70
sea ice	30–40
liquid water	
30° lat. summer	6
30° lat. winter	9
60° lat. summer	7
60° lat. winter	21

From William D. Sellers, *Physical Climatology*, Chicago: University of Chicago, 1974. Used by permission.

The solar energy absorbed by a surface, let us call it *solar gain*, is equal to incoming shortwave (S_i) less the amount reflected (S_o):

$$\text{Solar gain} = S_i - S_o$$

This quantity represents energy added to the absorbing material in the form of heat, and it in turn will produce a rise in the material's temperature. The actual amount of temperature rise for a given amount of energy added will vary with the type of material as a function of its thermal properties; that is, equivalent amounts of heat in two different materials, say, water and soil, will not yield the same temperature. (This is explained mainly by differences in their volumetric heat capacities, that of water being high compared to that of sand, for example (see Table 4.1 in Chapter 4).)

Taking into consideration the fact that solar radiation usually strikes surfaces in the landscape at an angle, it is necessary to combine the concept of incident radiation flux (flow over the receiving area) with albedo to determine the solar heating for a surface. This can be done computationally using just three variables: the ground sun angle (based on latitude, date, and surface inclination or slope), the intensity of solar radiation, and the albedo of the surface:

$$SH = S_i\,(1 - A) \sin SA_g$$

where:

SH = solar heating in cal/cm^2 · min or joules/m^2 · sec
S_i = incoming solar radiation in cal/cm^2 · min or joules/m^2 · sec
A = albedo ($1 - A$ gives the percentage absorbed)
SA_g = ground sun angle in degrees

It is instructive to see how important slope and albedo are in the solar heating of a varied landscape. For example, given the surfaces (at 45 degrees north latitude) represented by the profile in Fig. 2.7a, the rates of solar heating at noon on the equinox would be as follows:

Building Roof

- slope = 45°
- orientation = south
- albedo = 10%
- SA_g = 90°
- S_i = .78 cal/cm^2 · min

$$\begin{aligned} SH &= .78\,(1 - .10) \sin 90° \\ &= .78\,(.9)\,1.0 \\ &= .70 \text{ cal/cm}^2 \cdot \text{min} \end{aligned}$$

Concrete Wall

- slope = 30°
- orientation = north
- albedo = 27%
- SA_g = 15°
- S_i = .78/cm^2 · min

$$\begin{aligned} SH &= .78\,(1 - .27) \sin 15° \\ &= .78\,(.73)\,.26 \\ &= .15 \text{ cal/cm}^2 \cdot \text{min} \end{aligned}$$

Plowed Field

- slope = 5°
- orientation = south
- albedo = 22%
- $SA_g = 50°$
- $S_i = .78 \text{ cal/cm}^2 \cdot \text{min}$

$$SH = .78\,(1. - 22)\sin 50°$$
$$= .78\,(.78)\,.77$$
$$= .47 \text{ cal/cm}^2 \cdot \text{min}$$

Sandstone Slope

- slope = 25°
- orientation = south
- albedo = 40%
- $SA_g = 70°$
- $S_i = .78 \text{ cal/cm}^2 \cdot \text{min}$

$$SH = .78\,(1 - .40)\sin 70°$$
$$= .78\,(.60)\,.94$$
$$= .44 \text{ cal/cm}^2 \cdot \text{min}$$

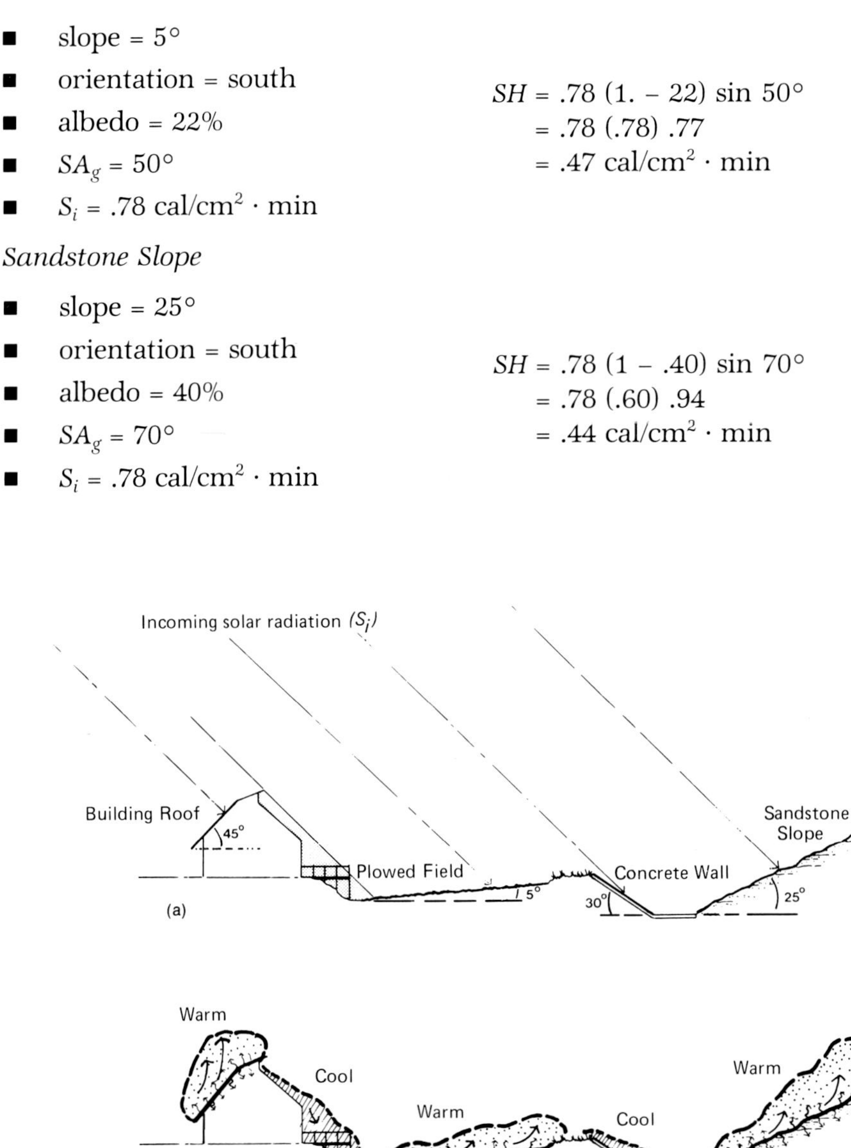

Fig. 2.7 (a) Variation in solar heating related to slope and surface materials. (b) The resultant differences in air heating and movement. The actual amount of air heating by surface materials depends not only on solar energy absorbed but on the volumetric heat capacity of the material. See Table 4.1 in Chapter 4.

The concrete wall gains the least energy, about one-third that of the field and slope and about one-fifth that of the roof.

The influence of these variations on ground-level climate, called *microclimate*, depends on many additional factors, including (1) how much of the solar energy absorbed is returned to the air over it as heat (either as sensible heat or by long-wave (infrared) radiation, which in turn may heat the air); (2) local wind conditions (which account for the rate of flushing of heated air from surfaces); and (3) the size of the area covered by each heating surface (which determines the relative balance of thermal influences among the different surfaces in an area). Under calm air conditions, the layer of air over surfaces such as these will begin to develop a pattern of temperatures roughly corresponding to the pattern of solar energy absorbed. This may then induce differential air movement with the warm air rising or sliding upslope and the cool air draining downslope (Fig. 2.7b). On some days this pattern may carry over well past the period of peak solar radiation and into the evening, given that regional weather systems do not obliterate it.

To gain an idea of the impact of land use change on the gain of solar energy by the landscape, one can compare the differences in slope (both angle and orientation) and surface materials before and after development. This would involve first mapping the predevelopment slopes of various angles, orientations, and compositions, measuring their areas, and then computing their total solar gain over some time period. These figures would be summed for the entire project area and compared to the parallel figure based on the same computation for the postdevelopment landscape. Although many additional factors would have to be taken into account to determine the climatic significance (that is, means and extremes in air temperature, wind, precipitation, humidity, and so on) of a change in solar gain, such a comparison does provide one measure of the relative impact of different land uses and development schemes on the environment at ground level.

CASE STUDY

Energy Management Planning and the Passive Solar Collector

Mitchell J. Rycus

In energy planning and management of parks and recreational areas one of the important concerns is that no one energy management strategy should be considered without first formulating a comprehensive energy plan. In a recent project for the United States Department of Interior this planning principle led to the use of a matrix designed to display an array of energy conservation strategies based on fuel conservation (gasoline, oil, electricity, etc.) and the type of facility involved (buildings, motor vehicles, ball parks, etc.). Included in the display of each strategy is the amount of fuel that could be conserved as a percentage of the total by fuel type consumed in the entire park. An estimate of the cost involved to achieve the saving for each strategy, such as the cost of retro-fitting buildings with solar collectors, is also displayed.

After all the strategies are placed in the matrix, summaries of the total cost of implementation, including fuel saved, can be calculated. Additional information on how to assess the political, economic, and institutional factors associated with different strategies is also placed in the matrix. At this point a park manager can examine the range of available strategies and from them develop an energy management program that would be most appropriate for his or her park. Once the manager decides on a particular set of strategies, he or she would then seek technical

(cont.)

CASE STUDY (cont.)

assistance, if needed, in computing actual costs and savings, or for the planning and designing of new facilities, as well as changing existing ones.

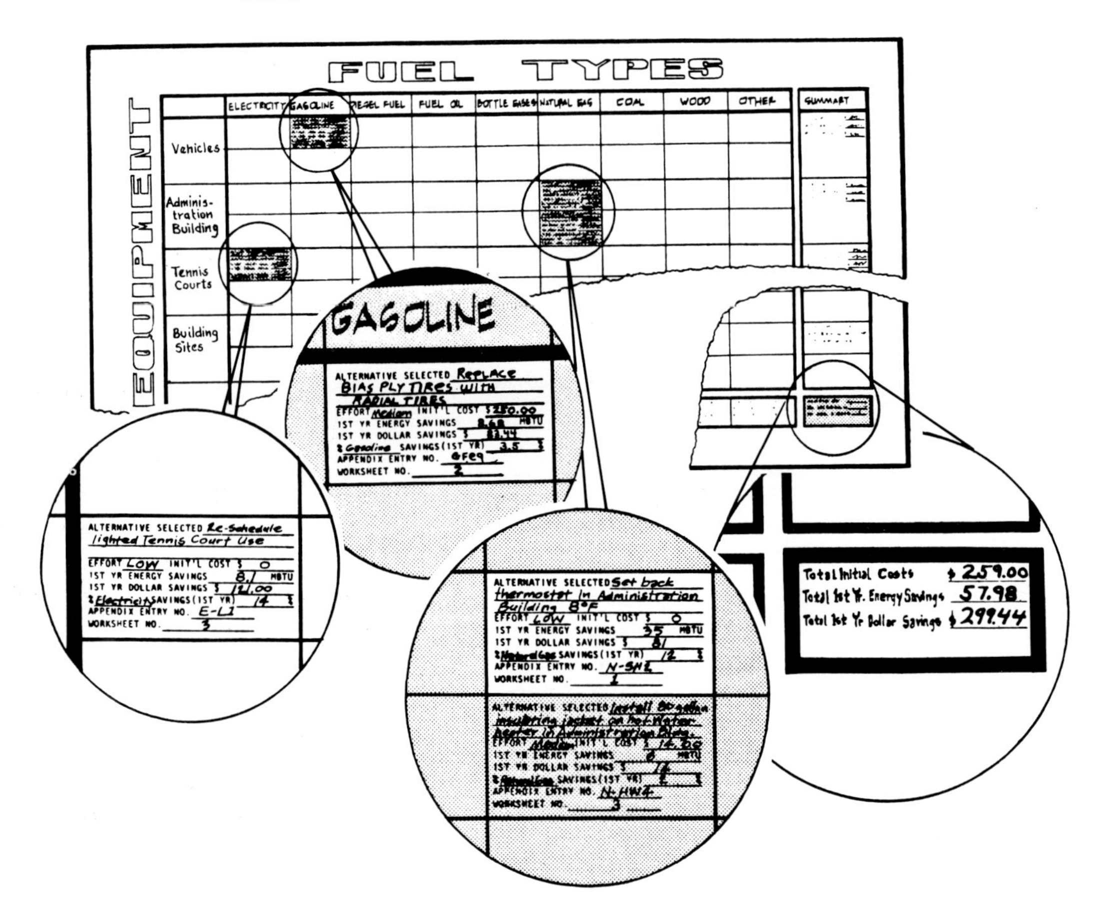

A number of the strategies deal with design criteria for new facility construction. The design information ranges from the use of insulating materials and light bulbs to building site locations and landscaping. Most of us are aware of sophisticated collectors that capture solar energy, convert it into heat, transfer it to a storage device such as a large tank of water, and then circulate the heated water through a building. However, a rational solar energy strategy for a park involves more than acquiring and setting up equipment, because careful consideration must also be given to building siting, landscaping, architecture, and engineering.

Passive solar energy is obtained through proper siting and landscaping, and can be achieved without active (powered-mechanical) apparatus such as pumps or tracking devices. For example, a large area of south-facing glass windows enclosing an attached room that collects the sun's energy is a passive system. Radiation enters the room and is absorbed by some material, the collector, which emits heat inside the room. Brick, concrete, or containers of water may serve as the collector and air warmed from the collector's heat convects naturally throughout the building. In the winter when the sun is low and deciduous trees are without foliage, the sun's rays pass easily into the facility, while in the summer when the sun is high an overhang can be used to shade the glass. Tree canopies can provide additional shading, keeping the facility from overheating. Actually,

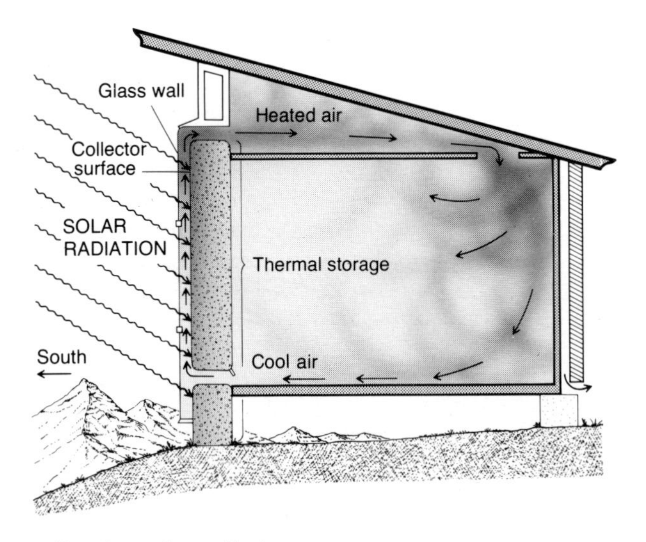

Passive solar collector.

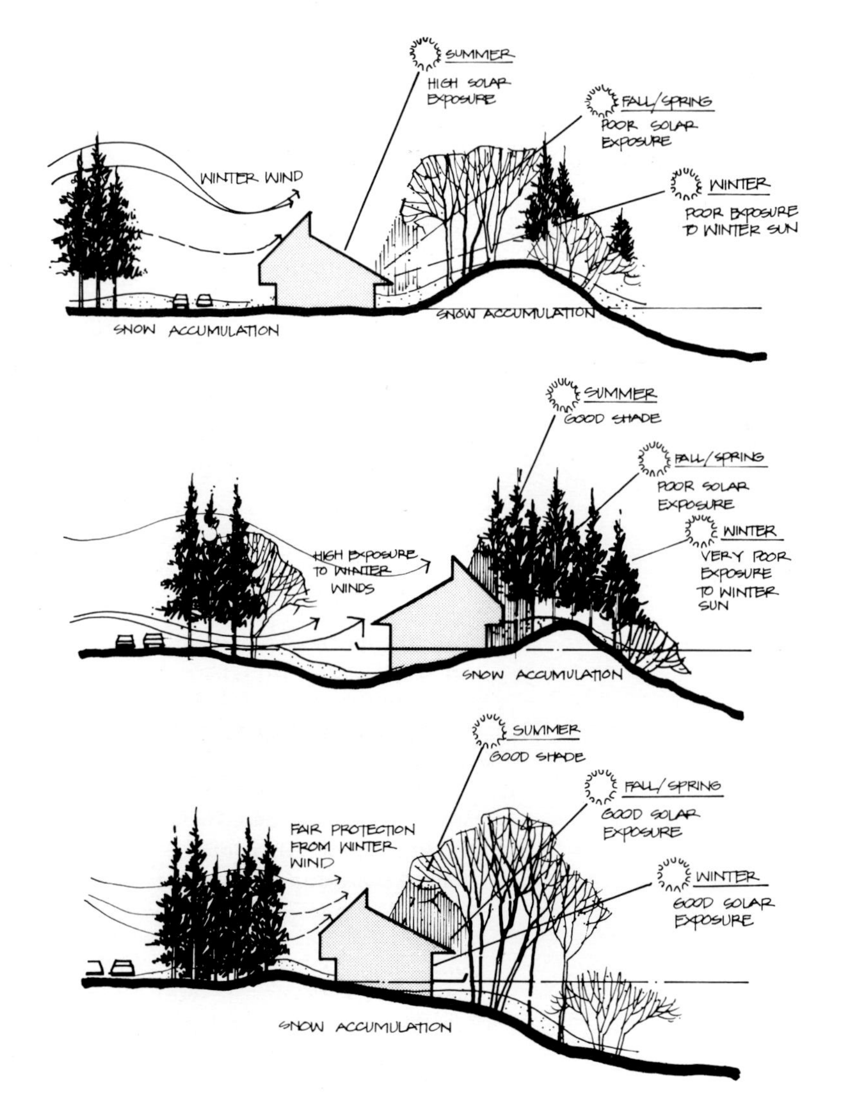

(cont.)

CASE STUDY (cont.)

some heat is useful in the summer to draw the cooler air from the north side of the building through the rooms and out roof or side vents.

If a passive solar energy strategy is selected for, say, a new park interpretation center, the following procedure might be used to plan the facility. After making an initial estimate of the savings, various sites having different terrain and vegetation characteristics would be examined. Each site would be analyzed according to the amounts of solar radiation available at different times of the year based on sun angles, topographic obstacles, and vegetation type and density. Other factors such as wind, fog, and snow drifting would also be considered with site analysis. Each site would then be evaluated for all these factors as well as for more detailed cost and savings data. Once completed, the preferred site would be selected and a design scheme would be worked out, taking into account elevations, building form and orientation, vegetation, and other factors.

Mitchell J. Rycus is a resource planner at the University of Michigan and a specialist in energy management.

IMPLICATIONS FOR VEGETATION AND SOIL

Variations in incident radiation due to differences in the orientation and inclination of slopes can have a profound influence on vegetation and ground conditions. In semi-arid mountainous areas, such as parts of Colorado, New Mexico, and California, the more direct sun angles on south-facing slopes result in greater surface heating and, in turn, higher rates of soil moisture evaporation and plant transpiration than on north-facing slopes. The resultant difference in soil moisture is often great enough to cause marked differences in vegetation on north- and south-facing slopes. The photograph in Fig. 2.8 shows one such example from southern Colorado, where moisture stress limits trees to north-facing slopes.

In even drier areas, where only herbs and shrubs can survive, the plant cover on south-facing slopes is often measurably lighter than that on north-facing slopes. Because more ground is exposed, erosion by runoff may also be higher on south-facing slopes, resulting in higher densities of gullies and lower slope angles. In addition, there may be a difference in the abundance of certain species, with the more drought-tolerant species making up a higher percentage of the plant cover on south-facing slopes.

Differences in species related to the influence of slope on incident radiation can also be found in humid regions, though the examples are rarely as obvious as those in dry areas. In the mid-latitudes, combinations of heat and light may ensure the survival of certain plants on extreme slopes. For instance, north-facing cliffs along the south shore of Lake Superior harbor species of ferns and mosses that are separated by hundreds of miles from the main bodies of their populations in arctic and subarctic regions (Fig. 2.9). Apparently the low light intensities and cool temperatures along these cliffs have favored the survival of these plants since the last continental glaciation.

Fig. 2.8 Differences in vegetation on north- and south-facing slopes. Cooler north slopes are able to sustain a forest cover because the moisture balance is better than on the south slopes, which are limited mainly to grasses.

Fig. 2.9 Shadow zone along north-facing cliffs of Lake Superior creates an environment conducive to certain arctic and subarctic plants.

IMPLICATIONS FOR BUILDINGS AND LIVING ENVIRONMENTS

The placement and size of buildings and trees in cities can seriously affect the reception of solar radiation. With rising concern over solar energy, this issue has gained new significance in urban planning and design. "Shadow corridors" and "solar windows" (or gaps) are two of the most common solar features of cities (Fig. 2.10). Solar windows are narrow spaces between tall buildings through which the solar beam passes to ground level. Depending on the orientation and spacing of the buildings, the shaft of light may illuminate a patch of ground for only a short time each day, making it difficult to maintain street plants and virtually impossible to utilize solar radiation as a source of energy.

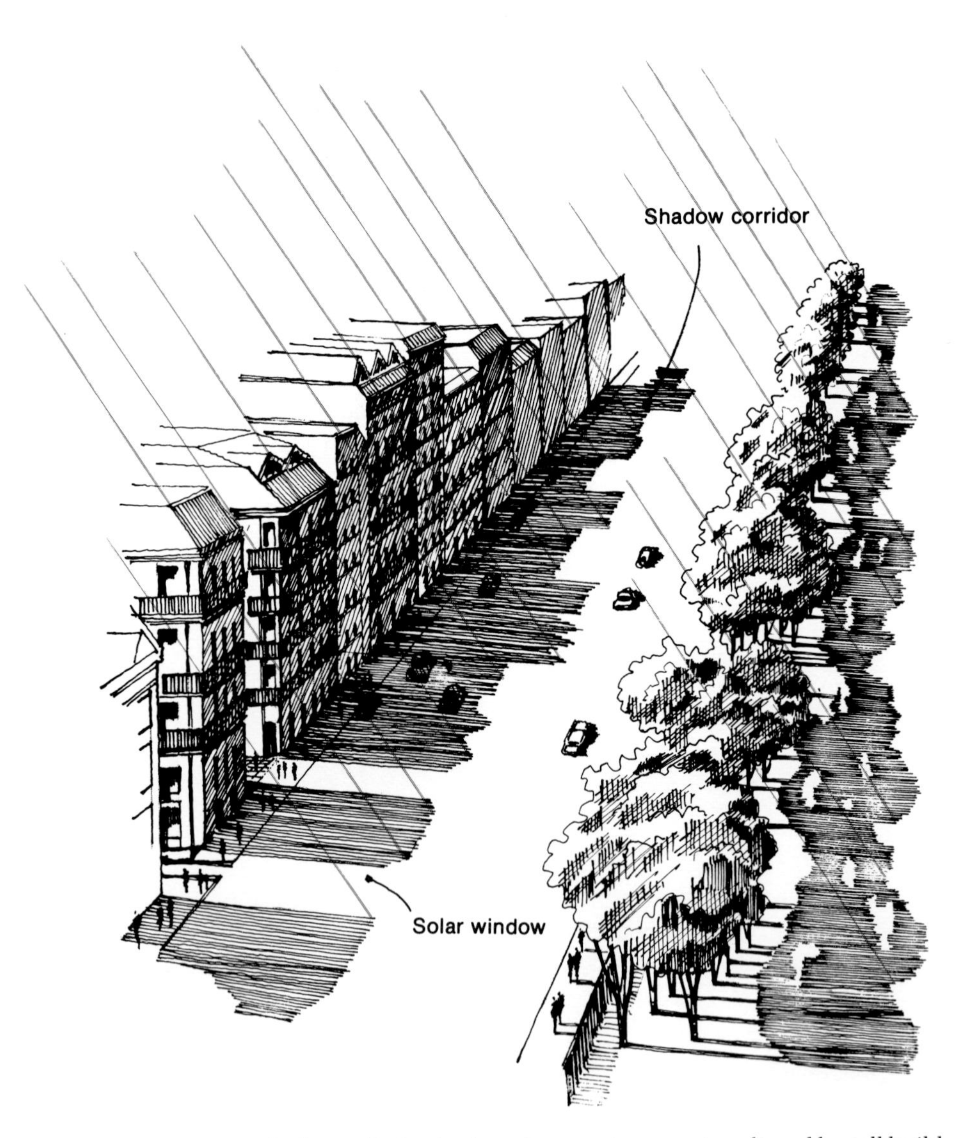

Fig. 2.10 The pattern of solar radiation in the urban environment as altered by tall buildings.

Afternoon shadows from the skyscrapers along Chicago's waterfront. (Photograph by Ann Clipson.)

Shadow corridors are elongated zones, bordered by a continuous ridge of tall buildings that block the sun. In the most extreme situations, direct (beam) solar radiation is never received in such environments; the only light comes from diffused sky radiation and radiation reflected from nearby buildings.

The length of a shadow cast by a building or tree is a function of the height of the object and the sun angle; computations can be made using the formula:

$$S_\ell = \frac{h}{\tan SA}$$

where:

S_ℓ = shadow length
h = height of the object
SA = sun angle

tangents (tan) for angles 5°–85°	
5° = .087	45° = 1.0
10° = .176	50° = 1.19
15° = .268	55° = 1.43
20° = .364	60° = 1.73
25° = .466	65° = 2.14
30° = .577	70° = 2.75
35° = .700	75° = 3.73
40° = .839	80° = 5.67
	85° = 11.43

This is traditionally used in site planning in areas of excessive heat and intensive solar radiation because it is necessary to provide for shade in pedestrian areas, parking lots, on building faces, plazas, and the like. (See the case study, "Heat Syndrome in Humans," in the following chapter.) The need for shade is generally greatest in the hours between 11 A.M. and 4 P.M. when high solar intensities are coupled with high air and ground temperatures (Fig. 2.11).

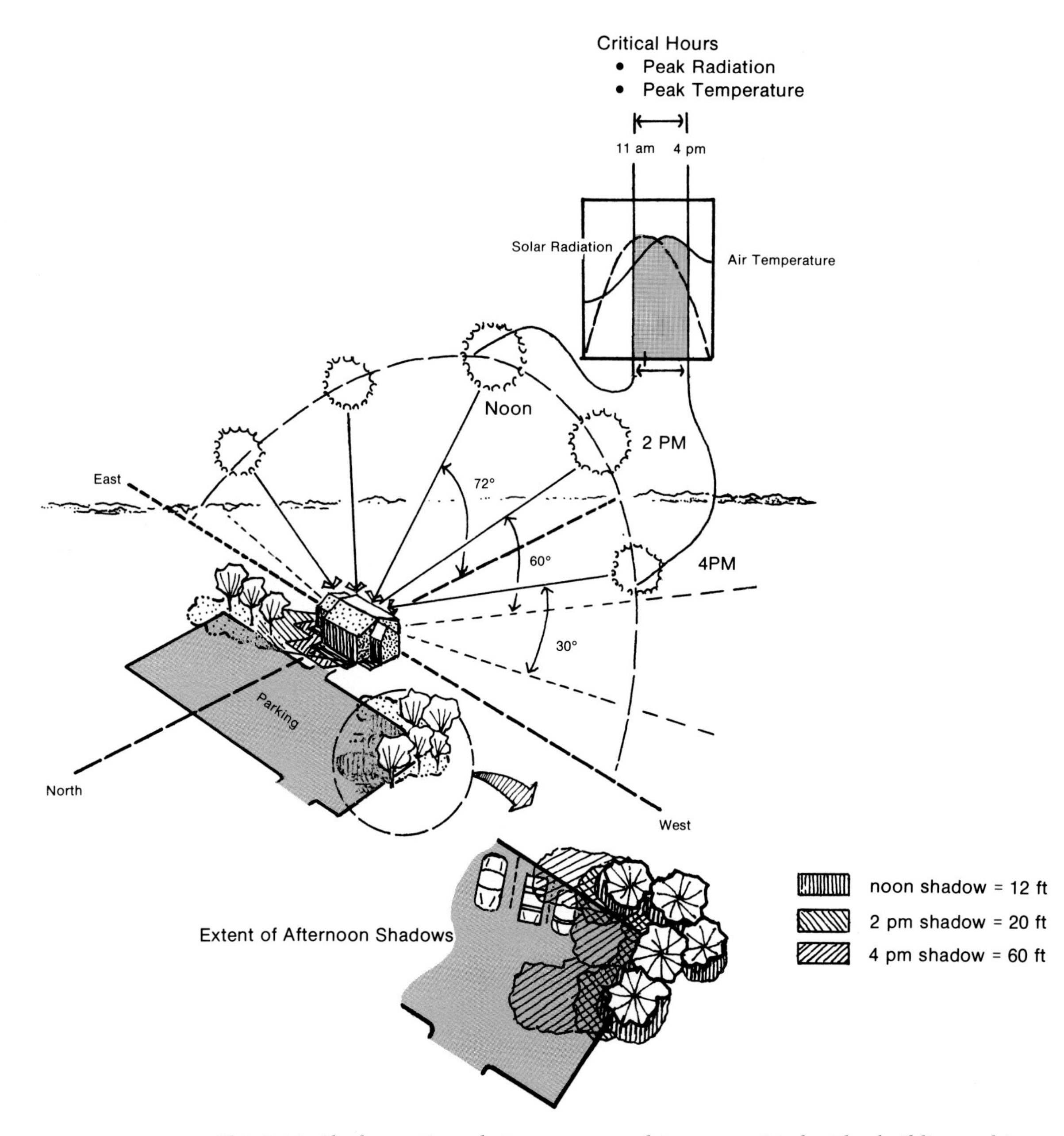

Fig. 2.11 Shadow patterns between noon and 4 P.M. associated with a building and trees near a parking lot. Shade is most critical between 11 A.M. and mid-afternoon when air and ground temperatures are highest.

While the shade can be a distinct advantage for local pedestrians and residents of cities prone to frequent heat waves, in northern cities, such as Minneapolis, Detroit, Toronto, and Montreal, shadow corridors encourage the buildup of ice and snow, making foot travel hazardous. Moreover, the solar gain is very poor in these zones for living units with northerly exposures, resulting in somewhat cooler room temperatures and higher heating costs. This is especially significant in light of recent findings in Great Britain and the United States concerning illness and death among the elderly caused by hypothermia.

Accidental hypothermia, a disorder characterized by low body temperature (near 90°F), slowed heart beat, lowered blood pressure and slurred speech, can be brought on in persons over seventy years old, by room temperatures as modest as 65°F, inadequate clothing, and prolonged periods of physical inactivity. Solar exposure may make a difference of several degrees in room temperatures, especially during cold spells, and in turn can tip the balance between hypothermia and a normal state of health in the elderly (Fig. 2.12). The United States National Institutes of Health estimate that as many as 2.3 million elderly people in the United States are vulnerable to accidental hypothermia. Undoubtedly, many of these persons inhabit buildings whose orientation, design, and neighborhood exclude or greatly restrict access to direct solar radiation in living spaces. On the other hand, these same conditions may be an advantage during the summer in many cities because they are not prone to excessive heating, a topic broached in the next chapter.

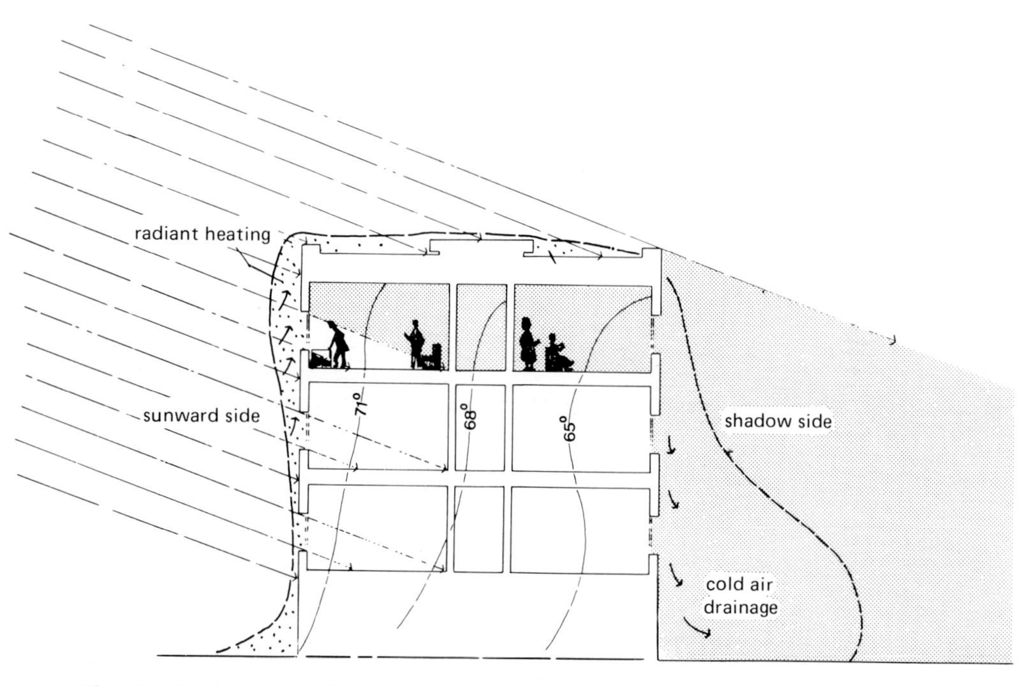

Fig. 2.12 Schematic diagram illustrating the effects of sun angle on living units in an apartment building.

PROBLEM SET

I. For each of the following locations, compute the mid-day sun angle for the date given.

1. Los Angeles (34°N), October 30; Kansas City, Mo. (39°N), March 1; Edmonton, Alberta, Canada (53.5°N), June 20; Fort Lauderdale (26°N), August 15.

2. Your own location for both solstices and equinoxes. In addition, use a protractor to construct these sun angles. Label the angles and dates and identify the north and south sides of the diagram.

II. The diagram below represents a north/south profile across a river valley at your location.

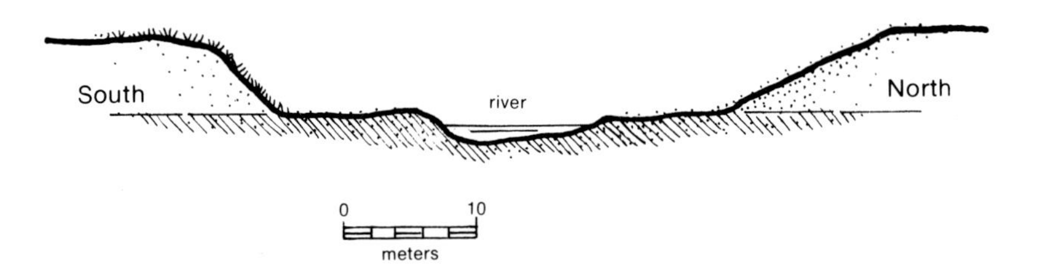

1. Determine the mid-day sun angles on the valley floor and on the north and south slopes during the winter solstice.
2. If the south slope is covered by brown grass, the valley floor by moist gray soil, and the north slope by dry sand, determine the mid-day solar heating of each if the incoming solar radiation (S_i) is 0.20 calories per square centimeter per minute (140 joules per square meter per minute). (If the surface is in shadow, give shadow as the answer.)
3. In order to improve the aquatic habitat of the river channel it is recommended that twenty-foot (6 m) high shade trees be planted along the south bank of the river. If the trees are planted six feet from the water's edge, how far over the water will the mid-day shadow extend on June 20–22, August 1, October 1, and September 1? (Assume that the bank will add another four feet (1.2 m) to the height of the trees.) Construct a diagram showing the stream bank, stream surface, trees, sun angle, and shadow.

III. Faced with the problem of designing a solar collector for your own residence:

1. Identify the optimum angle and albedo of the collector surface for the solstices and equinoxes, assuming your goal is to maximize radiation absorption in all seasons.
2. Draw a diagram showing the profile of the building and nearby obstacles, and indicate how you would take these factors into account in each season.

SELECTED REFERENCES FOR FURTHER READING

American Institute of Architects Research Corp. *Solar Dwelling Design Concepts.* Washington, D.C.: Dept. Housing and Urban Development, 1976.

Buffo, John, et al. "Direct Solar Radiation On Various Slopes from 0 to 60 Degrees North Latitude." *U.S.D.A. Forest Service Research Paper PNW–142,* 1972, 74 pp.

City of Davis (California). *A Strategy for Energy Conservation.* Davis, California: Energy Conservation Ordinance Project, 1974.

Feldt, Allan G.; Rycus, M. J.; and Hassett, M. L. *Energy Planning and Management for Parks and Recreation.* Washington, D.C.: Government Printing Office, 1981, 250 pp.

Land Design/Research, Inc. *Energy Conserving Site Design Case Study, Burke Center, Virginia.* Washington, D.C.: Dept. of Energy, 1979, 60 pp.

Marsh, William M., and Dozier, Jeff. "The Radiation Balance." In *Landscape: An Introduction to Physical Geography.* Reading, Mass.: Addison-Wesley, 1981, pp. 21–35.

National Institute On Aging. "A Winter Hazard for the Old: Accidental Hypothermia." Washington, D.C.: U.S. National Institutes of Health, Dept. of Health, Education and Welfare, 1981 (?), pub. no. (NIH) 78-1464.

Sizemore and Associates. *Methodology for Energy Management Plans for Small Communities.* Washington, D.C.: U.S. Dept. of Energy, 1978.

Schumm, S. A. "Application of Landform Analysis in Studies of Semiarid Erosion." *Circular 437,* U.S. Geological Survey, 1960.

Sterling, Raymond, et al. *Earth Sheltered Community Design.* New York: Van Nostrand Reinhold, 1981, 270 pp.

Tuller, S. E. "Microclimatic Variations in a Downtown Urban Environment." *Geografiska Annaler,* vol. 54A, 1973, pp. 123–135.

3

MICROCLIMATE AND THE URBAN ENVIRONMENT

- Introduction
- The Urban Heat Island
- Microclimatic Variations
- Applications to Urban Planning
- Climatic Criteria for Urban Planning and Design
- Problem Set
- Selected References for Further Reading

INTRODUCTION

Urbanization can cause significant changes in the atmospheric conditions near the ground. In extreme situations, such as in the heavily built up areas of larger cities, these changes extend hundreds of meters above the ground and are of such magnitudes that they produce a distinct climatic variant, the urban climate. Generally speaking, the urban climate is warmer, less well-lighted, less windy, foggier, more polluted, and often rainier than the region-wide climate.

Within the urban landscape microclimatic variations can also be considerable: air quality may be exceptionally poor along transportation corridors and in industrial sectors; certain neighborhoods may be warmer than average in summer; and small areas between tall buildings may receive little or no beam radiation, receiving much of their energy instead from the heat loss of buildings.

These variations can be important considerations in urban planning. Documentation of the desirable climatic effects of vegetated areas, for example, helps provide rationale for the inclusion of parks and greenbelts in master plans. Transportation planning today invariably includes air quality guidelines and goals, and proposals for industrial development must include forecasts on gaseous and particulate emissions and plume patterns under different atmospheric conditions.

THE URBAN HEAT ISLAND

Urbanization transforms the landscape into a complex environment characterized by forms, materials, and activities that are vastly different from those in the rural landscape. Not surprisingly, the flow of energy in the urban landscape is also different. As a whole, the receipt of solar radiation is substantially lower, while the generation of sensible heat at ground level is greater in cities compared to the neighboring countryside. Further, the rate of heat loss from the urban atmosphere through convective and radiant flows is lower. On balance, the increase in sensible heat coupled with lower rates of heat loss is more than enough to offset the thermal effects of the decrease in solar radiation, resulting in somewhat higher air temperatures in urban areas throughout most of the year and much higher temperatures on selected days. The spatial pattern of these temperatures is often concentric around the city center, producing a "heat island" in the landscape. The geographic extent and intensity of the urban heat island varies with city size (based on population) and with regional weather conditions. In general, large cities under calm, sunny weather produce the strongest heat islands.

The overall structure of the urban atmosphere can be envisioned as a large dome centered over the urban mass. This body of air is called the *urban boundary layer*. Because of a heavy particulate content, it is highly efficient in back-scattering solar radiation, often effecting a reduction of 50 percent in the lower 1,000–2,000 m above a city. A growing amount of research findings reveal that increased cloudiness and precipitation are associated with the urban atmosphere and that these trends carry downwind to neighboring areas in the urban region. Thus it appears that the dome-like structure of the boundary layer is pronounced only during relatively calm atmospheric conditions; whereas during a steady airflow across the region, the dome is tipped downwind and develops a plume (Fig. 3.1).

At ground level the causes underlying the formation of the urban heat island, and indeed the urban climate in general, are many and complex. To begin with, the materials of the urban landscape possess different thermal properties than those of the rural landscape. The volumetric heat capacities of street and building materials are appreciably lower than that of materials in the rural landscape (see Table 4.1 in Chapter 4). This means that urban surfaces generally reach a higher temperature with the absorption of a given quantity of radiation and, in turn, heat the overlying air faster. Second, the Bowen Ratio, which is a measure of the heat released from a surface in the sensible form relative to the latent form, is much higher in cities owing to the limited areas of open water, vegetation, and exposed soil. With a paucity of vapor sources, latent heat flux from the surface is relatively low; conversely, sensible is relatively high, giving rise to higher air temperatures. Added to this is the heat released from artificial sources (automobiles, buildings, etc.). In the mid-latitudes, these sources typically contribute more energy to a city in winter than the solar source does.

Fig. 3.1 Configuration of the urban "dust dome" during a cross wind. This plume may extend great distances downwind from the city, where it can produce a measurable increase in precipitation and cloudiness.

CASE STUDY

Heat Syndrome in Humans

NOAA

In the period 1950 through 1967, more than 8,000 persons were killed in the United States by the effects of heat and solar radiation. These were identified by health and medical authorities as direct casualties. How many deaths in the aged and infirm were encouraged by excessive heat or solar radiation is not known because medical records usually do not identify climatic factors as the cause of death in such cases.

Heat syndrome refers to several clinically recognizable disturbances of the human thermoregulatory system. The disorders generally have to do with a reduction or collapse of the body's ability to shed heat by circulatory changes and sweating, or a chemical (salt) imbalance caused by too

(cont.)

CASE STUDY (cont.)

much sweating. Ranging in severity from the vague malaise of heat asthenia to the extremely lethal heat stroke, heat syndrome disorders share one common feature: the individual has been subject to overexposure and/or overexercise for his age and physical condition and the thermal environment.

Two climatic conditions are associated with most epidemics of heat syndrome: regional heat waves and thermal microclimates. Heat waves are usually accompanied by an increased mortality, especially among the elderly; and this correlation can be expected without the complicating factors of high humidity or air pollution. Studies of heat syndrome show that it affects all ages of humans, but, other things being equal, the severity of the disorder tends to increase with age; heat cramps in a seventeen-year-old boy may be heat exhaustion in someone of forty years, and heat stroke in a person over sixty years of age.

There is evidence that heat waves are worse in the brick and asphalt canyons of the "inner cities" than in the more open and better vegetated suburbs. The July 1966 heat wave in St. Louis is a case in point. Most of the 236 deaths attributed to excessive temperatures occurred in the more heavily built-up areas of the city. Records also show that the death rate soared when the daily high temperature exceeded 100°F and that the highest tolls lagged behind temperature peaks by about a day.

For people with heart disease, thermal stress is worse than for others. In a hot, humid environment, impaired evaporation and water loss hamper thermal regulation, while physical exertion and heart failure increase the body's rate of heat production. The ensuing cycle is vicious in the extreme.

In a healthy person, the body acclimates to heat by adjusting perspiration-salt concentrations, among other things. In the mid-latitude conti-

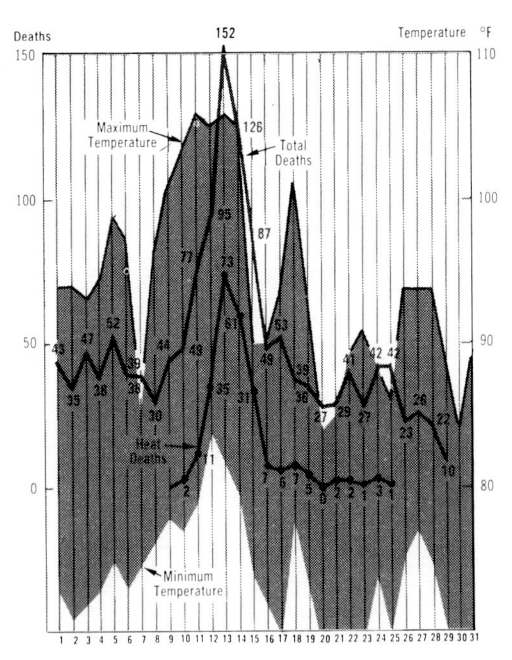

DAILY DEATHS AND TEMPERATURES—ST. LOUIS
JULY 1966

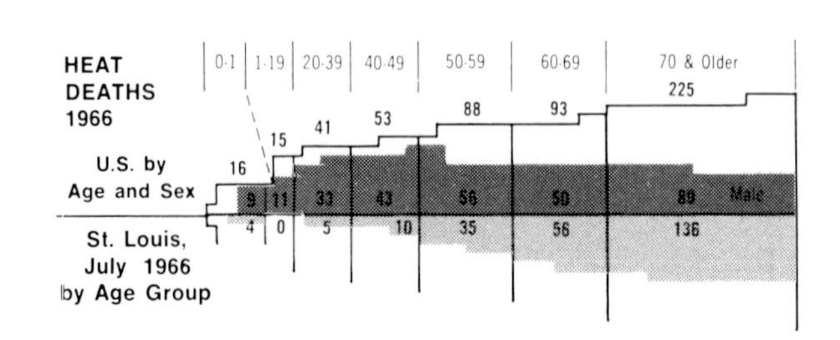

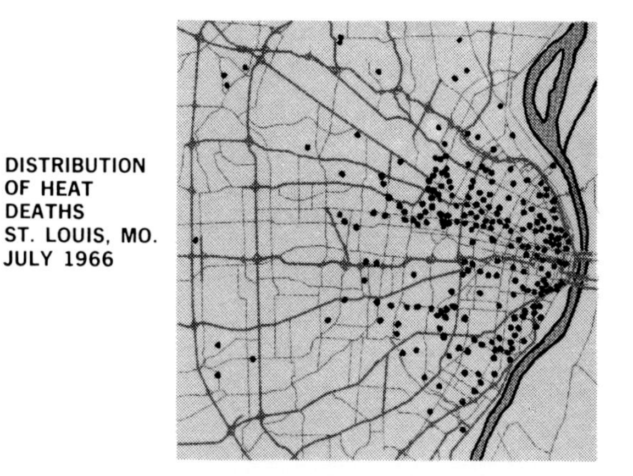

DISTRIBUTION
OF HEAT
DEATHS
ST. LOUIS, MO.
JULY 1966

nental climates, this concentration changes in winter and summer just as it does when one moves from Boston to Panama. The body seeks an equilibrium in which enough water is lost to regulate body temperature without upsetting its chemical balance. Females appear to be better at this than males, because females excrete less perspiration and so less salt; therefore, heat syndrome usually strikes fewer females.

NOAA, The National Oceanic and Atmospheric Administration, is the U.S. federal agency responsible for monitoring and forecasting weather and climate.

Heat inputs represent only one side of the system; the other is, of course, heat outputs, or losses, from the urban atmosphere. The principal consideration in this regard is wind speed. Overall, cities tend to have much lower wind speeds at ground level; therefore, heated air tends not to be flushed away as readily as it is in rural landscapes. Further, the urban atmosphere retains more heat because of a higher carbon dioxide content. On balance, then, the urban landscape yields and retains more heat, thereby accounting for the heat island effect.

MICROCLIMATIC VARIATIONS

While the climate of an entire city is an important issue for regional authorities, climatic variations within the city have become increasingly important issues to the urban planner, landscape architect, and architect. Perhaps the easiest variation to visualize is that associated with solar radiation around tall buildings, but other parameters including temperature, wind, fog, and pollution also show considerable variation within the urban landscape (Fig. 3.2). The nature and significance of these variations is not well documented; however, there seems to be little doubt among experts that extremes in these components of the urban climate do impair the health and safety of a significant number of people in most cities. Further, there is also agreement that improvement of the urban climate is possible through planning and management of land use activities and new approaches to urban design.

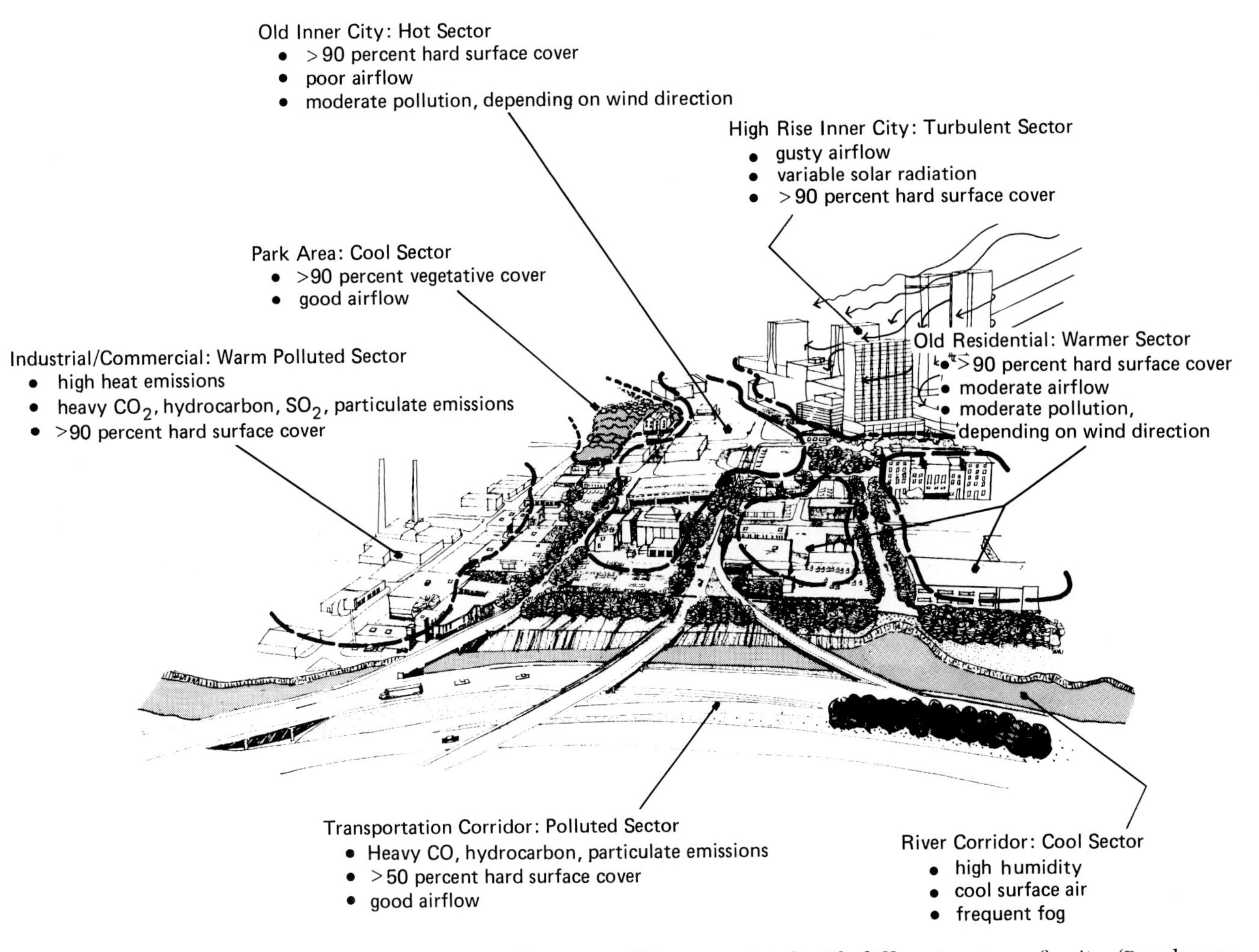

Fig. 3.2 Microclimatic conditions associated with different sectors of a city. (Based on an original drawing by Mary-Margaret Jones, Johnson, Johnson and Roy, Inc.)

Solar Radiation Beam radiation is intercepted by buildings, and, depending on sun angle and building height, a shadow of some size is created. Where buildings are closely spaced, a shadow corridor may form (see Fig. 2.10 in Chapter 2). For a single building site, incoming solar radiation varies with season, wall orientation, time of day, and the location and size of neighboring buildings. Pockets sheltered from beam radiation are illuminated by diffused and reflected radiation only, amounting to very small solar gains. (See Chapter 2 for more details on solar radiation in urban settings.)

Temperature Most cities are geographically diverse in surface materials, physical forms, and activities, and one would expect settings as different as people parks and industrial parks to develop markedly different temperature regimes. Studies show, however, that this is so only where thermal variations are not masked by strong regional weather systems or extreme local influences on climate. The latter is exemplified by a small park of vegetation in the midst of an inner city; whatever modification in temperature

Contrasting thermal environments within a typical urban landscape. The combination of full sunlight and dry, hard surface materials can result in higher ground level temperatures in built-up areas than in nearby parks.

is achieved by the park is masked by the thermal umbrella of the surrounding mass of buildings.

Thermal modification of the urban heat island by a large park or greenbelt can be significant, however. For example, in a ninety-acre Montreal park, a set of daytime temperature readings in summer showed the park interior to be 2°C cooler than the built-up area immediately surrounding it (Fig. 3.3). Other investigations have shown older residential areas with mature trees to be cooler than new residential areas and other urban surfaces. In Washington, D.C., it is not uncommon for the corridor of parks and water along the Potomac River Valley to be cooler during summer days and evenings than the heavily built-up areas on either side of it.

On the perimeter of the city the urban heat island may decline sharply where the urban landscape quickly gives way to the rural landscape. Pictured in a temperature profile, this sort of border is characterized by a "cliff" in the graph line. From a planimetric perspective, the border configuration appears to be very irregular in detail with cool inliers represented by parks and river corridors and warm outliers represented by large shopping centers and industrial parks (Fig. 3.4).

Wind and Convective Mixing The general influence of a city on airflow is to reduce wind speed at levels near the ground. This can be illustrated by comparing the profiles of wind speed over urban and rural surfaces. The elevated topography of the urban environment displaces the profile upward, leaving a thicker layer of slow-moving air near the ground (Fig. 3.5). At a more detailed level of observation, however, large variations in wind speeds can be found within relatively small areas. Much of this

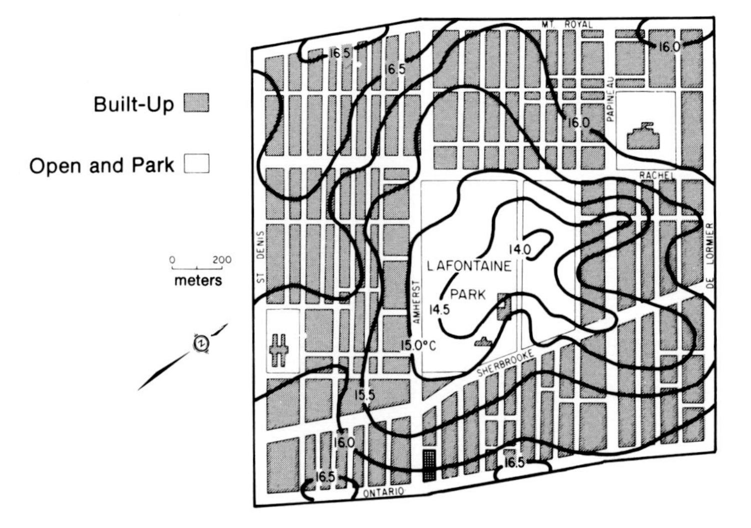

Fig. 3.3 A patch of cooler air associated with a large park in the midst of the Montreal heat island. The lower temperatures are related mainly to the vegetation and the change is produced in latent heat flux and volumetric heat capacity. (From T. R. Oke, "Evapotranspiration in Urban Areas and Its Implication for Urban Climate Planning," in *Teaching the Teachers on Building Climatology*, Stockholm: Swedish National Institute for Building Research, 1972, vol. 3. Used by permission of the author.)

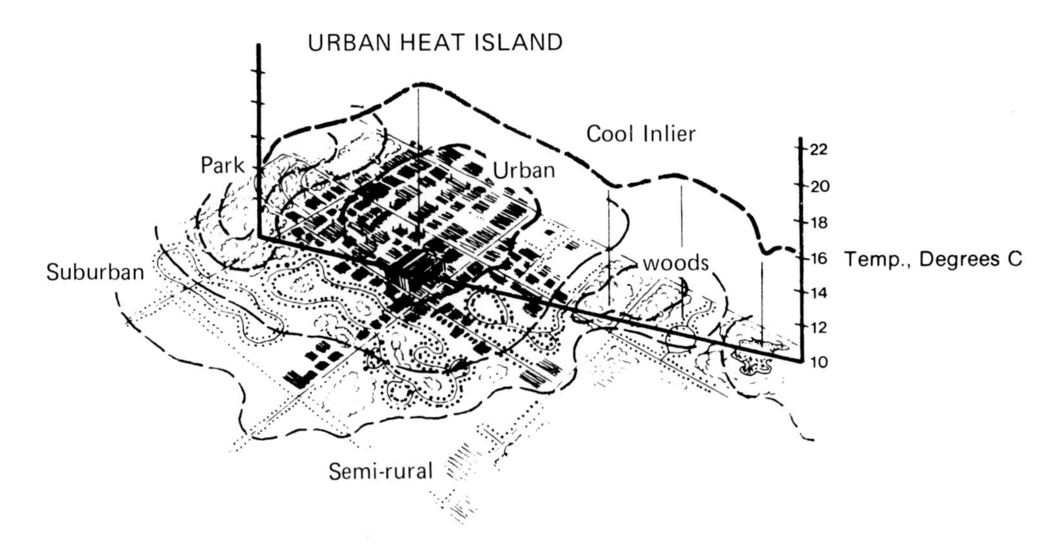

Fig. 3.4 A schematic diagram depicting the nature of the urban heat island and its boundary on the urban fringe.

variation is related to the size, spacing, and arrangement of buildings. Three examples are noteworthy. First, in the case of an individual building, the structure represents an obstacle to airflow and, in order to satisfy the continuity of flow principle, wind must speed up as it crosses the building. In a two-dimensional model the highest speeds are reached on the windward brow of the building and across the roof. Air is also deflected from the brow down the face of the building; on the leeward side, speeds decline and streamlines of wind spread out with some descending to the ground (labeled 'A' in Fig. 3.6).

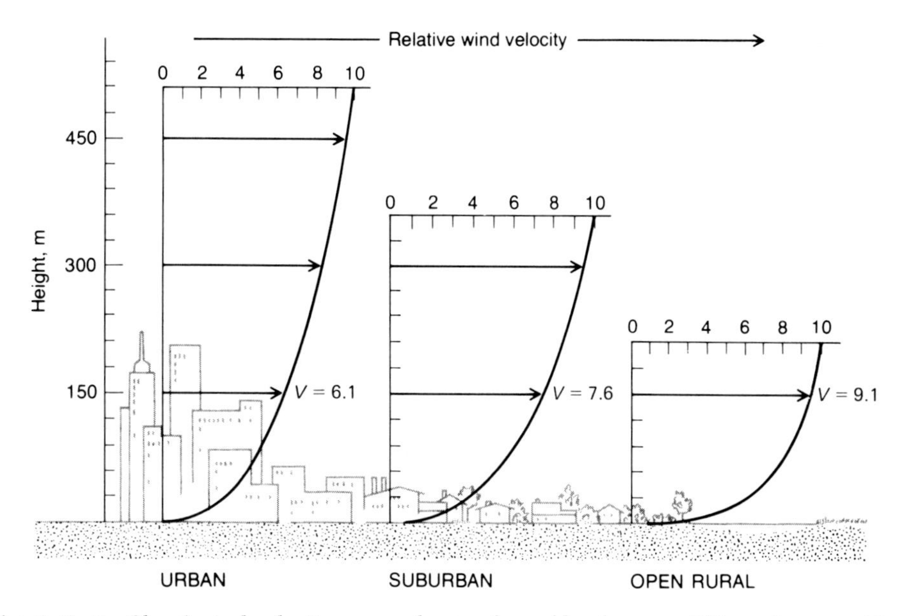

Fig. 3.5 Profile of wind velocity over urban and rural landscapes. Although ground level wind velocities are markedly lower in cities, turbulence tends to be higher because of tall buildings.

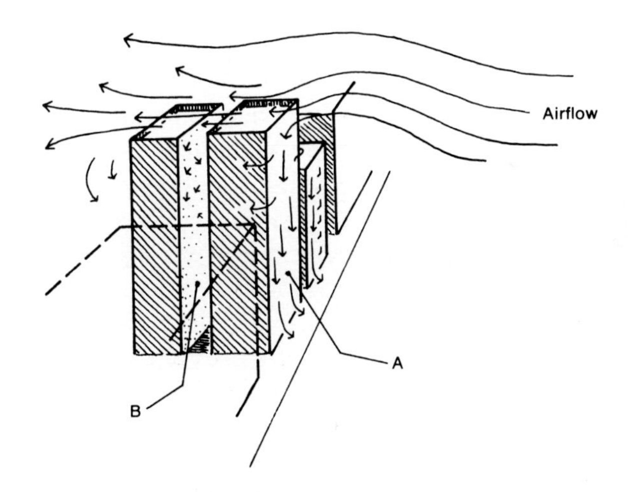

Fig. 3.6 Airflow over buildings. Highest velocities are reached on the windward brow and across the roof of the tallest building. A strong flow of air is also deflected down the building face (A) but a calm zone develops in the space between the buildings (B).

Where two tall buildings of similar heights are spaced close to each other, the streamlines of fast wind do not descend to the ground but are kept aloft by the roof of the second building. This gives rise to a small pocket of calm air between the buildings where mixing with the larger atmosphere of the city is limited (labeled 'B' in Fig. 3.6). Depending on local conditions, the air in such pockets may be measurably different than the surrounding atmosphere (see Chapter 2).

The third example involves the alignment of buildings and streets. Streets bordered by a continuous mass of tall buildings have the topographic character of canyons and, if aligned in the direction of strong winds, tend to channel and constrict airflow. This produces higher wind velocities at street level, especially during gusts, and increased turbulence along the canyon walls.

Fog The incidence of fog in cities may be twice that of surrounding country landscape. Most of this is usually attributed to the abundance of condensation nuclei from urban air pollution. Local concentrations of fog are common in selected areas, especially under calm atmospheric conditions coupled with strong night-time cooling at the surface. Several contributing factors can be identified besides air pollution, and one is related to the availability of water vapor near the ground. In low-lying coastal areas and in river valleys, the concentration of vapor may be appreciably higher than elsewhere in the city. In addition, cold air drainage into low-lying areas promotes fog development; conversely, heated buildings and hard surfaces may locally reduce fog development because they tend to limit the normal rate of fall in night-time air temperatures.

Air Pollution Although a body of heavily polluted air may blanket an entire urban region under certain atmospheric conditions, pollution levels are on the average higher in the inner city than in surrounding suburban areas (Table 3.1). In addition, pollution levels on many days vary sharply from one quadrant or sector of an urban area to another. Two factors account for this: (1) the site-specific nature of many pollution sources such as power plants, highway corridors, and industrial plants; and (2) short-term changes in the mixing and flushing capacity of the urban boundary layer. During windy and unstable weather, pollutants are mixed into the larger mass of air over the city and flushed away, thereby limiting heavy concentrations, if any, to relatively small zones downwind of discharge points. During calm and stable conditions, however, pollutants tend to build up over source areas, and if these conditions are prolonged, the concentrations coalesce to form a composite mass over the urban region. An intermediate condition might be characterized by a light crosswind which draws plumes of polluted air from discharge points and areas. The behavior of individual pollution

Table 3.1 Air Quality and Location in the Urban Region

Pollutant ($\mu g/m^3$)	*Inner City*	*Suburb*	*Rural*
Suspended particulates	102 (260)	40	21
Sulphur dioxide	65–80 (372)	60	40
Oxidants	125 (na)	na	na
Lead	0.21 1.11	0. 09	0.02
Nitrate ion	1.4 2.4	0. 8	0.4
Sulphate ion	10.0 10.1	5. 3	2.5

$\mu g/m^3$ = micrograms per cubic meter
na = not available
(260) = peak value
From U.S. Council on Environmental Quality, *Environmental Quality: Third Annual Report*, Washington, D.C., 1972.

plumes depends on the thermal structure of the receiving atmosphere, wind direction and speed, and the height of release (Fig. 3.7).

The mass balance of pollutants for a given volume of atmosphere can be estimated based on the total rate of pollutant emission and the rate of removal by airflow. Removal includes both lateral and vertical components; therefore, it is easy to imagine how heavy the buildup of pollution can become during a prolonged thermal inversion when airflow in all directions is negligible. Moreover, stagnation of polluted air increases the prospects for oxidation and photochemical processes involving sulfur dioxide, nitrogen oxides, and hydrocarbons, leading to the formation of sulfuric acid, nitric acid, and noxious gases such as ozone. Under severe episodes of air pollution, the only realistic management option (other than regulating people's activities) is to reduce the rate of emission. In several instances, officials in American cities have actually restricted automobile traffic and industrial activity to avert a health disaster. Such decisions depend not only on the gross level of air pollution, but on the levels of critical pollutants, especially hydrocarbons, oxides of nitrogen, sulfur dioxide, and airborne particles (Table 3.2).

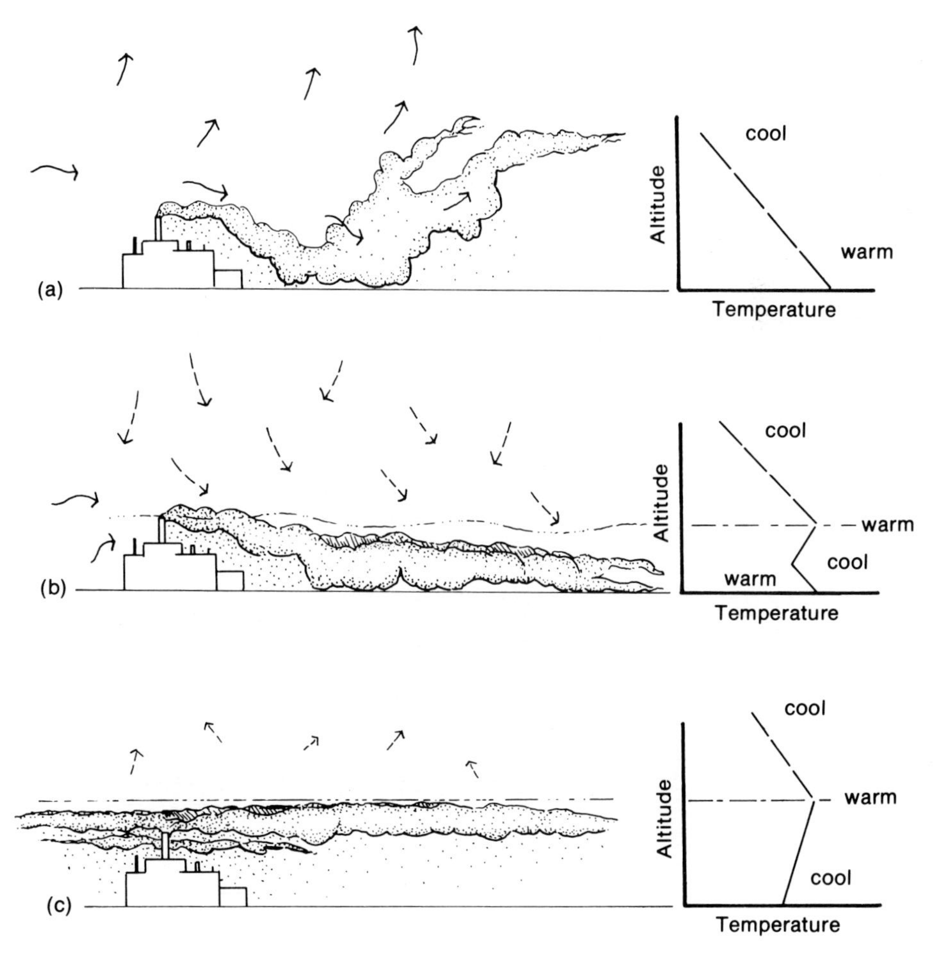

Fig. 3.7 Three basic types of stack exhaust plumes under different atmospheric conditions: (a) looping plume associated with unstable air; (b) fumigation plume associated with descending air; and (c) fanning plume associated with a thermal inversion.

Table 3.2 Major Air Pollutants and Their Sources

Pollutant	*Source*	*Effects*
Carbon monoxide	• gasoline-powered vehicles • industry using oil and gas • building heating using oil and gas	• enters human bloodstream rapidly, causing nervous system dysfunction and death at high concentrations
Sulfur oxides (sulfur dioxide and sulfur trioxide)	• industry using coal and oil • heating using coal and oil • power plants using coal, oil, and gas	• irritates human respiratory tract and complicates cardiovascular disease • damages plants, especially crops • promotes weathering of building skin materials
Nitrogen oxides (nitric oxide and nitrogen dioxide)	• gasoline-powered vehicles • building heating using oil and gas • industry and power plants	• irritates human eyes, nose, and upper respiratory tract • damages plants • triggers development of photochemical smog
Hydrocarbons (compounds of hydrogen and carbon)	• petroleum-powered vehicles • petroleum refineries • general burning	• toxic to humans at high concentrations • promotes photochemical smog
Particulates (liquid or solid particles smaller than 500 micrometers)	• vehicle exhausts • industry • building heating • general burning • spore- and pollen-bearing vegetation	• some are toxic to humans • some pollens and spores cause allergic reactions in humans • promotes precipitation formation

The plumes of polluted air generated from metropolitan areas are known to extend tens, hundreds, and in extreme cases, thousands of miles beyond their source areas (see Fig. 3.1). The effects on regional climate are not well documented, but it is known that they are more pronounced in certain regions and seasons and are characterized by increased cloudiness, precipitation, and turbulent weather. Recently added to this is another regional effect—acid rain in southeastern Canada and northeastern United States. This is caused by the formation of sulfuric acid from the combination of atmospheric moisture and sulfur trioxide in polluted air. Because of the regional flow of weather systems across the Midwest, the acidic moisture is carried from industrial areas and precipitated in Ontario, Quebec, and New England where the biota of thousands of lakes and ponds have been damaged by the increase in water acidity.

APPLICATIONS TO URBAN PLANNING

Climate is widely recognized by planners and designers as an important element in urban planning, but few have been able to incorporate climatic variables effectively into the information base for decision making. Several factors account for this, and one of the most important is the level of scientific understanding of microclimate in the built environment. Although architects and engineers understand many of the influences of climate *on* a building, for example, wind stress, solar exposure, and corrosion of skin materials, comparatively little is known about the influence of buildings on climate. In contrast to urban hydrology, for instance, technical planning is able to provide fewer models and less accurate forecasts to guide the urban planner in setting the heights of buildings, the balance between hard surfaces and vegetative surfaces, the widths of streets, and the like.

A second factor is the general lack of planning regulations pertaining to climate. Although it is widely recognized that urban climate affects the health and well being of people, resulting among other things in greatly increased medical costs, few ordinances have been enacted establishing climate performance standards for residential areas. The exceptions to this are in the area of air quality: national regulations on industrial and automotive emissions are aimed at improving the living conditions in cities. In local transportation projects involving federal funds, a transportation master plan is required that takes air quality into consideration. In locales that are subject to severe episodes of air pollution, such as Los Angeles County, local agencies are responsible for regulating the outdoor activity of school children, and, under emergency conditions, reducing automotive and industrial activity. Beyond examples related to air quality, however, little attention is paid by planning agencies to climatic parameters such as temperature, airflow, fog, and radiation. Prospects for change in this state of affairs are not good where a direct relationship to human safety or to capital costs is not apparent; that is, where a direct savings to individuals, companies, agencies, or institutions are not evident. A case in point is the emergence of solar energy-oriented communities where ordinances on "solar rights" are beginning to appear because access to the sun can be given some economic value.

Climate is invariably addressed in environmental impact statements, but it usually consists of a description of existing conditions with some "forecasts" about potential changes given a proposed action. Only cases involving air quality changes are a source of serious concern and may be the basis for recommending against or altering a proposed action. As for changes in physical components of climate, no guidelines or performance standards have been established to aid planners in formulating plans and reviews. As a result, in most cases no one is quite sure how much importance should be ascribed to a suspected change in some aspect of physical climate, and thus the issue is often relegated to the bin of unused information.

CLIMATIC CRITERIA FOR URBAN PLANNING AND DESIGN

Modern cities in most countries appear to be undergoing almost constant physical change mainly in response to economic and political forces. This demands that the urban planner and designer be on constant watch for opportunities to improve pedestrian movement, commuter traffic, land use, air quality, wastewater disposal, and so on. The overriding challenge is to achieve a proper balance between the economic functions that are necessary to the city's existence and an environment that allows for the health and well-being of the populus. Planning to improve climate and the quality of the atmosphere is a good example of this challenge, judging from the struggle in the United States in the past decade over whether to relax pollution emission standards in order to improve the industrial economy.

Five climatic factors influence the comfort and health of people in most urban environments: air temperature, humidity, solar radiation, wind, and air pollution. While little can be done in designing cities to combat regional atmospheric conditions, measures can be taken to minimize thermal extremes and high levels of air pollution associated with microclimates within the city. Basically, only four types of climatic controls or changes are possible through urban planning and design, given the goal of improving living conditions in cities prone to excessive heat and air pollution:

1. Reduce solar radiation by shading critical surfaces, e.g., pedestrian walks, waiting areas, and busy streets.
2. Reduce the abundance of concrete and asphalt, and increase the amount of vegetation and open water. This will create higher volumetric heat capacities, greater rates of latent heat flux, and thereby lower air temperatures.
3. Increase airflow at ground level to flush heated and polluted air away from the city.
4. Reduce pollution by decreasing emission rates, improving flushing rates, and locating discharge points to minimize impact on heavily populated sectors.

In applying climatic factors to urban design, it is important to first consider the scale of the problem. For problems of city-wide scope, the location, structure, and layout of streets, building masses, and industrial parks must be weighed against airflow patterns, sources of pollution (such as existing traffic corridors), and the ratio of open to developed space. In inland cities the maintenance of ground level airflow in summer is very important; therefore, street corridors should be wide, aligned with prevailing winds, and kept free of major obstacles.

CASE STUDY

Microclimate and Landscape Design

Carl D. Johnson

One of the primary objectives in residential and urban design is mitigation of the climatic extremes in spaces occupied by humans. In architecture the focus is on the internal climate of buildings, and this is achieved through air conditioning, light control, and so on. In landscape architecture the primary concern is with outdoor spaces, and climatic modification is attained through the use of vegetation, siting of buildings, the use of different ground materials, and topographic features, either as they exist or as they could be constructed.

In the continental mid-latitudes, discomfort from the cold poses a major restriction to the use and enjoyment of outdoor space. Therefore, in the design of modern residential complexes, it is desirable to achieve some modification of microclimate to encourage greater use of patio and yard space. Understandably, the level of modification that can be expected

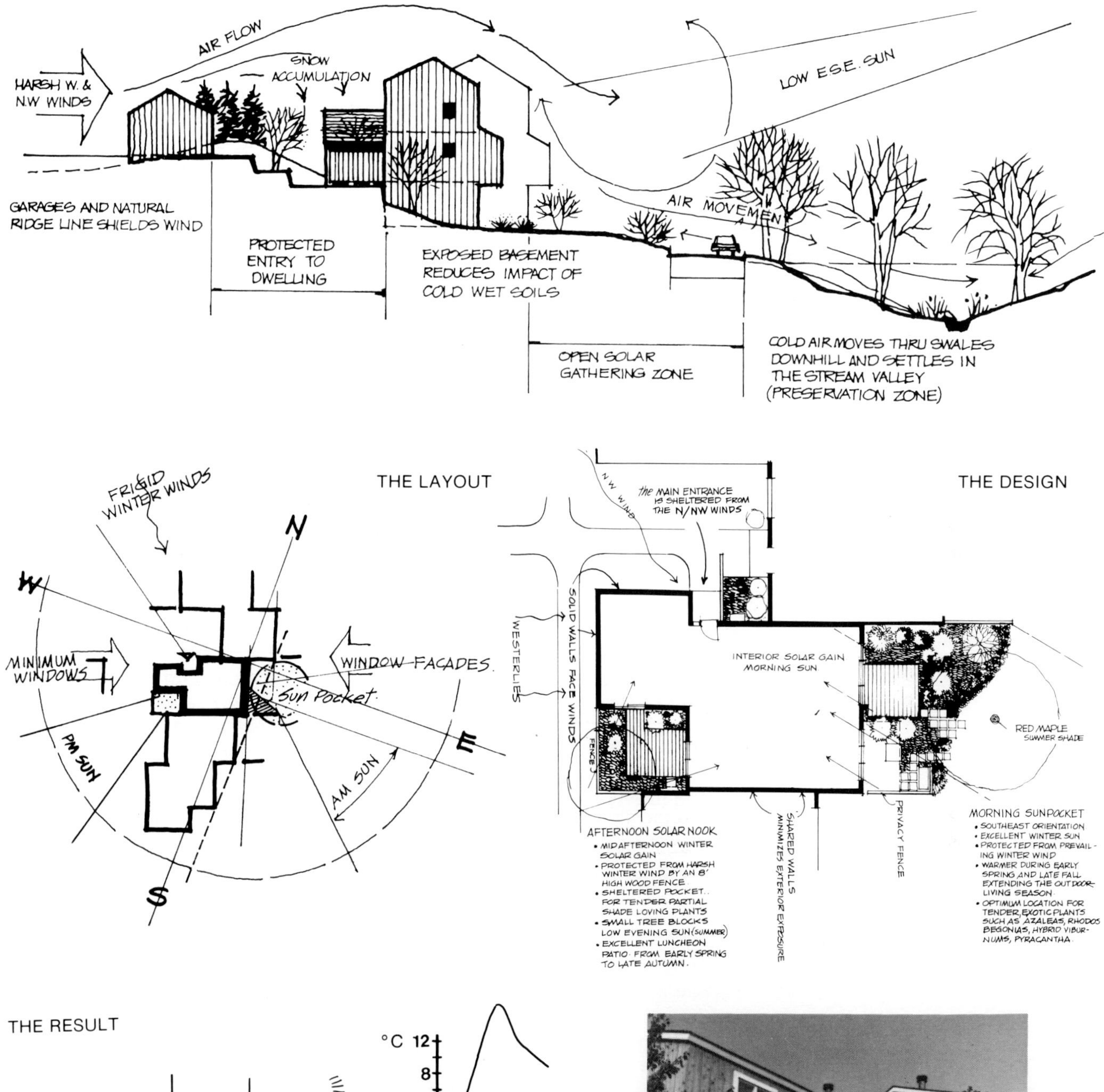

THE RESULT

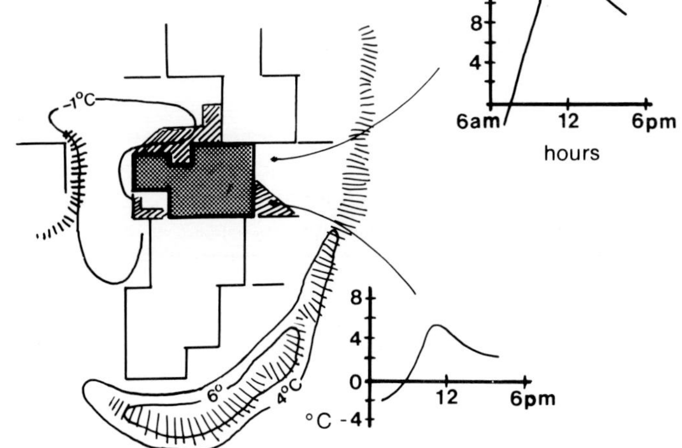

Air temperature near the ground on a sunny March day

(cont.)

CASE STUDY (cont.)

through landscape planning is relatively modest, particularly when set into a Minnesota or Quebec winter. On the other hand, small modifications of marginally cold or cool weather, such as that of spring and fall, are indeed possible, and days that would otherwise be uncomfortably cool can in fact be made to be quite pleasant through sensitive planning and design.

Newport West is a low density townhouse development located on the northeast edge of Ann Arbor, Michigan. The development site is characterized by south- and east-facing slopes leading down to a swale along one border. The northwest slope of the swale, an old farmfield fringed by trees, was selected for the building site because it offered the greatest opportunity to optimize microclimate conditions and conserve building energy.

The housing units were arranged in a series of clusters to form solar pockets and provide protective buffers from the cold, windy northern exposures. The sun pockets were designed to provide comfortable outdoor spaces in fall and spring, and to afford habitats for exotic plants such as azaleas and rhododendrons. The townhouses were constructed with large south-facing windows to receive solar radiation and augment interior heating in fall and winter. On the southwest sides of the units, facades were protected from excessive heating by afternoon sun with full crown deciduous trees.

After construction and landscaping were completed and the units were occupied, a set of temperature readings were taken in early March to determine the effectiveness of the design in the modification of microclimate. Ground temperature varied substantially, depending on solar exposure, and were 7°C higher at the 5 cm depth on inclines near south-facing walls compared to surfaces near north- and northwest-facing walls. In the patio spaces, air temperatures varied with shade and beam radiation receipt; in the southeast-facing sun pocket the daily high temperature (at surface level) on one bright day was nearly 10°C higher than in permanently shaded areas nearby. On cloudy days and windy days the difference was negligible, however.

An examination of the climatic records shows that in the Midwest a total of ten to twenty days in fall and spring can be classed as calm and sunny with uncomfortably cool ambient air temperatures. This brief study suggests that these sorts of conditions can be improved in near-building spaces through climate-oriented building design and siting. It also suggests that conditions for exotic plants are more favorable in sun pockets, although it should also be recognized that summer heat may be excessive and protective shading with deciduous trees may be necessary. The study also implies that residential units that offer both warm and cool outdoor exposures are preferable to those with single exposures in the warm/cold climates because they increase the opportunities for the seasonal use of outdoor space. This is growing increasingly important as people are faced with smaller residences situated in community or neighborhood clusters with limited yard space.

Carl D. Johnson is a principal in Johnson, Johnson and Roy, a planning and landscape design firm, and a professor of landscape architecture at the University of Michigan.

Where vacant land is available in and around built-up areas, it should be converted to vegetation and, to as great an extent as possible, vegetation should be expanded into inner city areas along streets, in pocket parks, and on rooftops. Where pollution is a problem, seasonal patterns of airflow and weather should be considered in locating industry, power plants, and the like. In the mid-latitudes, winter weather often produces plume patterns that direct polluted air toward the ground; where this is known to occur, the location of polluting activities should be situated to minimize impact to residential areas.

At the scale of individual blocks of buildings, attention must be given to orientation with respect to airflow and solar radiation and to building sizes and forms. Building heights must be taken into account; to minimize the nuisance and danger of gusts to pedestrians, studies have shown that the taller of two adjacent buildings should not exceed the shorter by more than twofold.

In considering the vertical dimension of urban climate, it is important to ask what level (elevation) is most appropriate for different human activities? Clearly, ground level in the inner city has several distinct drawbacks, including severe heat, pollution, as well as competition with automotive traffic. Similarly, high elevations pose the hazard of high speed winds that can damage structures and impair human safety. At the middle level, however, in the four- to ten-story range, air is generally cleaner than that at ground level but substantially less windy than that at higher elevations, providing a somewhat healthier and more comfortable climate. Therefore, where heat and pollution at ground level are a problem it seems that rooftop spaces and balconies in this zone offer promise for expanded human use if appropriate landscaping could be introduced to ensure shade and safety. This concept is similar to the model of bioclimatic zonation in large, tropical forests where the middle level of the canopy is the optimum climatic zone for many creatures, the top being too windy and hot and the floor too humid and shaded.

Rooftop storage of stormwater may also be desirable in modifying the urban climate.. Upon evaporation, large quantities of sensible heat are taken up and released with the water vapor, thereby cooling the roof surface and the air over it. In Texas, for example, as much as eighty inches of water can be evaporated from open water surfaces in the average year, twice the average annual precipitation for most of the state. Therefore, it is easily possible to dissipate the total annual quantity of stormwater if the water could be held on rooftops.

At the street scale, consideration must be given to the potential for thermal stress on people in waiting and walking spaces. This is especially critical in cities that record official temperatures above 90°F on many days per month in summer, because in thermal microclimates such readings translate into temperatures above 100°F (Fig. 3.8). In addition to high temperatures, intensive solar radiation, poor airflow, high humidity, and physical exertion also contribute to heat syndrome. Thus along pedestrian corridors with high solar exposures, poor air circulation, and long walking distances, the potential is great for heat syndrome among walkers. To avoid this, shaded rest stops with good ventilation should be provided at appropriate locations. The distribution and location of stops should be based on origin and destination patterns for different walkers, for example, elderly, disabled, and youth.

Finally, it is necessary to evaluate the performance of completed urban design projects based on microclimatic factors related to health (heat and pollution), safety (e.g., wind), energy costs, and aesthetics. With respect to heat syndrome, performance standards can be based on bioclimatic criteria

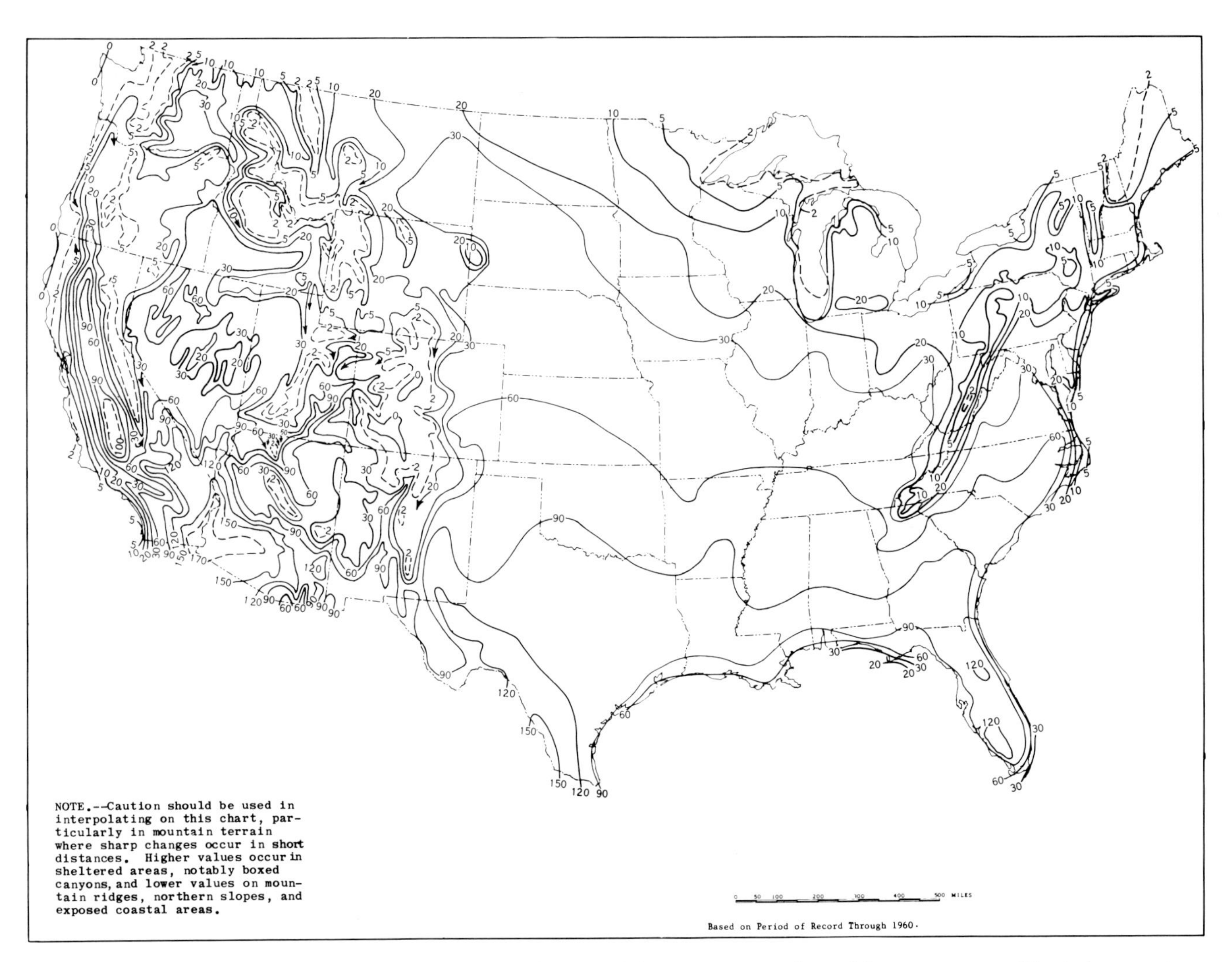

Fig. 3.8 The distribution of the average number of days per year with air temperature reaching 90°F or more. (From *U.S. Climatic Atlas*, Washington, D.C., GPO).

such as those in Olgyay's bioclimatic chart, and would require field measurements of humidity, wind, temperature, and solar radiation in various types of spaces occupied by people (Fig. 3.9). Following design evaluation, modifications should be made to improve performance.

PROBLEM SET

I. Consult an atlas, a recent source of United States population data, and the map in Fig. 3.8.

1. Identify those cities of 500,000 people or more that can expect ninety or more days per year with a high temperature of 90°F or more.
2. Name several factors in the urban environment that will drive this temperature up by 5°F, 10°F, or even 15°F.

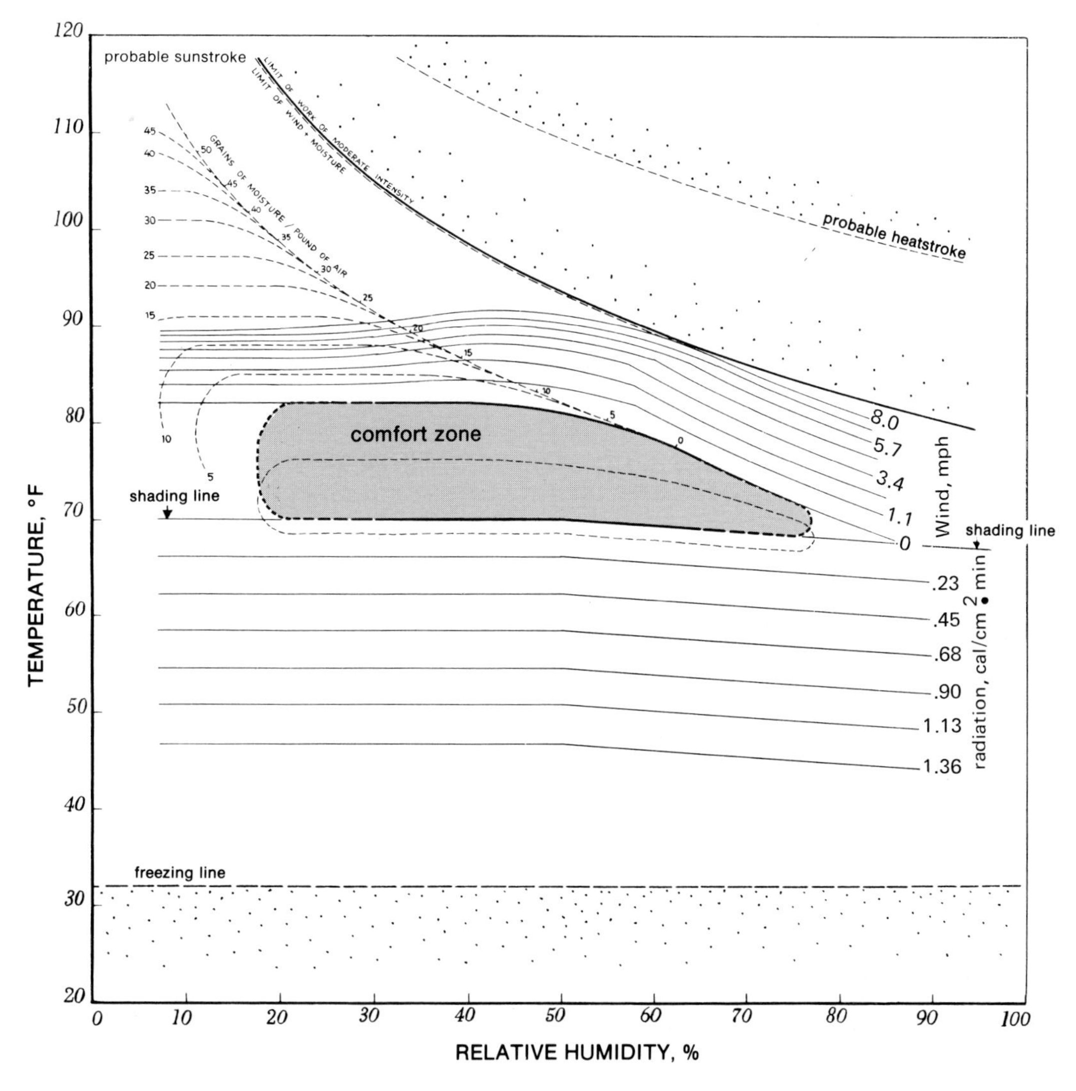

Fig. 3.9 Climatic comfort chart that can be used to test urban environments for their suitability for human habitation. (From Victor Olgyay with Aladar Olgyay, *Design with Climate: Bioclimatic Approach to Architectural Regionalism.* Copyright © 1963 by Princeton University Press. Fig. 45, p. 22, adapted by permission of Princeton University Press.)

3. Describe the kind of urban residential setting that is most susceptible to very high temperatures and the threat of heat syndrome.

II. The following data represent air temperature readings at ground-level taken along a transect through a city.

Distance	*Temperature*	*Site*
0	22°C	residential
175 m	22°C	residential
725 m	26°C	shopping center
1200 m	24°C	old residential
1650 m	28°C	inner city
1900 m	23°C	park/river corridor
2450 m	27°C	industrial/commercial
3100 m	21°C	vacant

1. Construct a graph showing the change in air temperature with distance along the transect.
2. At each of the peaks and troughs in the profile identify the type of land use and list the factors that are likely to contribute to each condition.

III. Construct a wind rose (direction and frequency diagram) for your city or town and identify where within the city a new coal-burning power plant should be located to have least effect on residential areas. Also give consideration to seasonal weather conditions, such as the passage of cyclonic storms, as they relate to plume behavior.

SELECTED REFERENCES FOR FURTHER READING

American Society of Landscape Architects Foundation. *Landscape Planning for Energy Conservation.* Reston, Va.: Environmental Design Graphics, 1977, 224 pp.

Berry, Brian J. L., and Horton, F. E. *Urban Environmental Management: Planning for Pollution Control.* Englewood Cliffs, N.J.: Prentice-Hall, 1974, 425 pp.

Chandler, T. J. *The Climate of London.* London: Hutchinson and Co., 1965, 292 pp.

Ellis, F. P. "Mortality from Heat Illness and Heat-aggravated Illness in the United States." *Environmental Research*, vol. 5, no. 1, 1972, pp. 1–58.

Federer, C. A. "Trees Modify the Urban Microclimate." *Journal of Arboculture*, vol. 2, 1976, pp. 121–127.

Landsberg, H. E. "The Climate of Towns." In *Man's Role in Changing the Face of the Earth.* Chicago: University of Chicago Press, 1956, pp. 584–606.

Marsh, William M., and Dozier, Jeff. "The Influence of Urbanization on the Energy Balance." In *Landscape: Introduction to Physical Geography.* Reading, Mass.: Addison-Wesley, 1981, 636 pp.

Oke, T. R. "Inadvertent Modification of the City Atmosphere and the Prospects for Planned Urban Climates." *Symposium on Meteorology as Related to Urban and Regional Land Use Planning.* Asheville, N.C.: World Meteorological Organization, 1975.

Oke, T. R. *Boundary Layer Climates.* New York: Halsted Press, 1978, 372 pp.

Olgyay, Victor. *Design With Climate.* 4th ed. Princeton, N.J.: Princeton University Press, 1973, 190 pp.

4

SEASONAL GROUND FROST, PERMAFROST, AND LAND DEVELOPMENT

INTRODUCTION

Practically everywhere in the landscape we can see the direct or indirect influences of ground heat. The germination of many seeds depends on ground temperature. Evaporation of soil moisture is also influenced by soil heat. Permafrost, which occupies 25 to 30 percent of the land area of this planet, is a form of ground frost that can place severe stress on most modern land uses. Though less serious than permafrost, seasonal ground frost can be an important consideration in the planning and engineering of facilities in the mid-latitudes. In particular, water pipes and sewer lines must be laid below the frost line, and building foundations and roadbeds must be designed to minimize disruption and damage from frost.

DAILY AND SEASONAL VARIATIONS IN SOIL HEAT

Soil heat does not exist in a static state in the ground, but is almost constantly changing in response to changes in heat at the surface. When the surface is relatively warm and the soil cool, then heat flows into the soil. When the soil is warmer than the atmosphere, heat flows out of the soil. Hardly ever are the soil and atmosphere at the same temperature, because the atmosphere is subject to such rapid and large temperature changes. The soil, on the other hand, is slow to change temperature, especially at depth.

We can examine soil heat flow in various time frames, beginning with a day/night period. On a summer day, for example, the soil surface may heat to a temperature of 35° to 45°C, while just 20 cm or so below, the ground temperature is only 20° to 25°C. The heat flow is downward, of course, but because soil is not a good conductor, it does not reach far into ground before the sun sets and the surface heat is lost. Later in the night the surface cools to a temperature even lower than that of the underlying soil, and the heat flow reverses as the heat gained during the day flows upward. The maximum depth at which the variability of day/night change is negligible is called *diurnal damping depth* (Fig. 4.1).

The ground temperature also varies with the seasons. If we examine the average surface temperatures for summer, fall, winter, and spring, it is apparent that from winter to summer the soil should be heating up, while from summer to winter it should be cooling down. The depth of the seasonal change is much greater than that of the daily change, so the seasonal damping depth is much greater, on the order of 3 m in the mid-latitudes. However, owing to the time it takes heat to reach to a depth of 3 m, the soil does not reach its maximum temperature until a month or more after the surface has. Thus, the heat seasons in the soil do not coincide with the heat seasons on the surface; ground heat and surface heat are always out of phase with each other. In addition, because the soil acts as an insulation layer, it is always cooler in summer and warmer in winter than the surface. These facts have some important implications for building architecture and energy conservation. In regions with hot summers and cold win-

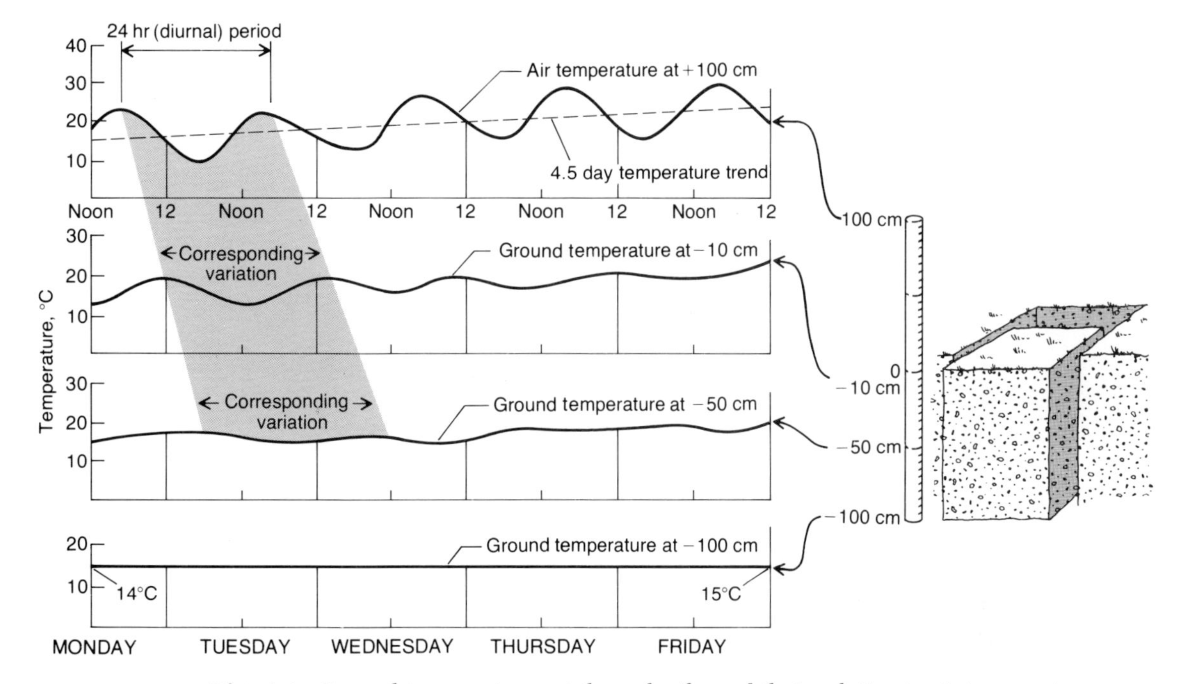

Fig. 4.1 Ground temperatures at three depths and their relation to air temperature over a 4.5-day period. The diurnal damping depth appears to lie close to 50 cm, for beyond that depth the daily variation in surface temperature is not apparent. (From W. M. Marsh and Jeff Dozier, *Landscape: An Introduction to Physical Geography*, © 1981, Addison-Wesley, Reading, Massachusetts. Fig. 3.6. Reprinted with permission.)

ters, subterranean structures have a distinct thermal advantage over above-ground structures. Basements, for example, are cooler in summer, and in winter, a basement is less expensive to heat than a comparable structure above ground (Fig. 4.2).

INFLUENCES ON SOIL HEAT AND GROUND FROST

The rate at which heat flows into and out of the soil depends on two main factors: (1) the temperature differential between the soil at some depth and the surface; and (2) the composition of the soil, which determines its *thermal conductivity* (Table 4.1). In the first column of Table 4.1, the thermal conductivities are given for nine different earth materials. Notice that sand and clay conduct heat better than organic material, and conductivity increases with soil moisture content. Organic matter is a very poor heat conductor; as a result, it often serves as an effective thermal insulator, which helps explain why permafrost is particularly prominent and lasting in areas of muck and peat soils.

The rates at which a given temperature, such as the 0°C line, actually moves into the soil is somewhat different than is suggested by the conductivity value. This rate, called *diffusivity*, is a product of the volumetric heat capacity, given in the second column of Table 4.1, and the conductivity of the soil. Diffusivity is highest at moisture contents between 8 and 20 per-

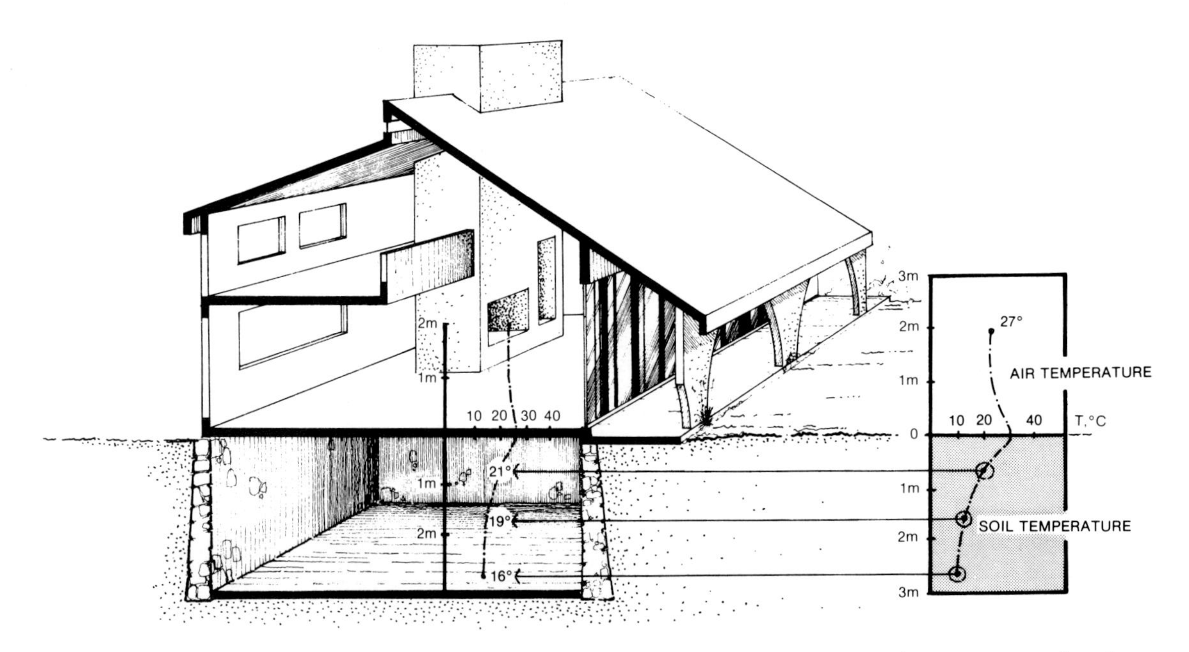

Fig. 4.2 Summer ground temperatures at depths of 1 m, 2 m, and 3 m, compared to those at the same depths in the basement of a house.

Table 4.1 Thermal Properties of Some Common Earth Materials

Substance	*Thermal Conductivity*[1]	*Volumetric Heat Capacity*[2]
Air		
Still (at 10°C)	0.025	0.0012
Turbulent	3,500–35,000	0.0012
Water		
Still (at 4°C)	0.60	4.18
Stirred	350.00 (approx.)	4.18
Ice (at −10°C)	2.24	1.93
Snow (fresh)	0.08	0.21
Sand (quartz)		
Dry	0.25	0.9
15 percent moisture	2.0	1.7
40 percent moisture	2.4	2.7
Clay (nonorganic)		
Dry	0.25	1.1
15 percent moisture	1.3	1.6
40 percent moisture	1.8	3.0
Organic soil		
Dry	0.02	0.2
15 percent moisture	0.04	0.5
40 percent moisture	0.21	2.1
Asphalt	0.8–1.1	1.5
Concrete	0.9–1.3	1.6

[1]Heat flux through a column 1 m² in W/m when the temperature gradient is 1°K per meter.
[2]Millions of joules needed to raise 1 m³ of a substance 1°K.

cent. Thus, in saturated soils, frost penetration is usually not as great as it is in damp soils, owing to the higher heat capacity of wet soil (Fig. 4.3).

Other factors also play a part in ground frost penetration, in particular, vegetation, snow cover, and land use. Snow cover and vegetation tend to reduce soil heat loss in winter, whereas land use has a variable effect; for instance, a building reduces heat flow from the soil, whereas a barren highway usually serves to increase it. The combined effects of land use, vegetation, and snow cover can be dramatic. Where forest cover has been removed for agriculture or urban development, winds are able to blow most snow off the ground, which in turn facilitates soil heat loss. The graphs in Fig. 4.4 illustrate the influence of snow cover for a grass-covered site in Minnesota.

For most parts of North America, the depth of frost penetration is not well documented, and is usually estimated from climatic records. The standard map of frost depth in the coterminous United States is based on representative winter temperatures, and does not take into account factors such as snow cover or vegetation (Fig. 4.5). As a result, field measurements of frost penetration will often show appreciable deviation from this map for any winter. A snow-covered swamp in Maine may receive no ground frost, whereas a nearby airfield may receive 2 m or more.

When snow cover *is* taken into account, the following formula and graph can be used to estimate the depth of frost penetration in northern United States and southern Canada. The formula combines snow depth and heating degree days to give degrees temperature per inch (or centimeter). Taking the result of a computation using this formula, the maximum

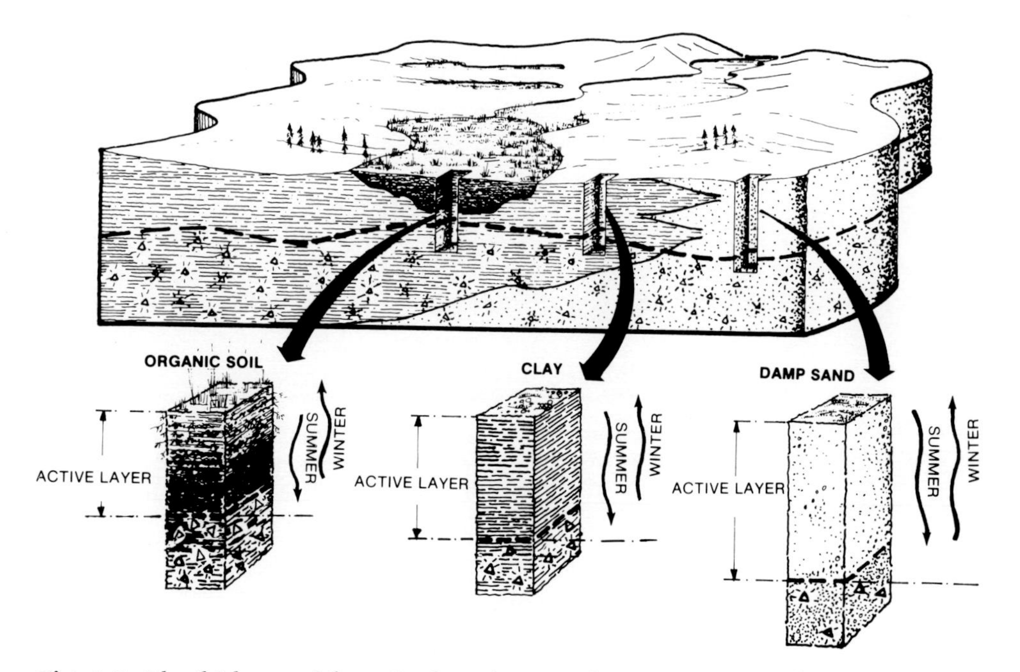

Fig. 4.3 The thickness of the active layer in permafrost regions can reflect the influence of soil composition and moisture content on seasonal heat flow. The directions and depths of seasonal heat flows are shown in the soil sections.

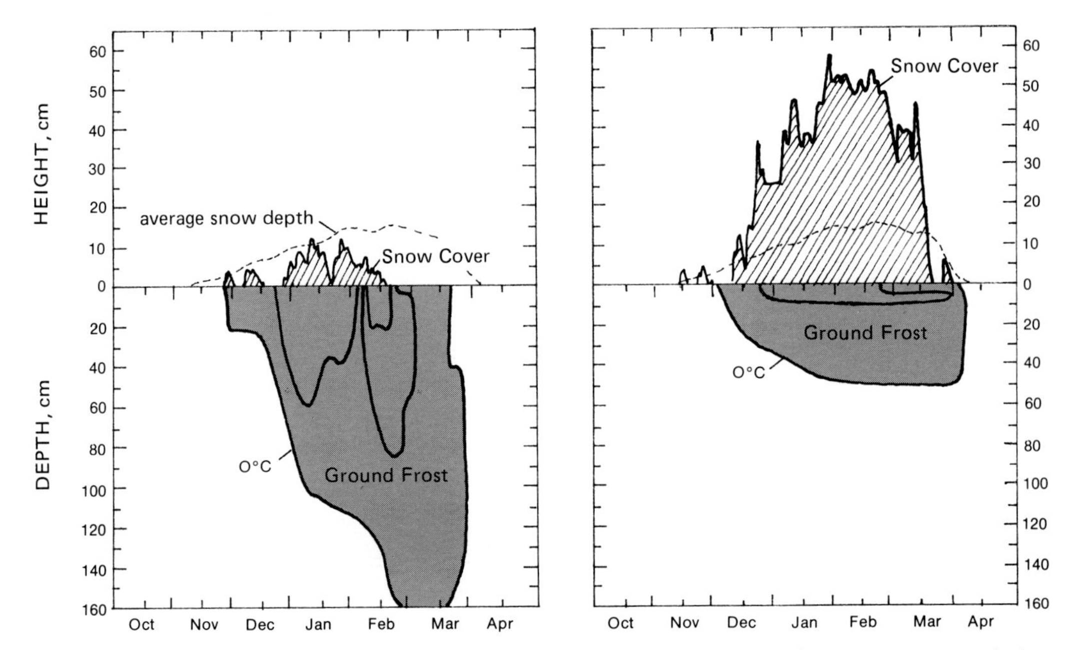

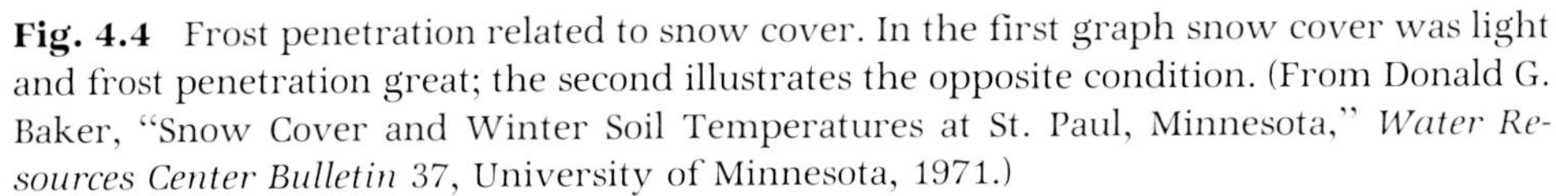
Fig. 4.4 Frost penetration related to snow cover. In the first graph snow cover was light and frost penetration great; the second illustrates the opposite condition. (From Donald G. Baker, "Snow Cover and Winter Soil Temperatures at St. Paul, Minnesota," *Water Resources Center Bulletin* 37, University of Minnesota, 1971.)

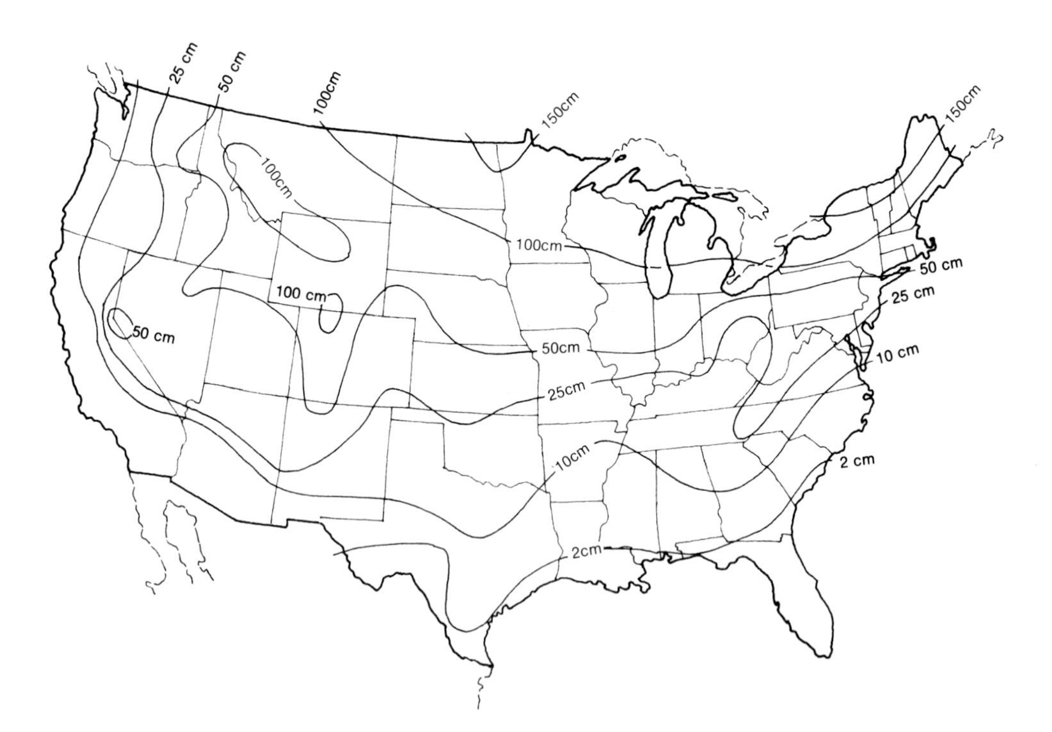

Fig. 4.5 Expected ground frost penetration by the end of winter. Departures from these values may be considerable, depending on the year and local snow, soil, and land use conditions.

depth of the zero-degree isotherm for the winter is read from the graph, in the manner illustrated for 20°F per inch (4.4°C/cm). For this example, the depth of the zero-degree isotherm would be about 105 cm.

$$D = \frac{\Sigma HDD_{\text{OCT-MAR}}}{\Sigma(d_1 \cdot d_2)_{\text{NOV-MAR}}}$$

where:

D = degrees temperature per inch (or cm) (to find the frost depth, this figure is read into the vertical scale of the graph)
HDD = heating degree days, summed Oct. through Mar.
d_1 = average monthly snow depth, inches
d_2 = number of days with snow cover of 0.5 inch or more ($d_1 \cdot d_2$ is computed month by month and summed Nov. through Mar.)

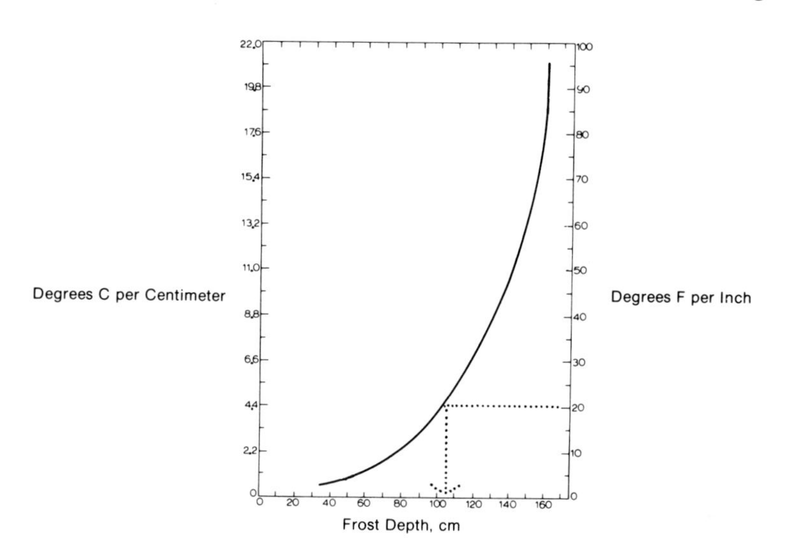

In addition, the orientation and exposure of the ground can be critical; on barren slopes a southward exposure may make an appreciable difference in radiation receipts (see Fig. 2.7a in Chapter 2), whereas north-facing slopes may not only receive less radiation but also lose ground heat more rapidly because of exposure to northerly winds (Fig. 4.6).

SUMMARY OF FACTORS INFLUENCING GROUND FROST DISTRIBUTION

On balance, then, many factors must be taken into account when one attempts to forecast the pattern of ground temperature fluctuations and frost penetration, especially in areas of varied terrain. Unfortunately, mathematical models that integrate many variables are difficult to manipulate and require data from the field in order to be set up. In environmental inventories for impact studies, master planning, or constraint studies, we must instead often turn to simpler and less expensive methods in identifying areas susceptible to heavy frost penetration. One of these is the map overlay method in which individual maps showing the influences of topog-

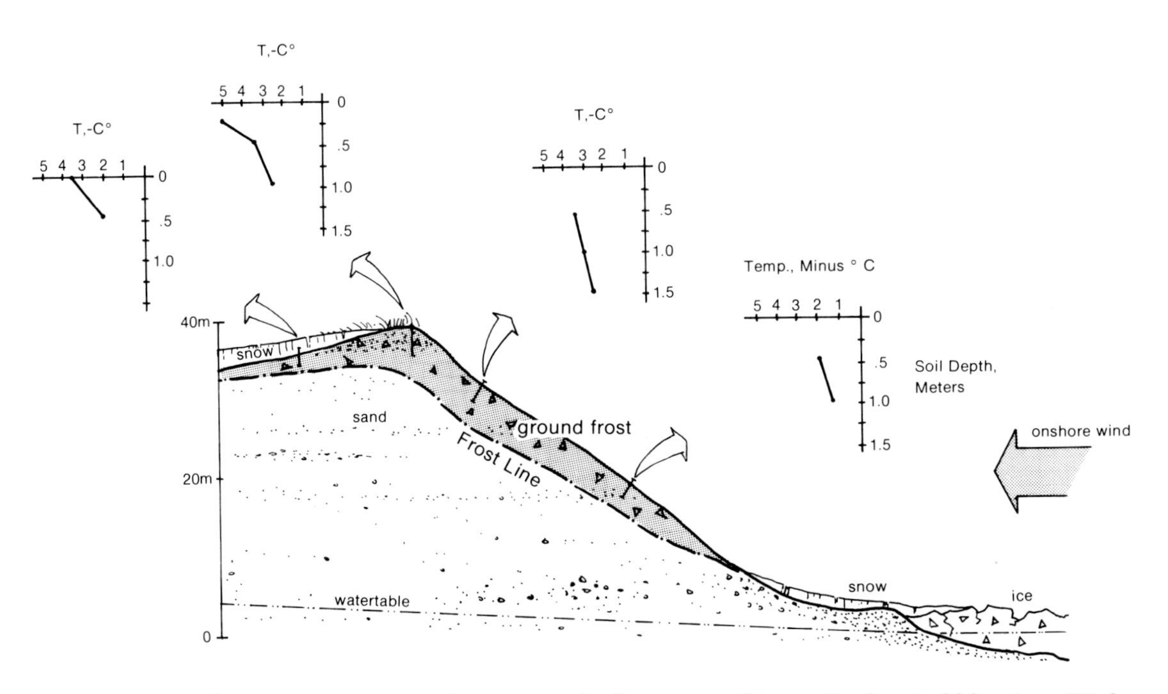

Fig. 4.6 Frost penetration on a sandy slope exposed to northerly wind blowing off Lake Superior. Wind velocity increases upslope, and heat loss increases, with wind velocity producing greatest frost penetration on the upper slope. (Data by William M. Marsh and Mark L. Hassett.)

raphy, vegetation, snow cover, exposure, and soils are superimposed on one another and the resultant combinations are designated high, medium, or low susceptibility. Usually each of the maps is broken down into categories that are numerically coded prior to overlaying; for example, soils might be classed as moist organic (1), moist mineral (2), well-drained mineral (3), three being most susceptible to seasonal frost penetration. The results of this method do not indicate how deep frost penetration should be; rather, they indicate only the relative penetration, and may be used to isolate areas where more detailed analysis could be carried out (Table 4.2).

PERMAFROST

From the southern border of the United States, the depth of ground frost penetration increases northward to a point in Canada where the inflow of summer heat is inadequate to melt the winter frost completely away. The layer of frozen ground that remains is permafrost. In varied terrain, permafrost first appears in isolated pockets on north-facing slopes where solar heating is weakest. In North America, such pockets of permafrost are reported as far south as 50° north latitude; in Asia they extend to 45° north latitude and beyond, into the Tibetan plateau and neighboring highlands. These areas mark the southern fringe of the *discontinuous zone* of permafrost.

Northward the pockets of permafrost grow much broader and thicker and are overlain by a layer of soil called the *active layer*, which freezes and thaws seasonally. Near the Arctic Circle, the discontinuous zone gives way

Table 4.2 Ground Frost Susceptibility

	Low	*Medium*	*High*
Soil Type	organic	wet mineral (clay, loams, sand)	well-drained mineral (loams, sand)
Soil Moisture	saturated	moist (near field capacity)	damp (less than field capacity)
Vegetative Cover	heavy forest	grass	barren
Wind Exposure	low exposure to fast cold wind (usually S, SW, SE facing slopes)	intermediate exposure (such as E, NE, W facing slopes)	high exposure to cold, fast wind (usually N, NW facing slopes)
Snow Cover	>50 cm (Nov.–Mar.)	10–50 cm (Nov.–Mar.)	<10 cm; (intermittent cover throughout winter)
Solar Exposure	south-facing slope >20%	flat ground or locally irregular terrain	north-facing, shaded

to the *continuous zone* where permafrost extends uninterrupted over vast areas of land and reaches depths as great as 1000 m. The thickness of the active layer also changes with latitude. In the discontinuous zone it is usually several meters thick, but poleward declines to a very thin layer or disappears altogether in the continuous zone (Fig. 4.7).

Nowhere is the seasonal flow of soil heat more apparent than in permafrost regions. In summer, the active layer develops with the penetration of heat from the surface (Fig. 4.8a). With the onset of cold weather in fall, the heat flow reverses in the upper active layer, and frost begins to penetrate the soil from the surface. Since the lower active layer is still thawed, heat flows both upward and downward from this relatively warm zone (Fig. 4.8b). The temperature profile assumes a spoon shape at this time; but in the ensuing months of winter, when the active layer freezes out completely and surface temperatures fall far below 0°C, the thermal gradient is fully reversed from that of summer, and the heat flow is upward. Paradoxically, the permafrost layer is the primary source of heat for the landscape during winter (Fig. 4.8c).

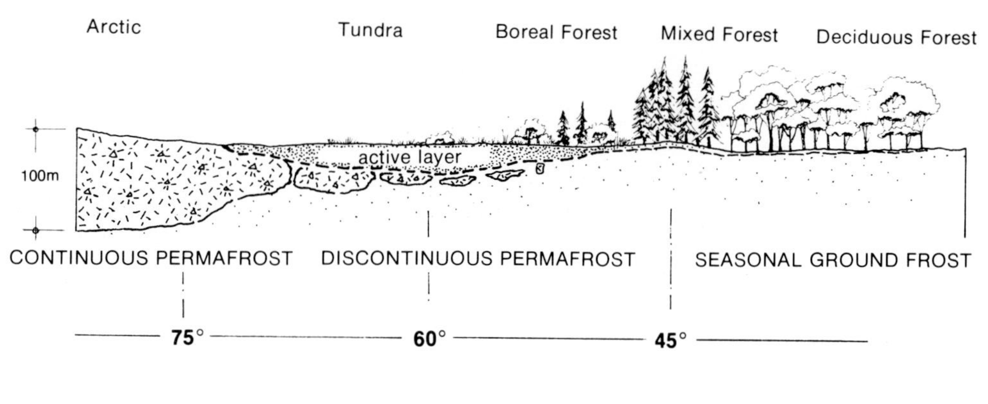

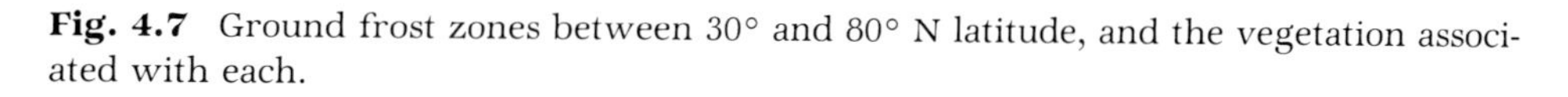

Fig. 4.7 Ground frost zones between 30° and 80° N latitude, and the vegetation associated with each.

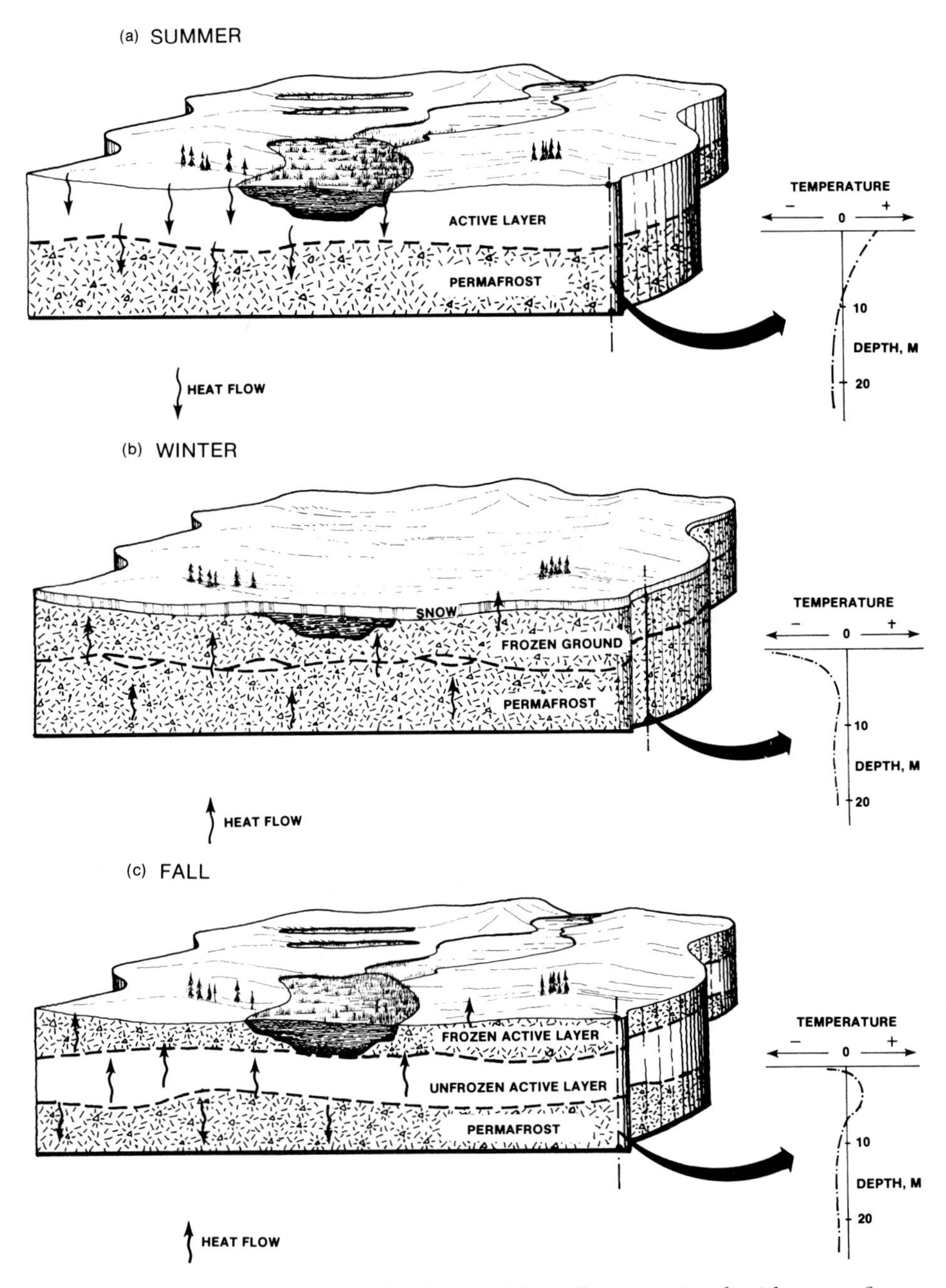

Fig. 4.8 The three seasonal models of ground heat flow associated with permafrost.

LAND USE AND FROZEN GROUND

Modern land uses have proven to be problematic in most permafrost regions. In the past half-century many military installations, railroads, highways, and communities have been built in Alaska, northern Canada, and Russia, and they have provided ample evidence to illustrate the nature of the problem. The energy flow at ground level is altered first with the clearing of vegetation and surface grading. When a foreign material such as concrete or asphalt is placed on the ground, the thermal regime of the active layer is altered further, making it grow colder or warmer. In soils where ice comprises part of the soil bulk (in amounts greater than the volume of interparticle spaces at dry weight), thawing can cause subsiding in the permafrost (because of the volume reduction with change from ice to liquid water) and with it subsiding of the ground surface as well. When heated buildings, utility lines, or oil lines are set on the ground without adequate insulation to check the flow of heat into the active layer, subsiding can be dramatic (Fig. 4.9).

Fig. 4.9 Subsiding caused by melting of the permafrost from the heat generated in this building. The structure was built in 1951 on a concrete basement heated by a furnace. (Photograph by Troy L. Péwé. Used by permission.)

CASE STUDY

Permafrost and the Trans-Alaska Pipeline

Peter J. Williams

Remarkable as it now seems, the earliest proposals in [the Alyeska Oil Pipeline Project] were to construct the pipeline in a conventional manner, burying it in the ground for virtually the entire distance [of 800 miles from Prudhoe Bay to Valdez]. In fact, the most basic problem was that the warm pipeline might thaw the underlying permafrost. Wherever the permafrost contained quantities of ice in addition to that within the soil pores, settlement or subsidence would inevitably follow thawing. At one planning

(cont.)

CASE STUDY (cont.)

Air view of service road and Alaska Pipeline under construction, 1975. (Photograph by T. L. Péwé, April 5, 1975. Used by permission.)

stage serious consideration was given to using a chilled pipeline: cooling the oil so that the pipe could be buried without thawing the permafrost. This proposal was rejected because of the effect of low temperatures on the oil, which would not flow satisfactorily. The interaction of a chilled pipeline with the soil, which was to present problems for subsequently-proposed gas pipelines, was apparently not foreseen.

Probably three-quarters of the proposed route overlay permafrost, and at least half of this was estimated to contain ice, which on melting would cause settlement. The effect of the warm pipeline was such that up to 10 m of soil could be expected to thaw in the first year. The amount would vary, of course, depending on the pre-existing ground temperatures and the type of soil.

Near Prudhoe Bay the permafrost is some 600 m thick. Southward it becomes generally thinner although the thickness is very variable. Thawing around a buried, warm pipeline would progress downward, although at a decreasing rate, for many years. The degree of consequent settlement, or subsidence, would depend on the amount of "excess" ice in the thawed

Vertical support members with thermal devices (on top) to keep footings frozen in the ground. (Photograph by T. L. Péwé, July 17, 1977. Used by permission.)

layer, but quite often there would be several metres displacement. Obviously, such effects left unchecked would cause great disruption of the pipeline. . . .

As soil surveys proceeded much more ground ice was discovered than was initially predicted. This, coupled with findings as to the amount of thaw that would occur, led gradually to the decision to build more and more of the pipeline above ground elevated on pile supports, rather than buried in the ground as first envisaged. The air passing beneath an elevated line would dissipate most of the heat from the pipe and greatly reduce the thawing of the permafrost. Furthermore, the pile supports, or "vertical support members" (known as VSMs) could be designed to permit lateral movements of the pipe as it expanded or contracted with temperature changes.

Raising the pipe above ground on the VSMs did not ensure that there would be *no* thawing of the ground. The disturbance inflicted on the ground during the course of construction was sufficient to initiate temperature changes in the soil which could result in thawing to a signifi-

(cont.)

CASE STUDY (cont.)

cant depth, in all except the coldest, most northern areas. Climatic change too, might in places initiate a continuing thawing.

Ultimately, the solution of the problem lay in the so-called thermal VSM. The thermal VSMs are equipped with devices known as heat pipes. These are sealed 2-inch diameter tubes within the VSMs. They extend below the surface and contain anhydrous ammonia refrigerant. In the winter months this evaporates from the *lower* end of the tube and condenses at the top where there are metallic heat exchanger fins. The evaporation process occurs because, during the winter months, the ground is *warmer* than the air outside. The evaporation process itself cools the lower end of the pipe and the surrounding ground, and this is the point of the device: by cooling the permafrost in winter its temperature is sufficiently lowered to prevent thaw during the summer. The mean ground temperature falls, the heat pipes preventing the warming that would otherwise occur following disturbance of the ground surface. About 610 km of pipe were built above ground, and about 80 per cent of this length had thermal VSMs.

From *Pipelines and Permafrost: Physical Geography and Development in the Circumpolar North* (Longman, 1979) by Peter J. Williams, Professor of Geography and Director of the Geotechnical Science Laboratories at Carleton University, Ottawa, and a specialist in problems of development in cold environments. (Used by permission of the author.)

Other problems experienced by land use in permafrost regions include inadequate drainage in summer, mass movement (such as mudflows and landslides) of surface material, and difficulties in procuring groundwater for water supplies. Faced with these problems, urbanization has been very limited in permafrost regions. In North America, the northern limit of urban development roughly coincides with the southern fringe of the permafrost zone.

Frozen ground is also a problem outside permafrost areas. In northern Europe, southern Canada and the northern United States, as well as in many other areas of the world, ground frost causes highway buckling, damage to building foundations, and freezing and breakage of water pipes (Fig. 4.10). In the United States and Canada, building codes generally recommend foundations to be set below frost depth (usually given as four to five feet in the northern tier of states and in southern Canada) to minimize heaving and damage. In highway construction, frost heaving caused by the growth of ice lenses in the roadbed is a serious problem. To alleviate this, gravel-based roadbeds are required because gravel does not transmit capillary water upward from the underlying soil fast enough to allow ice lenses to form under the cold concrete or asphalt.

PLANNING APPLICATIONS

With the exception of engineering design standards, such as those previously mentioned, little formal attention has been given to ground frost in community planning outside permafrost regions. Within permafrost re-

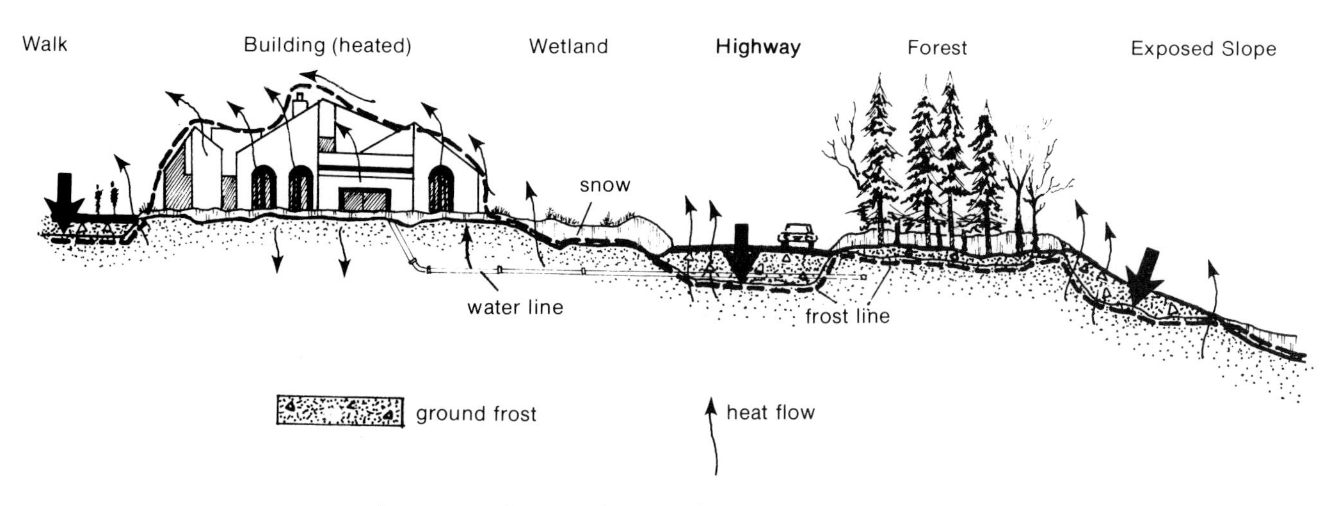

Fig. 4.10 Schematic diagram showing relative frost penetration as a function of soil, vegetation, land use, snow cover, and slope exposure.

gions the picture is quite the opposite, though permafrost is by no means universally recognized and addressed in planning methodology and practice, even in the most hostile settings.

In the Fairbanks, Alaska, planning region, the Soil Conservation Service (SCS) reviews development proposals with an eye to potential permafrost problems. The first level of evaluation involves checking the location of the proposed development against the distribution of soils known to have permafrost problems. Drawing on the results of permafrost research in the Fairbanks area, the SCS has been able to classify the soils of the Goldstream, Saulich, Ester, and Lemeta series as those with greatest susceptibility to permafrost (Fig. 4.11). For projects that would involve these soils, the review may be taken to a second level of evaluation and call for an examination of the types of activities and facilities actually proposed. In some cases the project may be compatible, because it is judged to be neither prone to damage from the environment nor itself of a significant threat to the environment. In other cases, modifications may be recommended, such as changes in building sites or the use of special engineering technology for footings and utility lines (Fig. 4.12); and in still other cases, the project may be viewed as incompatible with the environment and not recommended for approval.

The SCS recommendation is then passed on to the staff of the planning agency in charge, where it is combined with recommendations from other technical fields and interest groups to form a general recommendation on the proposal. This statement is then studied and discussed by a decision-making body such as a planning commission, county board, or city council; a vote is taken; and the proposal is approved, denied, approved with specified reservations, asking the applicant to agree to certain changes, or sent back to the applicant or the reviewer for further work and documentation.

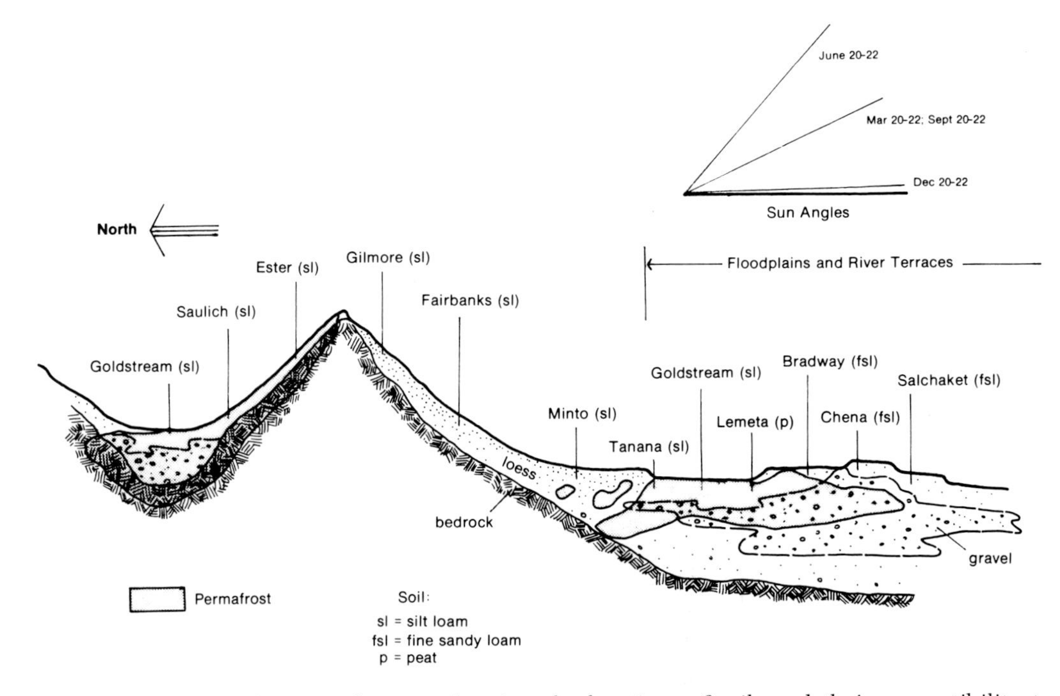

Fig. 4.11 Schematic diagram showing the locations of soils and their susceptibility to permafrost in the Fairbanks, Alaska, region. (From U.S. Soil Conservation Service, *Soil Survey: Fairbanks Area, Alaska*, 1963; after Troy L. Péwé, "Effect of Permafrost on Cultivated Fields, Fairbanks Area, Alaska," in *Mineral Resources of Alaska, Geological Survey Bulletin* 989, 1951–1953.)

Fig. 4.12 Design modifications in residential development in response to permafrost, Norman Wells, Northwest Territory, Canada. Homes, water lines, and sewage lines are built above ground to minimize thermal disturbance of the permafrost. (Photograph by Troy L. Péwé. Used by permission.)

PROBLEM SET

I. The topographic map and cross-section of the tract of land below represent a north-south alignment through an area of irregular terrain located at latitude 50 degrees north. The noon sun angles for December, January, and February range between 22 and 32 degrees and the strongest and coldest winds in winter are from the north and northwest. In the average year, permanent snow cover in this region is established by November 1 and lasts until the end of March, but locally the snow depths vary radically because new snow is redistributed by wind. In protected sites, typical midwinter snow depths range from .4 m to .6 m.

1. Taking wind direction, topography, and vegetation into consideration, estimate the spatial pattern of snow accumulation and map the

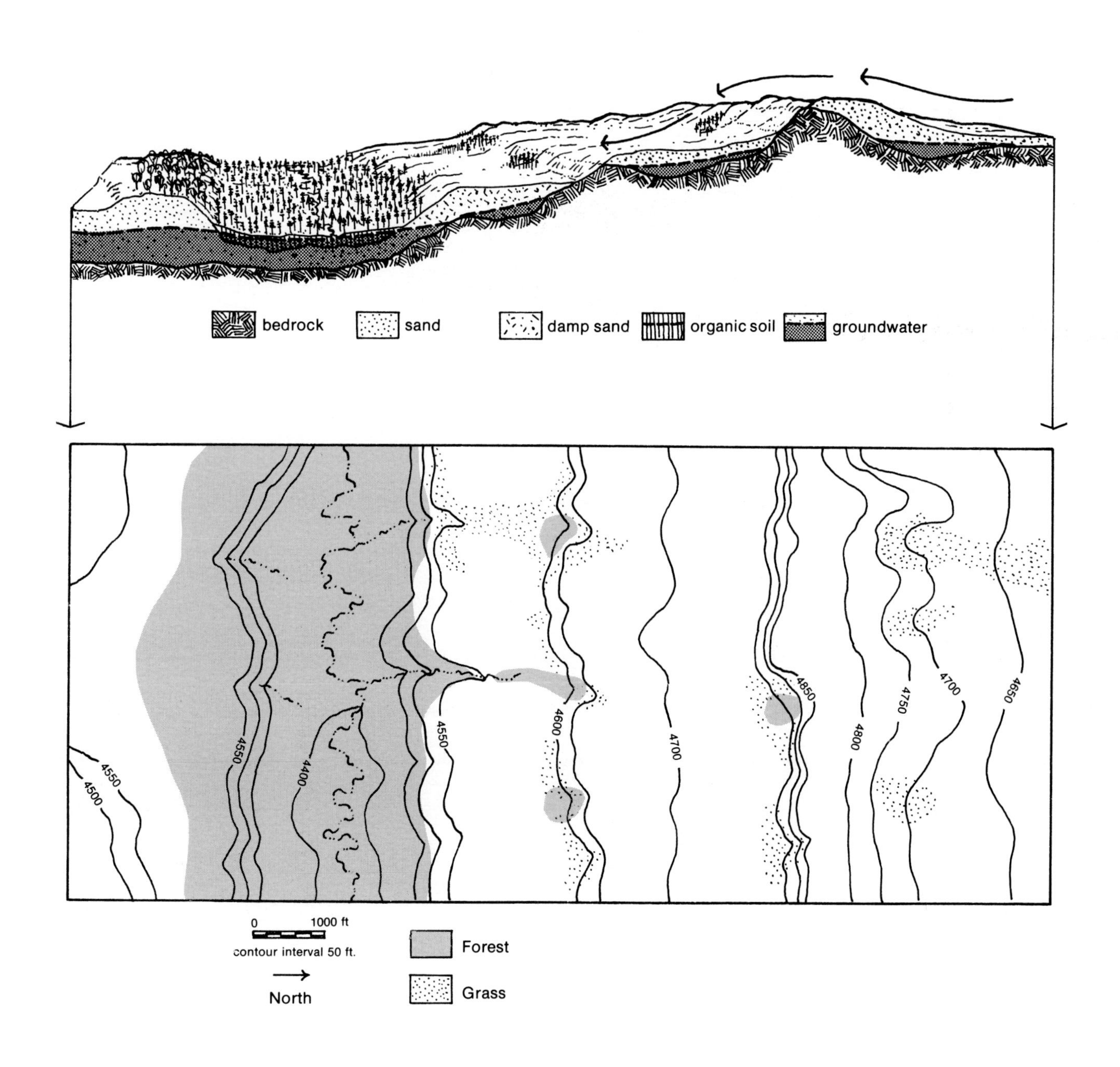

areas where you would expect snow cover to be heavy, intermediate, and light (see Table 4.2 for snow cover definitions).

2. Using the criteria given in Table 4.2, subdivide the area shown in the topographic map according to susceptibility to ground frost. Use four categories: high, intermediate, low, and negligible. (One way of doing this is to assign numerical values to the susceptibility classes given in Table 4.2 (high, intermediate, low) and then build a map for each criteria, that is, soil, solar exposure, etc. When the maps are overlaid, the values can be combined and the totals used to determine the four categories. (See Table 6.2 in Chapter 6 for an example of how this technique is used in soil mapping for on-site wastewater disposal.)
3. Using the formula for a graph approximating the depth of the freezing temperature based on snow cover, determine the freezing depth for sites A, B, and C given the following:

	Site A					*Site B*					*Site C*				
• Heating degree days (Oct.–March)	7100 (°F)(3910°C)					7100 (°F)					7100 (°F)				
	N	D	J	F	M	N	D	J	F	M	N	D	J	F	M
• Average monthly snow depth, in. (Nov.–March)	4	8	16	18	8	0	2	5	7	2	2	8	12	9	3
• Days with measurable snow cover (Nov.–March)	12	18	31	23	12	0	3	9	6	3	4	14	22	18	5

SELECTED REFERENCES FOR FURTHER READING

Allen, L. J. *The Trans-Alaska Pipeline.* Alyeska Pipeline Service, 1977, 2 vols.

Baker, Donald G. "Snow Cover and Winter Soil Temperatures at St. Paul, Minnesota." *Water Resources Research Center Bulletin 37,* University of Minnesota, 1971, 37 pp.

Brown, R. J. E. "Influence of Climate and Terrain on Ground Temperatures In The Continuous Permafrost Zone of Manitoba and Keewatin District, Canada." *Third Conference of Permafrost Proceedings,* Edmonton, vol. 1, 1978, pp. 16–21.

Brown, R. J. E. *Permafrost In Canada,* Toronto: University of Toronto Press, 1970, 234 pp.

Ferrians, O. J., et al. "Permafrost and Related Engineering Problems in Alaska." U.S. Geological Survey Professional Paper 678, 1969, 37 pp.

French, H. M. *The Periglacial Environment.* New York: Longman, 1976, 309 pp.

Péwé, Troy L. "Effect of Permafrost on Cultivated Fields, Fairbanks Area, Alaska." In *Mineral Resources of Alaska, Geological Survey Bulletin* 989, 1951–1953, pp. 315–351.

Smith, M. W. "Microclimatic Influences on Ground Temperatures and Permafrost Distribution in the Mackenzie Delta, Northwest Territories." *Canadian Journal of Earth Science,* vol. 12, no. 8, 1975, pp. 1421–1438.

U.S. Soil Conservation Service. *Soil Survey: Fairbanks Area, Alaska.* Washington, D.C.: U.S. Government Printing Office, 1963, 41 pp.

Washburn, A. L. *Periglacial Processes and Environments.* New York: St. Martin's Press, 1973, 320 pp.

Williams, Peter J. *Pipelines and Permafrost: Physical Geography and Development in the Circumpolar North.* New York: Longman, 1979, 98 pp.

5

SOIL AND DEVELOPMENT SUITABILITY

INTRODUCTION

The relationship between soil composition and agriculture is apparent at practically any scale of observation. The relationship between soil composition and other types of land uses, however, is often not so apparent. At least part of the reason for this in North America is that precious little attention has been given to soils by developers, especially in the last several decades. Among the factors contributing to this state of affairs is a sense of complacency about soils that has been brought on by a popular impression that modern engineering technology can overcome problems of the soil. Granted, the technology does exist to build practically any sort of structure in or on any environment; however, the costs involved in terms of both dollars and environmental damage are often prohibitive. Increasingly, decision-makers are turning to soil surveys to help guide in site selection for residential, industrial, and other forms of development that involve surface and subsurface structures.

SOIL PROPERTIES

Several features, or *properties,* are used to describe soil for problems involving land development. Of these, texture and composition are generally the most meaningful; from them we can make inferences about bearing capacity, internal drainage, erodibility, and slope stability.

Composition refers to the materials that make up a soil. Basically, there are just four compositional constituents: mineral particles, organic matter, water, and air. Mineral particles comprise 50 to 80 percent of the volume of most soils and form the all-important skeletal structure of the soil. This structure, built of particles lodged against each other, enables the soil to support its own mass as well as that of internal matter such as water and the overlying landscape, including buildings. Sand and gravel particles generally provide for the greatest stability, and if packed solidly against one another, will usually yield a relatively high bearing capacity. *Bearing capacity* refers to a soil's resistance to penetration from a weighted object such as a building foundation. Clays tend to be more variable in stability with loosely packed, wet particle masses having a tendency to compress and slip laterally under weight (stress) (Table 5.1).

The quantity of organic matter varies radically in soils, but is an extremely important component inasmuch as its skeletal structure of organic particles is usually weak. Deep organic soils thus pose a serious limitation to land uses involving surface structures; on the other hand, the organic matter in top soil is important in general soil fertility, moisture absorption and retention, and landscaping.

The water content of soil varies with particle sizes, local drainage and topography, and climate. Most water in the soil occupies the spaces between the particles; only in organic soils do the particles themselves actually absorb measurable amounts of water. Two principal forms of water occur in both mineral and organic soils: capillary and gravity. Capillary

Table 5.1 Bearing Capacity Values for Rock and Soil Materials

Class		Material	*Allowable Bearing Value, tons per square foot*
1	rock	Massive crystalline bedrock, e.g., granite, gneiss	100
2	rock	Metamorphosed rock, e.g., schist, slate	40
3	rock	Sedimentary rocks, e.g., shale, sandstone	15
4	soil materials	Well compacted gravels and sands	10
5	soil materials	Compact gravel, sand/gravel mixtures	6
6	soil materials	Loose gravel, compact coarse sand	4
7	soil materials	Loose coarse sand; loose sand/gravel mixtures, compact fine sand, wet coarse sand	3
8	soil materials	Loose fine sand, wet fine sand	2
9	soil materials	Stiff clay (dry)	4
10	soil materials	Medium-stiff clay	2
11	soil materials	Soft clay	1
12	soil materials	Fill, organic material, or silt	(fixed by field tests)

Source: *Code Manual*, New York State Building Code Commission.

water is a form of molecular water, so-called because it is held in the soil by the force of cohesion among water molecules. Under this force, water molecules are mobile and can move from moist spots to dry spots in the soil. In the summer, most capillary water transfer is upward toward the soil surface as water is lost in evaporation and transpiration.

Gravity water is liquid water that moves in response to the gravitational force. Its movement in the soil is preponderantly downward, and it tends to accumulate in the subsoil and underlying bedrock to form groundwater. Groundwater completely fills the interparticle spaces; thus below the watertable the soil is largely devoid of air.

DRAINAGE

References to "drainage" in soil reports usually refer to gravity water and a soil's ability to transfer this water downward. Three terms are used to describe this process: (1) *infiltration capacity*, which is the rate at which water penetrates the soil surface (usually measured in cm or inches per hour); (2) *permeability*, which is the rate at which water within the soil moves through a given volume of material (also measured in cm or inches per hour); and (3) *percolation*, which is the rate at which water in a soil pit or pipe within the soil is taken up by the soil (used mainly in wastewater absorption tests and measured in inches per hour). Reference to poor drainage, for example, means that the soil is frequently or permanently saturated and may often have water standing on it. Terms such as "good drainage" and "well drained" mean that gravity water is readily transmitted by the soil and that the soil is not conducive to prolonged periods of saturation. Soil saturation may be caused by the local accumulation of surface water (because of river flooding or runoff into a low spot, for example), or by a rise in the level of groundwater within the soil column (because of the raising of a reservoir or excessive application of irrigation water, for example), or because the particles in the soil are too small to transmit infiltration water (because of impervious layers within the soil or clayey soil composition).

SOIL TEXTURE

The mineral particles found in soil range enormously in size from microscopic clay particles to large boulders. The most abundant particles, however, are sand, silt, and clay, and they are the focus of examination in studies of soil texture (Fig. 5.1). *Texture* is the term used to describe the composite sizes of particles in a soil sample, say, several representative handfuls. To measure soil texture, the sand, silt, and clay particles are sorted out and weighed. The weights of each size class are then expressed as a percentage of the sample weight.

Since it is unlikely that all the particles in a soil will be clay or sand or silt, additional terms are needed for describing various mixtures. Soil scientists use twelve basic terms for texture, at the center of which is the class *loam,* an intermediate mixture of sand, silt, and clay. According to the United States Soil Conservation Service (SCS), the loam class is comprised of 40 percent sand, 40 percent silt, and 20 percent clay. Given a slightly heavier concentration of sand, say 50 percent, with 10 percent clay and 40 percent silt, the soil is called *sandy loam.* In agronomy, the textural names and related percentages are given in the form of a triangular graph (Fig. 5.2). If you know the percentage by weight of the particle sizes in a sample, you can use this graph to determine the appropriate soil name.

MEASURING SOIL TEXTURE

In the field, soil texture can be estimated by extracting a handful of soil and squeezing it into three basic shapes: (1) *cast,* a lump formed by squeezing a sample in a clenched fist; (2) *thread,* a pencil shape formed by rolling soil

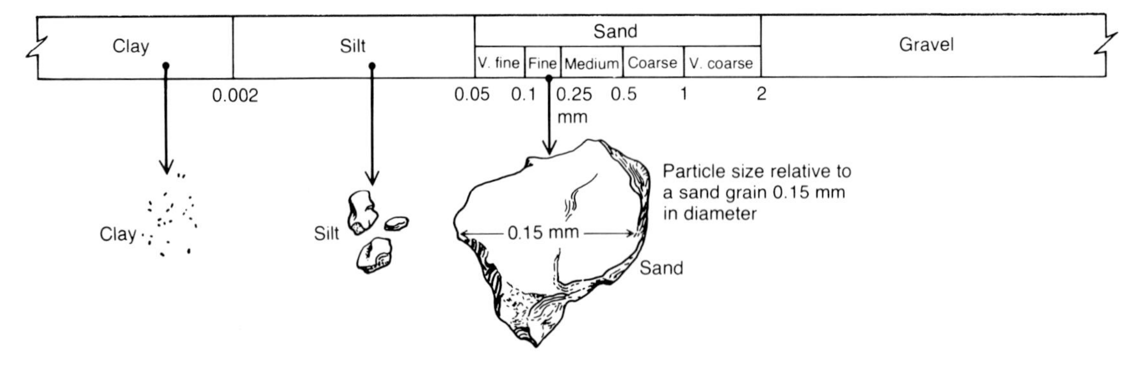

Fig. 5.1 A standard scale of soil particle sizes; the illustrations give a visual comparison of clay and silt with a sand particle 0.15 mm in diameter.

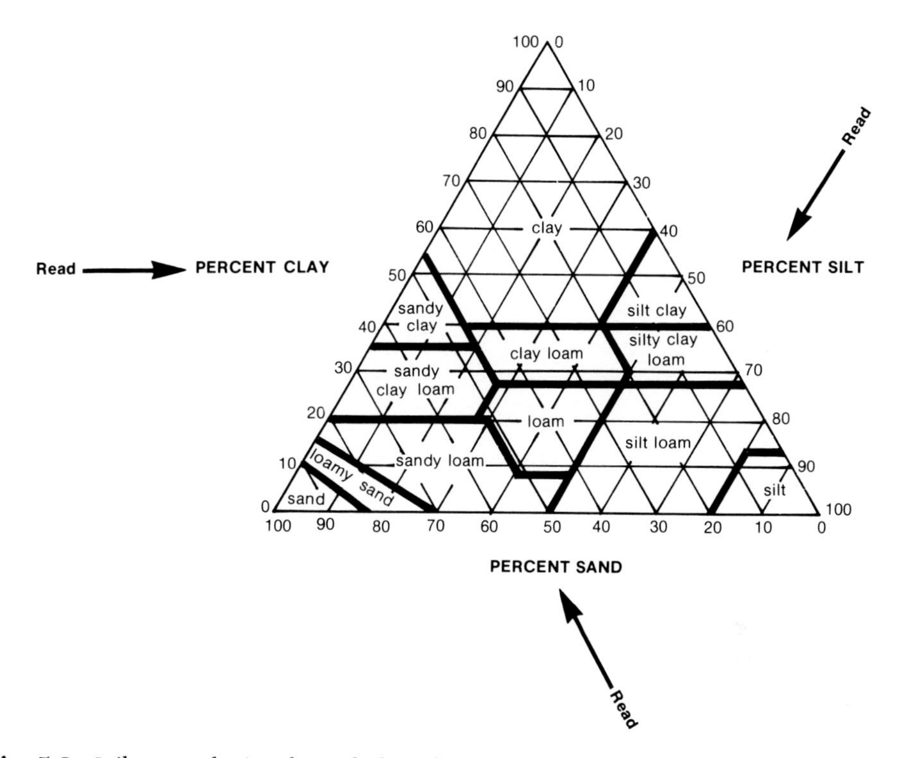

Fig. 5.2 Soil textural triangle, including the major subdivisions, used by the U.S. Department of Agriculture.

between the palms; and (3) *ribbon*, a flattish shape formed by squeezing a small sample between the thumb and index finger (Fig. 5.3). The behavioral characteristics of the soil when molded into each of these shapes, if they can be formed at all, provides the basis for a general textural classification. The sample should be damp to perform the test properly.

Behavior of the soil in the hand test is determined by the amount of clay in the sample. Clay particles are highly cohesive and, when damp-

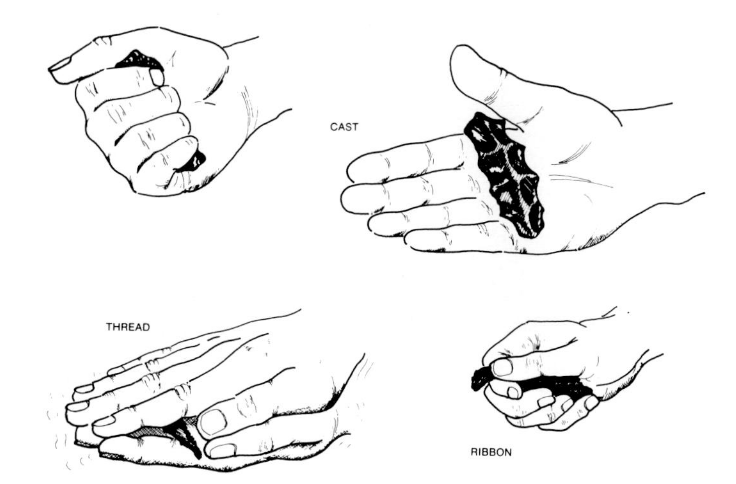

Fig. 5.3 The basic shapes used in the field hand test.

ened, they will behave as a plastic; therefore, the higher the clay content in a sample, the more refined and durable the shapes into which it can be molded. Table 5.2 gives the behavioral traits of five common soil textures.

Another method of determining soil texture involves the use of devices called *sediment sieves*, screens that have been built with a specified mesh size. When the soil is filtered through a group of sieves, each with a different mesh size, the particles become sorted in corresponding size categories (Fig. 5.4). Each category can be weighed and the weight expressed as a percentage of the total sample weight, in order to make a textural determination.

Although sieves work well for silt, sand, and larger particles, they are not appropriate for clay particles. Clay is too small to sieve accurately; therefore, in clayey soils the fine particles are measured on the basis of their settling velocity when suspended in water. Since clays settle so slowly, they are easily segregated from sand and silt. The water containing the clay can be drawn off and evaporated, leaving a residue of clay, which can be weighed.

SOIL, LANDFORMS, AND TOPOGRAPHY

Nearly all soils are formed in deposits laid down by geomorphic processes of some sort; for example, wind, glacial meltwater, ocean waves, river floods, and landslides. Once in place, the surface layer of these deposits is subject to alteration by processes associated with climate, vegetation, surface drainage, and land use. Depending on the type of deposit (e.g., sand, rock rubble, marine clay, or wind-blown silt), bioclimatic conditions, and

Table 5.2 Behavioral Characteristics of the Basic Shapes Used in the Field Hand Test and the Soil Type Represented by Each

Field Test (Shape)	*Soil Type*				
	Sandy Loam	*Silty Loam*	*Loam*	*Clay Loam*	*Clay*
Soil cast	Cast bears careful handling without breaking	Cohesionless silty loam bears careful handling without breaking; better-graded silty loam casts may be handled freely without breaking	Cast may be handled freely without breaking	Cast bears much handling without breaking	Cast can be molded to various shapes without breaking
Soil thread	Thick, crumbly, easily broken	Thick, soft, easily broken	Can be pointed as fine as pencil lead that is easily broken	Strong thread can be rolled to a pinpoint	Strong, plastic thread that can be rolled to a pinpoint
Soil ribbon	Will not form ribbon	Will not form ribbon	Forms short, thick ribbon that breaks under its own weight	Forms thin ribbon that breaks under its own weight	Long, thin flexible ribbon that does not break under its own weight

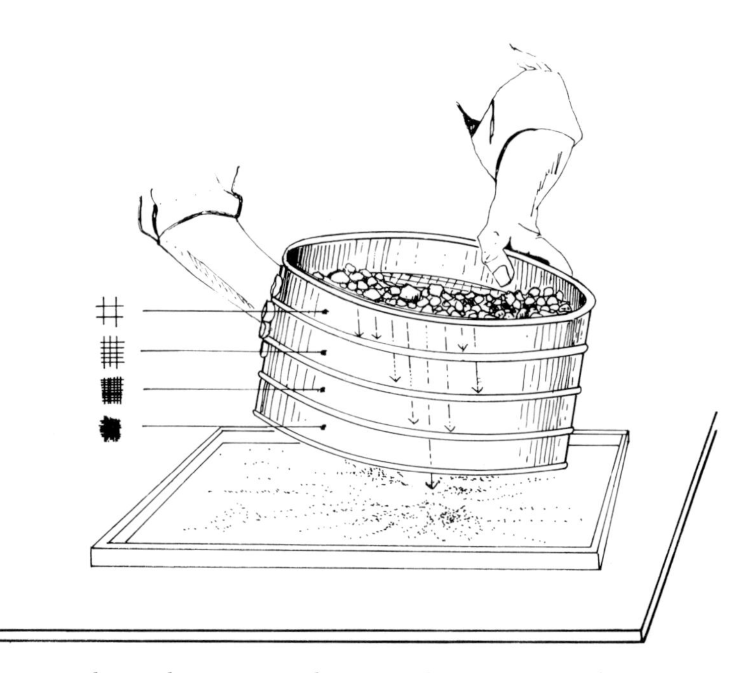

Fig. 5.4 Sieves can be used to measure the coarse fraction of particles in a soil sample.

local drainage, these processes in time produce a complex medium, usually 1 to 3 m deep, known as soil or the "solum." Below the solum, the bulk of most deposits, however, retain the basic properties imparted to them during their formation. This fact is especially significant in land development problems, because building footings, basements, roadbeds, slope cuts, and the like are built in these materials.

Geomorphologists have developed a basic taxonomy of the landforms associated with various types of deposits, making it possible to infer certain things about soil materials based on a knowledge of landforms. Any attempt to correlate landforms and soils, however, must recognize the limitations of geographic scale. Generally, the extremes of scale provide the least satisfactory results. At the scale of individual landform features, such as floodplains, sand dunes, or outwash plains, the correlation can be quite good for planning purposes (Fig. 5.5). Table 5.3 (page 98) lists a number of landform features and the soil composition and drainage associated with each.

Within many depositional features there are also predictable trends in soil make-up related to slope and the patterns of the processes responsible for the feature. Such trends are called *toposequences*, and identification of them can lead to valuable information for site planning. In the case of an alluvial fan, for example, texture tends to grade coarse to fine from the top to the toe (Fig. 5.6). In addition, alluvial fans are structurally complex, being comprised of many layers of sediment, some of which may be saturated and unstable for building. Another type of deposit, talus slopes, is comprised of rock fragments that have fallen into place from an upper slope. In contrast to alluvial fans, particle size tends to increase downslope, and drainage is good at all levels. On vegetated hillslopes, toposequences

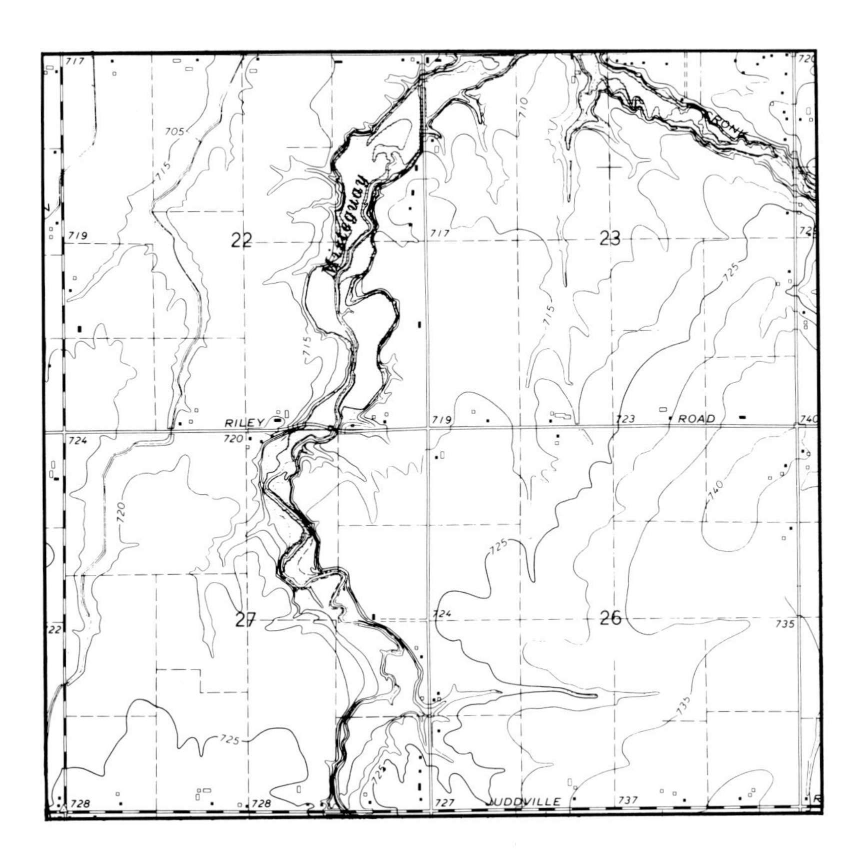

Fig. 5.5 The relationship between soils and landforms is illustrated vividly between the river valley and upland surfaces, but within the upland the correlation is not so sharp. Scale is 1:32,000; approximately 1 inch to 0.5 mile. ►

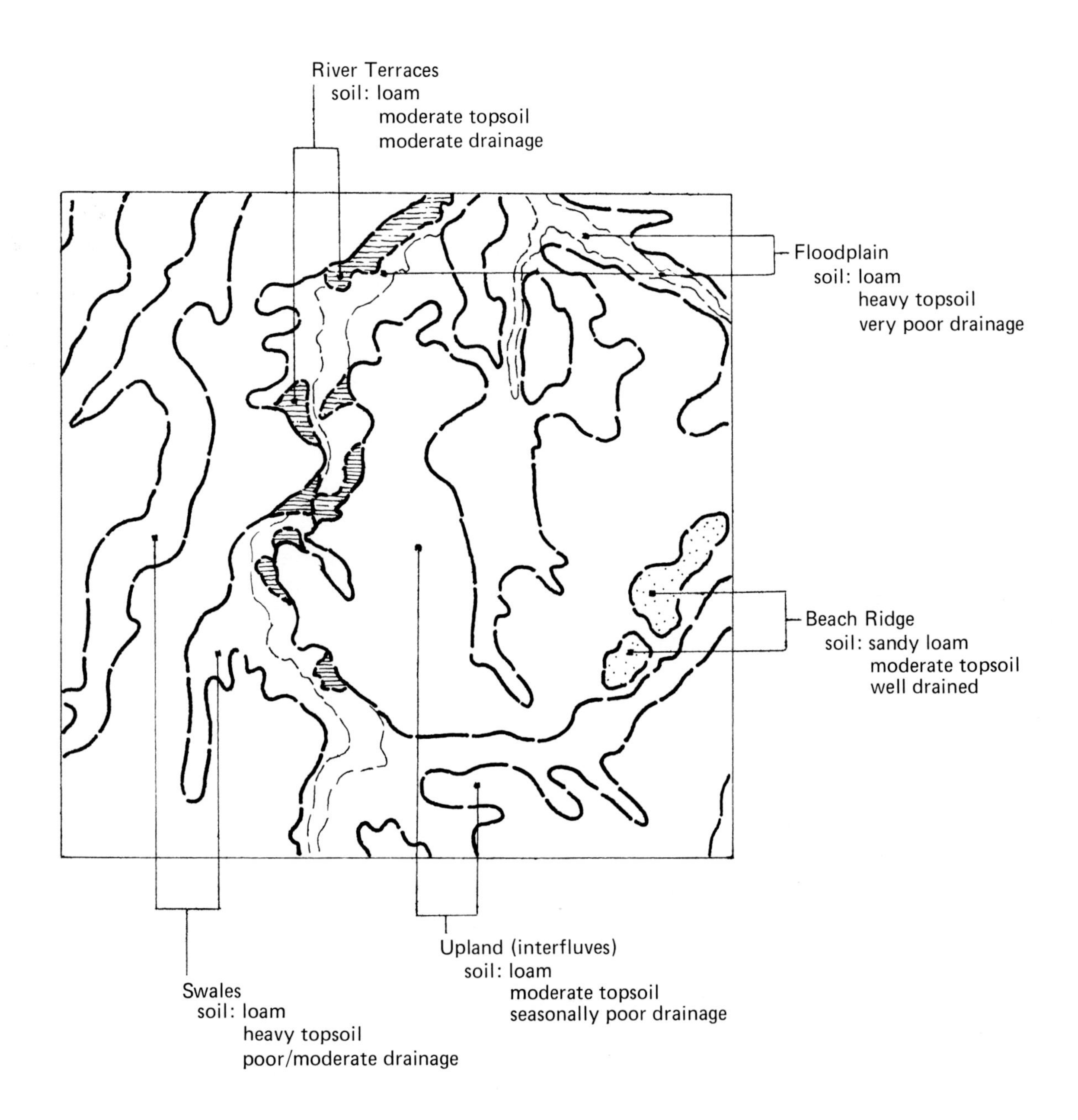
River Terraces
soil: loam
moderate topsoil
moderate drainage
Floodplain
soil: loam
heavy topsoil
very poor drainage
Beach Ridge
soil: sandy loam
moderate topsoil
well drained
Upland (interfluves)
soil: loam
moderate topsoil
seasonally poor drainage
Swales
soil: loam
heavy topsoil
poor/moderate drainage

Table 5.3 Landforms and Soil Materials

Feature	*Composition*	*Drainage*
Alluvial fan	Complex: sand, silt, clay with fraction pebbles and cobbles; markedly stratified and highly heterogeneous	Variable: upper portions may be well drained, lower portions may be very poor owing to groundwater seepage
Arroyo (also gulch or wash)	Complex: silt, sand, and pebbles in bed and valley of stream in arid setting	Poor; subject to seasonal and flash flooding
Barrier beach	Sand and pebbles	Good, but watertable often within several feet of surface
Beach	Variable; typically sand and pebbles but may also be clayey and silty or bedrock and rock rubble	Good if sandy, but watertable usually within several feet of surface
Beach ridge	Mainly sand, but pebbles common in lower portion	Excellent, especially in those with high elevation
Bog	Organic (muck, peat) with fraction mineral clay	Very poor
Cuesta	Bedrock often with partial coverage of thin soil and talus footslope	Good, but groundwater seepage common along footslope
Cusp or cuspate foreland	Sand and pebbles	Good, but watertable often within several feet of surface
Delta	Complex: usually clay, silt, and sand with local concentrations of organic material in stratified mass	Very poor to poor; high watertable; subject to frequent flooding
Drumlin	Clayey with admixture of coarser fractions as large as boulders	Good to poor
Escarpment (*see* Cuesta)		
Esker	Stratified sand and pebble mixture (gravelly) in the form of a sinuous ridge	Excellent
Floodplain	Complex: all varieties of soil possible including organic; assortment of stratified channel deposits with fraction flood and colluvial deposits	Poor to very poor; subject to high watertables and flooding
Ground moraine	Often sand, silt, clay admixture, but may be highly variable ranging from compacted clays to sand, pebbles, cobbles, boulders; usually gently rolling	Good to poor
Kame	Mainly stratified sand and gravel in the form of a conical-shaped hill	Excellent
Lake plain	Clayey with local concentrations of beach and dune sand	Poor to fair
Lake terrace	Usually sand and pebbles but may be bedrock or clay and silt	Excellent to good
Levee	Sand, silt, and clay deposits resting on floodplain (channel) sediments	Poor; slightly better than adjacent floodplain
Marsh	Organic (muck, peat) with fraction mineral clay	Very poor
Moraine	Often sand, silt, clay mixture, but may be highly variable ranging from compacted clays to sand, pebbles, cobbles, boulders; usually in form of irregular hilly terrain	Good to poor

Table 5.3 (cont.)

Feature	*Composition*	*Drainage*
Outwash plain	Sandy	Usually excellent, but high watertable in some locales
Pediment	Thin layer of sand and gravel over bedrock	Good, but infiltration capacity may be poor
River terrace	Variable; stratified clays, silts, sand	Excellent to fair
Sand dune (barchans, seifs, parabolic, hairpin, transverse, or coastal)	Pure sand	Excellent
Scarp (*see* Cuesta)		
Scree slope	Cobbles and pebbles (30°–40° slope)	Excellent
Spit	Sand and pebbles	Good, but watertable often within several feet of surface
Swamp	Organic (muck, peat) with fraction mineral clay	Very poor
Talus slope	Boulders in slabs, sheets or blocks (30°–40° slope)	Excellent
Tidal flat	Sand, silt, or clay with local concentrations of organic material	Very poor
Till plain	Often sand, silt, clay admixture, but may be highly variable ranging from compacted clays to sand, pebbles, cobbles, boulders (usually gently rolling)	Good to poor

Adapted from William M. Marsh, *Environmental Analysis for Land Use and Site Planning*, New York: McGraw-Hill, 1978.

Fig. 5.6 Typical textural variations with slope on an alluvial fan (left) and a talus slope (right).

are usually more subtle and limited largely to the surface layer. Topsoil, in particular, will vary with slope steepness, with the weakest development near midslope where the inclination is greatest and runoff removes most organic litter. Near the toe of the slope, topsoil thickens as runoff slows down and organic matter is deposited.

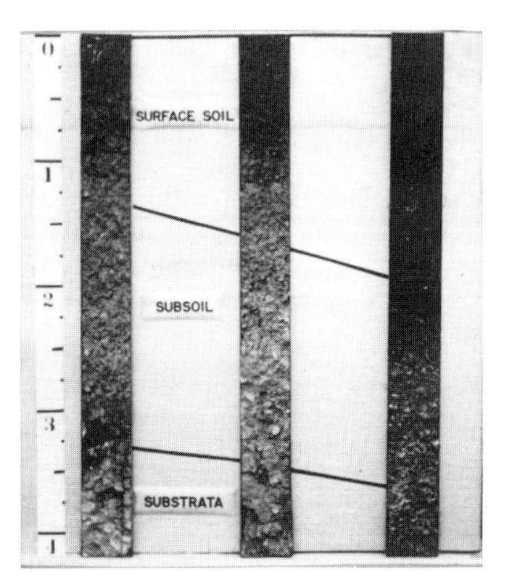

A typical example of soil variation associated with a toposequence across a hillslope. Organic matter increases markedly near the toe slope (right).

APPLICATIONS TO LAND PLANNING

In both the United States and Canada, extensive surveys have been conducted to map and classify soils. In the United States, surveys conducted by the Soil Conservation Service have traditionally focused on rural areas, and urban areas have generally been omitted. As a result, soil analysis for urban planning is usually done on a site-by-site basis by privately contracted firms. In addition, the SCS investigates soils only to a depth of 40 to 60 inches or so; therefore, in problems concerning heavy structures and underground facilities, additional examination of the soil to depths of many meters is often necessary.

In planning problems dealing with residential and industrial development, soil composition is one of the first considerations. Organic soils are at the top of the list because they are highly compressible under the weight of structures, and tend to decompose when drained. Where the organic mass is shallow, it is possible to excavate and replace the soil with a better material, but this is often financially expensive and ecologically harmful. Excavation and fill costs generally range from \$3.00 and \$6.00 per cubic yard; and from the ecological standpoint, one must recognize that organic soils are often indicative of the presence of wetlands that may support valued communities of plants and animals. On balance, then, organic soils should generally be avoided as development sites.

For mineral soils, texture and drainage are the important considerations. Coarse-textured soils, such as sand and sandy loam, are preferred for most types of development because support (bearing) capacity and drainage are usually excellent (see Table 5.1). As a result, foundations are structurally stable and usually free of nuisance water. Clayey soils, on the other hand, often provide poor foundation drainage, whereas bearing capacity may or may not be suitable for buildings. In addition, certain types of clays are prone to shrinking and swelling with changes in soil moisture,* creating stress on foundations and underground utility lines (Fig. 5.7). Therefore, development in clayey soils may be more expensive because special footings and foundation drainage may be necessary. Accordingly, site analysis often requires detailed field mapping followed by engineering tests to determine whether bearing capacity, drainage, and other problems may be posed by clayey soils.

In preparing soil maps, the scale of analysis is often as important as the types of analysis. For regional-scale problems, SCS maps are generally suitable and indeed are widely used along with topographic maps and other published sources of data to identify suitable locations for various land uses. At the site scale, however, SCS reports and maps are usually too general for most projects, especially those involving structures. In these cases, a finer grain of spatial detail is called for, which usually requires additional field sampling for texture, composition, and relations to topography and vegetation. On occasion, special laboratory analysis, such as Atterberg limits tests, may also be required for selected soils. The maps and data generated from such an effort can be instrumental in defining buildable land units, building locations, road alignments, and the like. Once a site plan is selected, analysis must be taken to a further level of detail and this usually involves making soil borings at critical points.

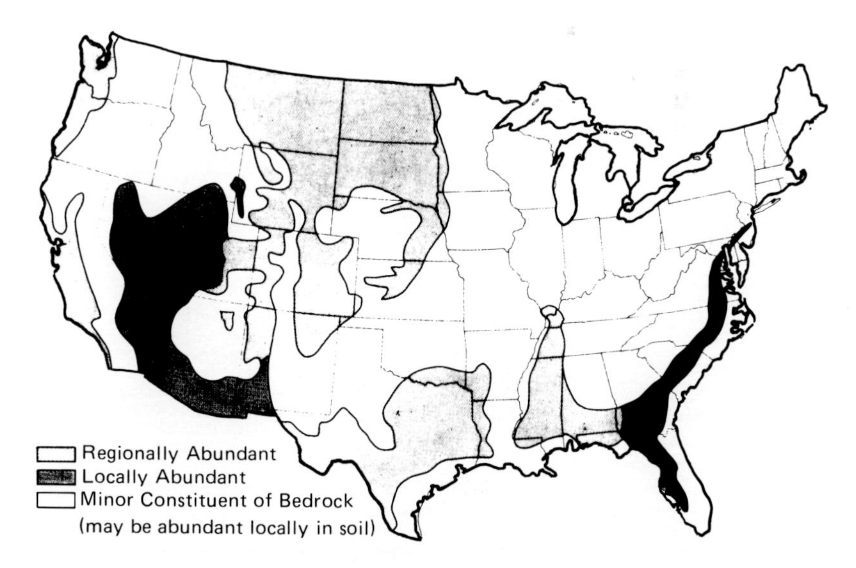

Fig. 5.7 The distribution of near-surface rock containing clay prone to swelling and shrinking. In soil, this clay (montmorillonite) is responsible for stress and cracking of foundations and utility lines. (From U.S. Geological Survey, 1978, based on H. A. Tourtelot, "Geologic Origin and Distribution of Swelling Clays," *Association of Engineering Geologists Bulletin*, vol. 11, no. 4, 1974.)

*Soils belonging to the *vertisol* order (of the U.S. Department of Agriculture soil classification scheme) are especially prone to expansion and contraction with changes in moisture.

CASE STUDY

Mapping Soils at the Site Scale for Private Development

W. M. Marsh

The aim in mapping soils for planning and architectural projects is to provide meaningful information vis-á-vis the objectives of the project. This requires a clear understanding of the development program; that is, what is to be built, how much floor space is called for, what ancillary facilities will be needed (parking and the like), what utility systems are necessary, and so on. These elements of the program must be known to determine: (1) the scales at which mapping must be conducted; (2) the soil features (parameters) that should be recorded; and (3) the depth to which soil must be examined.

The results of the soil survey, along with the results of related studies such as drainage, slope, and traffic, are used to formulate alternative plans for development of the site. In this particular project, the soil map was instrumental in defining buildable land units, which in turn served as a framework for the formulation of alternative development plans. Following the selection of an alternative, more detailed soil analyses were then conducted by engineers according to specific locations designated for buildings and facilities in the plan.

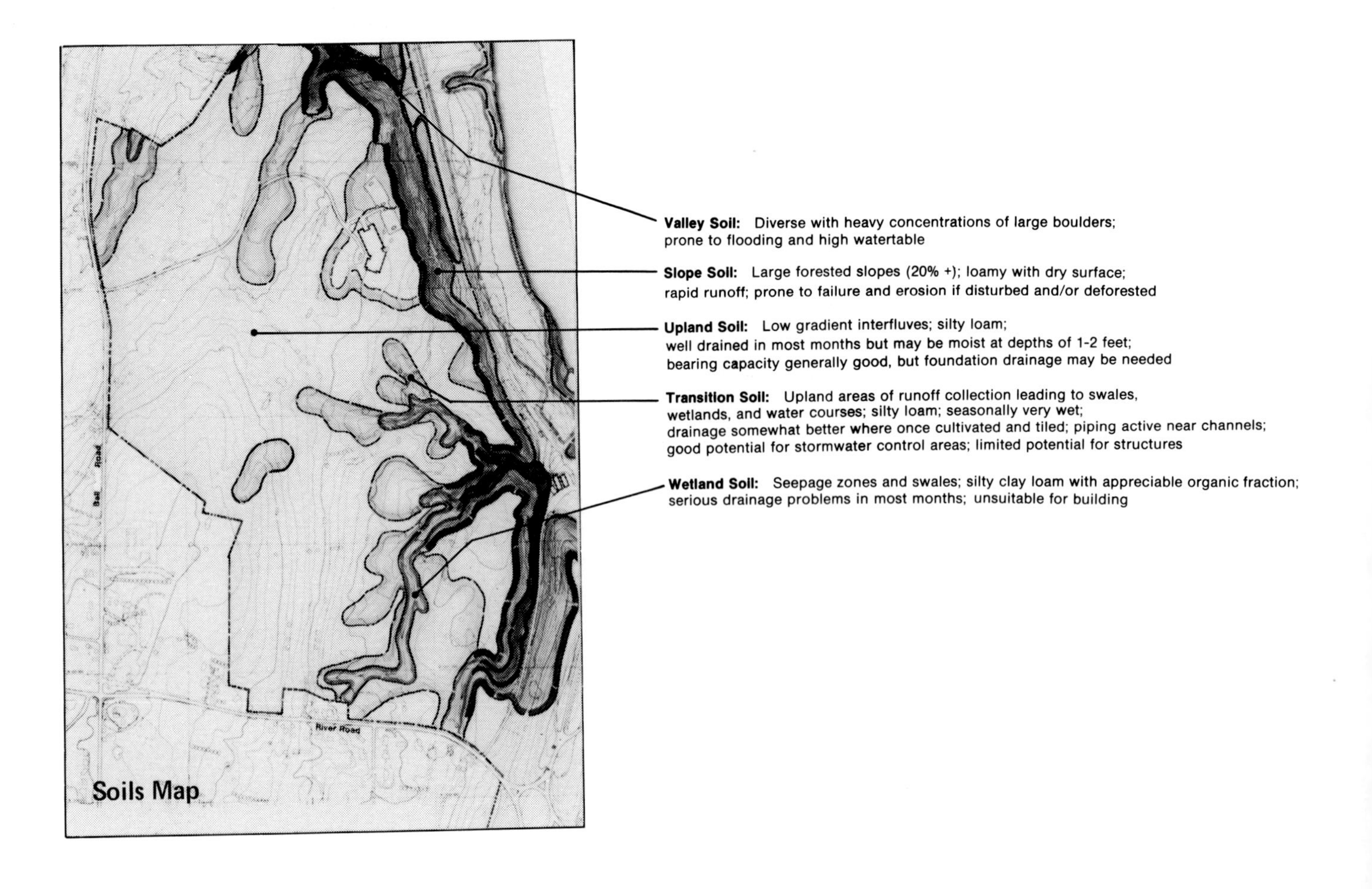

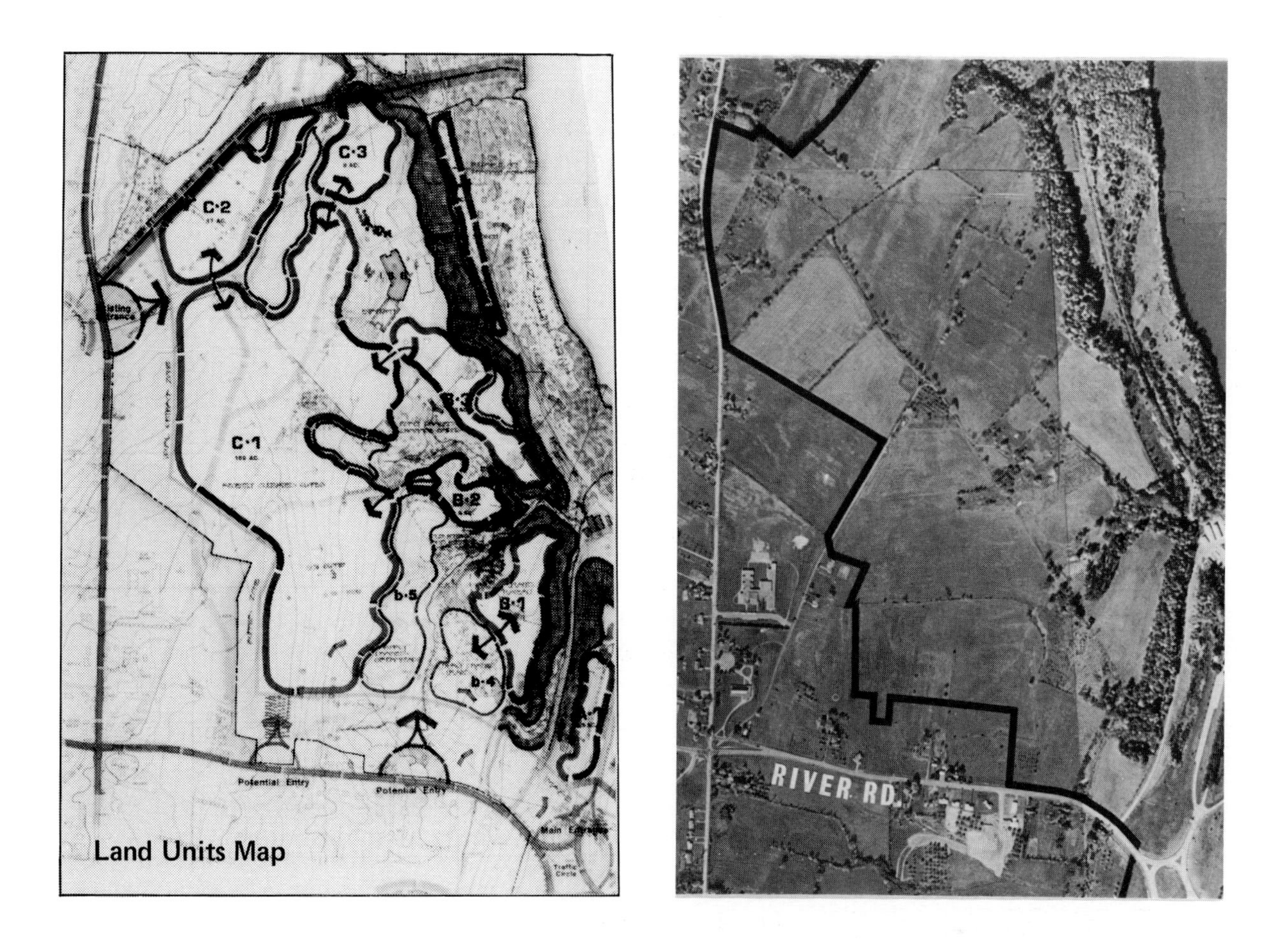

Note that the categories used actually describe the soil environment, as well as specific parameters such as texture, depth, and bearing capacity, and in this respect differ significantly from soil maps produced for other purposes such as agriculture.

Soil borings are test holes used to record the stratigraphy and composition of the subsoil. The arrangement of strata can be analyzed and samples can be extracted for laboratory analysis. The depth to which a boring is made depends on the building program requirements (for example, heavy buildings or light buildings), the types of materials encountered underground, and the depth to bedrock. If the soil mantle is relatively thin, say, less than 10 m, and the bedrock stable, the footings for large buildings will usually be placed directly on the bedrock. In areas of unstable soils, such as poorly consolidated wet clays, or weak bedrock, such as cavernous limestone, special footings are usually required to provide the necessary stability.

PLANNING CONSIDERATIONS IN SOLID WASTE LANDFILLS

One of the most pressing land use problems in urbanized and industrial areas today is the disposal of solid waste: municipal garbage, chemical residues from industry, rubble from mining and urban development, various forms of industrial and agricultural debris, and, most recently, nuclear residue from power plants and military manufacturing and development installations. Because of the large volumes involved and/or the composition or form of this material, it is neither economically nor technically feasible to dispose of this waste in conventional sanitary sewer systems; that is, those involving diffusion in water and transport through underground pipes to a treatment plant. Of the alternatives remaining, the practice of burial in the ground, called sanitary landfilling, is the preferred disposal method. Most other disposal methods are either environmentally unacceptable or too expensive for most waste; these include burning, open dumping, ocean dumping, and deep earth injection. Increasingly, however, with revelations that chemical and municipal landfills are resulting in contamination of soil and ground water, it has become more apparent that hazardous wastes such as chemicals and nuclear residues will have to be disposed of in more elaborate and expensive ways.

CASE STUDY

Hazardous Waste Disposal on Land

U.S. EPA

Groundwater supplies serve more than half of the United States population; for this reason, safeguarding the land from improper disposal of solid waste, especially hazardous waste, has become one of the U.S. Environmental Protection Agency's highest priorities. Under the Resource Conservation and Recovery Act of 1976 (RCRA), in 1979 the EPA proposed guidelines for landfill disposal of solid (nonhazardous) waste, and published criteria for classifying solid waste disposal facilities. In February 1981, the EPA announced temporary standards for four classes of new hazardous waste land disposal facilities: secure landfills, surface impoundments, land treatment facilities (land farms), and underground injection wells.

Since it became obvious that waste chemicals can cause harm, responsible producers of these chemicals have disposed of them in *secure landfills,* sometimes called secure chemical waste landfills. These are carefully constructed earthen excavations specifically designed to contain hazardous waste and to keep it from escaping into the environment, either by reaching into the groundwater or by evaporating.

Not too many years ago a "secure" landfill might have been simply a trench in the earth with drums of hazardous waste placed into it. The drums were then covered with clay to the height of the trench walls to keep out surface water. But clay is not 100 percent watertight, and hazardous waste could and sometimes did contaminate groundwater and surface water, as the waste corroded its way through the drums and leaked out.

Today's secure chemical landfill is technologically more sophisticated and must conform to RCRA standards. The owner or operator of a secure landfill must monitor surface and groundwater; the site must also be above the one-hundred-year floodplain, outside fault zones, or other likely disruptions. It must contain either natural or synthetic liners to restrict the loss of leachate, and provisions must be made for a network of pipes that collect any precipitation and leachate that accumulate in the landfill. Monitoring wells must also be installed to check the quality of groundwater within the area.

A *surface impoundment* is a basin designed to hold liquid wastes. It may be a natural or man-made depression, or a diked area, formed primarily from earthen materials. Such impoundments can be used for waste

CASE STUDY (cont.)

treatment, storage, or disposal, and are subject to most of the same standards that apply to secure landfills. There are some additional design modifications and standards aimed at preventing wind and water erosion and damage caused by plant roots and burrowing animals.

Surface impoundments offer a relatively inexpensive disposal alternative, but they are now known to present an environmental risk in the form of pollution from leaching and overflow. One EPA study has compiled some disturbing statistics on existing surface impoundments nationwide: 94 percent have no monitoring wells, 87 percent are located over underground drinking water supplies, 66 percent have no liners, and 38 percent are built in soils with high permeability.

For many hazardous wastes, especially nonchlorinated organic wastes, aerobic *"land farming"* is an acceptable alternative. In land farming, compatible wastes such as refinery sludge and petroleum-based solvents are mixed into soil, where soil microbes will degrade or purify the wastes. Working of the soil with earth-moving equipment hastens the purification process; air is mixed mechanically and circulated throughout the soil-waste combination so that atmospheric oxygen can support microbiological life. Land farms are usually located in industrial areas, safely removed from any source of drinking water. The disposal area is monitored continuously, and once treatment is complete the land may be put to other use—even turned into a park or a botanical garden, as has been done in California.

Deepwell injection is a disposal method that has been used by the petroleum industry for the disposal of salt brine since the late nineteenth century. It involves pumping liquid wastes through encased wells into porous rock formations deep in the earth, below underground sources of drinking water and isolated from mineral-bearing layers of rock such as coal beds. Today, underground injection is used for hazardous waste disposal by a wide variety of industries, especially refineries and petrochemical plants. The practice, however, is limited to certain geographic areas. Currently, commercial deep-well disposal services are available only in EPA Regions V (Midwest) and VI (Texas and bordering states). There are only sixty-five hazardous waste deep wells in Region V.

Until permanent regulations are developed under the Underground Injection Control program of the Safe Drinking Water Act, deep-well injections will be controlled under current RCRA regulations. These set strict construction, operating, and closure standards for injection wells that accept most types of liquid hazardous waste, except chlorinated hydrocarbons. Injection presents one of the cheapest forms of disposal, ranging from $15 to $100 per ton into established wells. If a well has to be constructed, however, it may cost more than $1,000,000 to sink a casing to a depth of 5,000 to 10,000 feet.

The United States Environmental Protection Agency is responsible for federal efforts in the control of air and water pollution, noise, radiation, and hazardous waste management, among other things.

Solid waste can be divided into four major types: mining debris, industrial refuse, agricultural waste, and urban garbage. Mining debris and agricultural waste account for 90 percent of the solid waste produced in the United States each year. Urban and industrial waste, though comparatively low in total output, present the most widespread and serious solid waste disposal problem, however. Several factors account for this, including the issue of public health and the growing concern over hazardous waste, conflict over aesthetic and real estate values, the limited availability of land around cities for disposal sites, and the high cost of garbage collection and hauling in urban areas. In addition, the production of urban and industrial waste has been rising steadily in most parts of the United States (Fig. 5.8).

The selection of a disposal site for urban and industrial solid waste is one of the most critical planning processes faced in suburban and rural areas today. Properly approached, it should be guided by three considerations: (1) *cost*, which is closely tied to land values and hauling distances; (2) *land use* in the vicinity of the site and along hauling routes; and (3) *site conditions*, which are largely a function of soil and drainage. In both urban and industrial landfills, the chief site problem is containment of the liquids that emanate from the decomposing waste.

These fluids are collectively called *leachate*, and are composed of heavy concentrations of dissolved compounds that can contaminate local water supplies. Leachates are often chemically complex and vary in makeup with the composition of the refuse; moreover, the behavior of leachates in the hydrologic system, especially in groundwater, is poorly understood. Therefore, the general rule in landfill planning and management is to restrict leachate from contact with either surface or subsurface water. Instances of groundwater contamination by leachates typically result in the loss of potable water for many decades.

Leachate containment may be accomplished in some cases by merely selecting the right site. An ideal fill site should be excavated in soil that is effectively impervious, neither receives nor releases groundwater, and is

Fig. 5.8 A modern sanitary landfill. The mass of garbage (right) is interlayered and covered with soil excavated from the area on the left. The dark material is organic soil that has been removed from the excavation.

not subject to contact with surface waters such as streams or wetlands. Dense (compact) clay soils and relatively high-level ground are the preferred site characteristics. In addition, the clay should not be interlayered with sand or gravel, not subject to cracking upon drying, and stable against mass movements such as landslides.

Where these conditions cannot be met, the site must be modified to achieve the satisfactory performance. If the soils are sandy and permeable, for example, a liner of clay or a synthetic substance such as vinyl must be installed to gain the necessary imperviousness. The same measure must be taken in stratified soils, wet soils, sloping sites, and sites near water features and wells. In many areas of the United States and Canada, however, the decision to modify a site is not left up to the landfill developer or operator because strict regulations often govern which sites can even be considered for landfills; thus, inherently poor sites are often eliminated from consideration by virtue of planning policy.

In addition to a formal site selection and preparation plan, a growing number of local and state/provincial governments are requiring the preparation of a management plan for the design and operation of landfills. In the case of sanitary landfills for municipal garbage, the planner is asked to address the following:

1. Compartmentalization of the filling into cells or self-contained units of some sort.
2. Phasing of the operation; for example, excavation and filling of a limited portion of area at any one time.
3. Landscaping, pest control, and protection of the site during operations.
4. Limiting the total thickness of refuse by interspersing layers of soil within the garbage.
5. Backfilling over the completed fill with a layer of soil with a provision for vent pipes to release gases, if necessary.
6. Preparation of a plan for grading, landscaping, and for future use of the site.

PROBLEM SET

I. The following data were produced from soil samples taken at the points shown on the map on page 108.

1. Based on the percentage of sand, silt, and clay in each sample, determine the appropriate texture for the sample using the textural triangle in Fig. 5.2. Plot these soil names on the topographic map next to their respective sample points, and note the relationship between topographic features and soil texture.
2. Based on observable similarities and trends in soil texture and topography, map the areas of different soil textures. For example, do valleys exhibit different soil textures than hillslopes? Where you are uncertain about a soil boundary, draw a broken line where you guess the boundary might lie.

Sample Pt.	*%Sa*	*Si*	*Cl*	*Sample Pt.*	*%Sa*	*Si*	*Cl*	*Sample Pt.*	*%Sa*	*Si*	*Cl*
1	71	19	10	17	41	43	16	33	organic		
2	67	23	10	18	36	30	34	34	75	21	4
3	72	15	13	19	organic			35	82	14	4
4	38	29	33	20	87	8	5	36	43	25	32
5	organic			21	organic			37	29	41	30
6	organic			22	88	9	3	38	43	38	19
7	94	4	2	23	29	41	30	39	72	18	10
8	organic			24	44	43	13	40	55	40	5
9	31	33	36	25	64	18	18	41	38	24	38
10	58	23	19	26	55	22	23	42	40	39	21
11	74	19	7	27	58	21	21	43	42	42	16
12	61	17	22	28	68	15	17	44	84	9	7
13	71	8	21	29	45	41	14	45	79	16	5
14	organic			30	42	29	29	46	74	22	4
15	62	19	19	31	81	15	4				
16	46	40	14	32	organic						

Sa = sand; Si = silt; Cl = clay

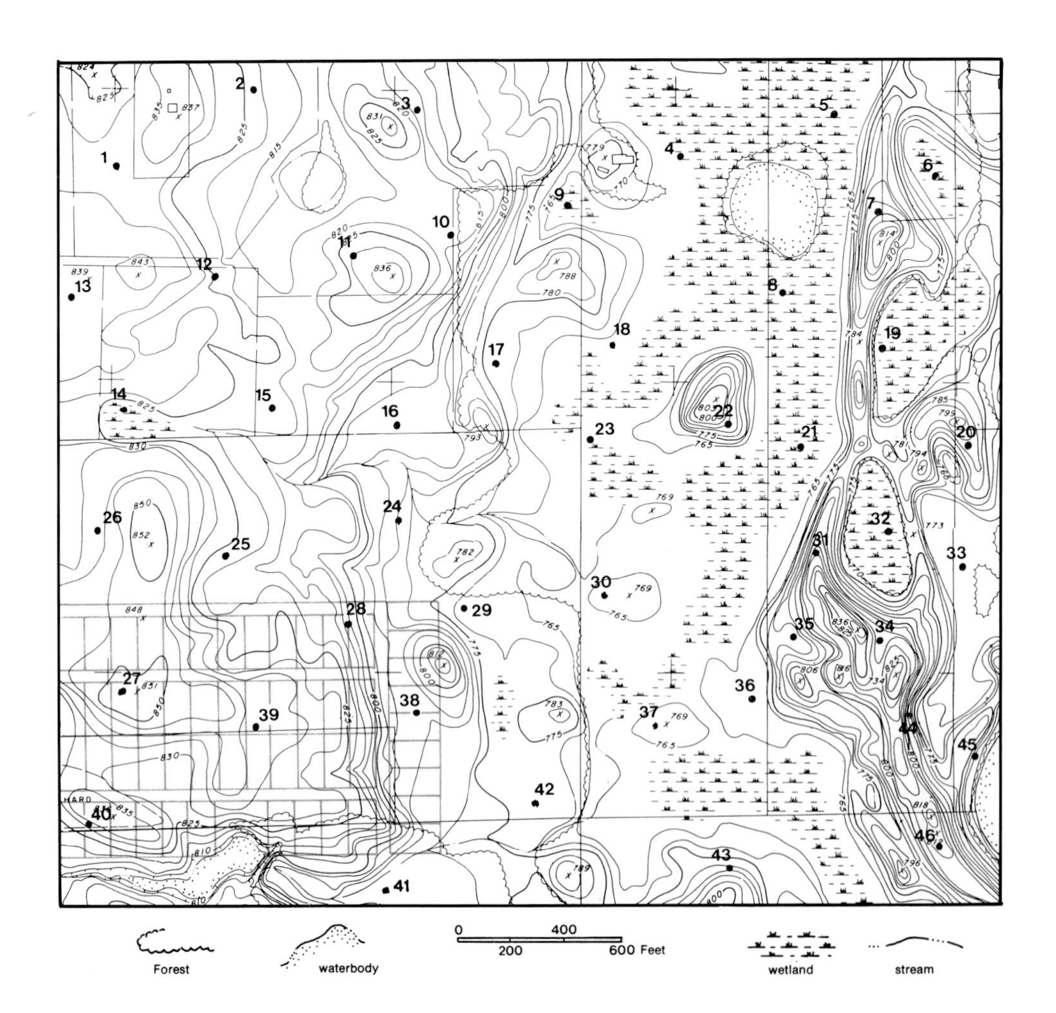

3. Color code the different categories of soils you have identified, and mark the distributions of these soils on the map. Provide a map legend for each category, and indicate the relative suitability of each for single-family residential development, assuming that little or no budget is available to cover major site improvement costs or special foundation construction.
4. A highway corridor with an alignment connecting points 3, 9, 8, 19, and 20 is proposed for the area shown in the accompanying map. Determine the cost of excavation and fill given the following data, and indicate what other factors should be taken into consideration besides costs in determining where this highway should be located.
   - width of roadbed = 125 feet
   - average depth of organic soil = 8 feet
   - cost of excavation = $2.00/yd^3
   - depth of fill = 12 feet
   - cost of fill = $3.25/yd^3

II. The map on page 110 shows an area near a town where a landfill site is being sought. The three zones represent hauling distances, the outer zone being the farthest and most expensive. According to the garbage haulers, only zones 1 and 2 are feasible for a landfill site given current trucking costs. Within these zones, a 160-acre site must be selected, and the local planning commission wants the area in zones 1 and 2 classified into just two categories: (1) suitable (no major restrictions); and (2) unsuitable (one or more major restrictions). The classification is to be based on four criteria: soil type, topography, nearness to water features, and nearness to residences, schools, churches, and airports according to the following limits:

	Soil	*Topography*	*Nearness to Water Features*	*Nearness to Communities*
Suitable	clay, silt clay, clay loam, silty clay loam, sandy clay	flat to gently rolling (0–15% slopes)	more than 1000 ft. (300 m)	more than 1000 ft. (300 m)
Unsuitable	sand, loamy sand, sandy loam, sandy clay loam, loam, silt loam, silt, organic, alluvial	hilly (>15% slopes) (4 contour intervals or more per 1000 ft.)	less than 1000 ft. (300 m)	less than 1000 ft. (300 m)

1. Map the unsuitable area using the above criteria and one of the following mapping techniques: map overlay or grid cell. The first technique involves building a separate map for each criterion and then combining the four maps; the second involves subdividing the entire area into grid cells, 160 acres or less in area, and then classifying each as suitable or unsuitable according to the four criteria.

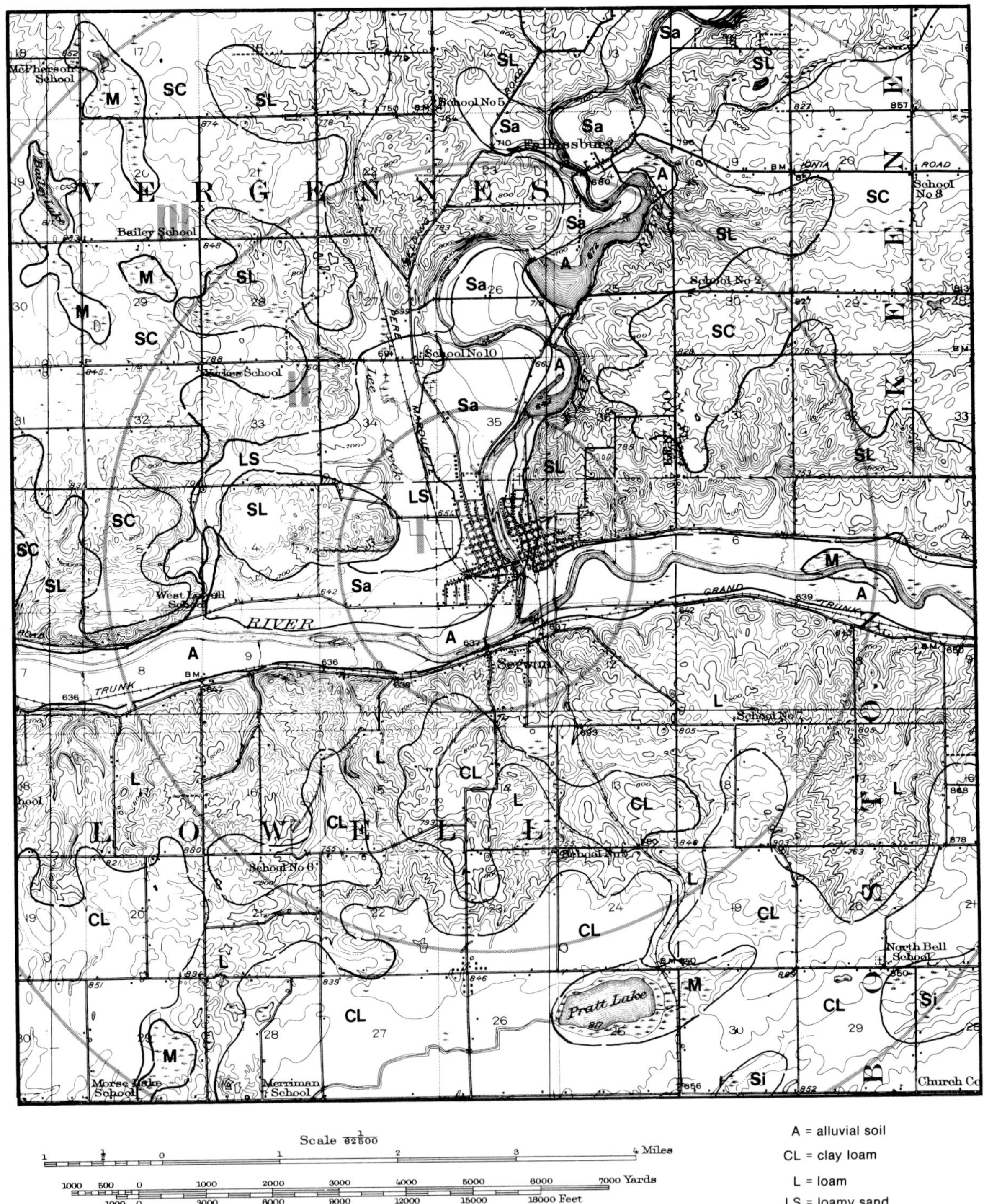

Scale $\frac{1}{62500}$

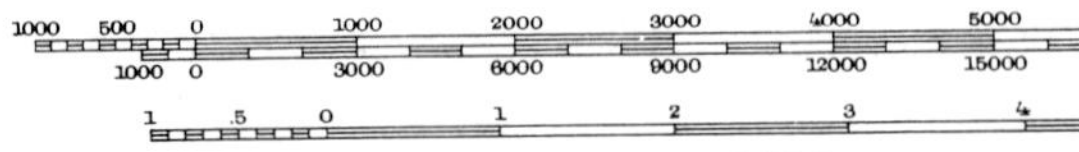

Contour interval 20 feet

A = alluvial soil
CL = clay loam
L = loam
LS = loamy sand
M= muck
Sa = sand
SC = sandy clay
Si = silt
SL = sandy loam

SELECTED REFERENCES FOR FURTHER READING

American Planning Association. *Soils and Land Use Planning.* Chicago: APA Press, 1966, 44 pp.

Briggs, David. *Soils.* Boston: Butterworths, 1977, 192 pp.

Colonna, Robert A., and McLaren, Cynthia. *Decision-Makers Guide to Solid Waste Management.* Washington, D.C.: Government Printing Office, U.S. Environmental Protection Agency, 1974, 157 pp.

Davidson, Donald A. *Soils and Land Use Planning.* New York: Longman, 1980, 129 pp.

Eckholm, Erik P. *Losing Ground.* New York: W. W. Norton, 1976, 223 pp.

Hills, Angus G., et al. *Developing a Better Environment: Ecological Land Use Planning in Ontario, A Study Methodology in the Development of Regional Plans.* Toronto: Ontario Economic Council, 1970, 182 pp.

Hopkins, Lewis D. "Methods of Generating Land Suitability Maps: A Comparative Evaluation." *Journal of the American Institute of Planners,* October 1977, pp. 388–400.

Metropolitan Area Planning Commission. *Halifax-Dartmouth Metro Area, Natural Land Capability.* Halifax, Nova Scotia: Nova Scotia Department of Development, 1973, 71 pp.

Pettry, D. E., and Coleman, C. S. "Two Decades of Urban Soil Interpretations In Fairfax County, Virginia." *Geoderma,* vol. 10, 1973, pp. 27–34.

Soil Society of America. *Soil Surveys and Land Use Planning.* Madison, Wis.: Soil Society of America, 1966.

6

SOILS AND WASTEWATER DISPOSAL

INTRODUCTION

Until this century, soil served as the primary medium for the disposal of most human waste. Raw organic waste was deposited in the soil via pits or spread on the surface by farmers. In either case, natural biochemical processes broke the material down, and water dispersed the remains into the soil, removing most harmful ingredients in the process. But this practice often proved ineffective where large numbers of people were involved, because the soil became saturated with waste, exceeding the capacity of the biochemical processes to reduce the concentration of harmful ingredients to safe levels.

The results were occasionally disastrous to a city or town. Water supplies became contaminated and rats and flies flourished, leading to epidemics of dysentery, cholera, and typhoid fever. In response to these problems, efforts were made in the latter half of the 1800s to dispose of human waste in a safer manner.

An outhouse privy near Sitka, Alaska, in the early twentieth century. Photographer G. K. Gilbert described this as evidence of suburban development.

In cities, where the problems were most serious, sewer systems were introduced, allowing waste to be transported through underground pipes to a body of water, such as a river or lake, beyond the city. Later this practice gave rise to critical water pollution problems, but it did solve the immediate problem of human health. Outside the cities, people continued to use pit-style privies until well into the twentieth century. But with the growth of suburban neighborhoods after 1930, the outdoor toilet proved unacceptable. In its place came the septic tank and drainfield.

THE SOIL-ABSORPTION SYSTEM

The septic drainfield method of waste disposal is one version of a soil-absorption system (SAS), so-called because it relies on the soil to absorb and disperse wastewater. The system is designed to keep contaminated water out of contact with the surface environment and filter chemical and biological contaminants from the water before they reach groundwater, streams, or lakes. The contaminants of greatest concern are nitrogen and phosphorus, which are nutrients for algae growth in aquatic systems, and certain forms of bacteria, which are hazardous to human health.

Most SAS systems are comprised of two components: (1) a holding or septic tank where solids settle out; and (2) a drainfield through which "gray" water is dispersed into the soil. The drainfield is made up of a network of perforated pipes or jointed tiles from which the fluid seeps into the soil (Fig. 6.1). Several different layout configurations are commonly employed.

Critical to the operation of a soil-absorption system is the rate at which the soil can receive water. Soils with high permeabilities are clearly preferred over those with low permeabilities. *Permeability* is a measure of the amount of water that will pass through a soil sample per minute or hour. In the health sciences, permeability is measured by the percolation rate, the rate at which water is absorbed by soil through the sides of a test pit (Fig. 6.2). The "perc" test is usually conducted *in situ*, meaning that it is conducted in the field rather than in the laboratory.

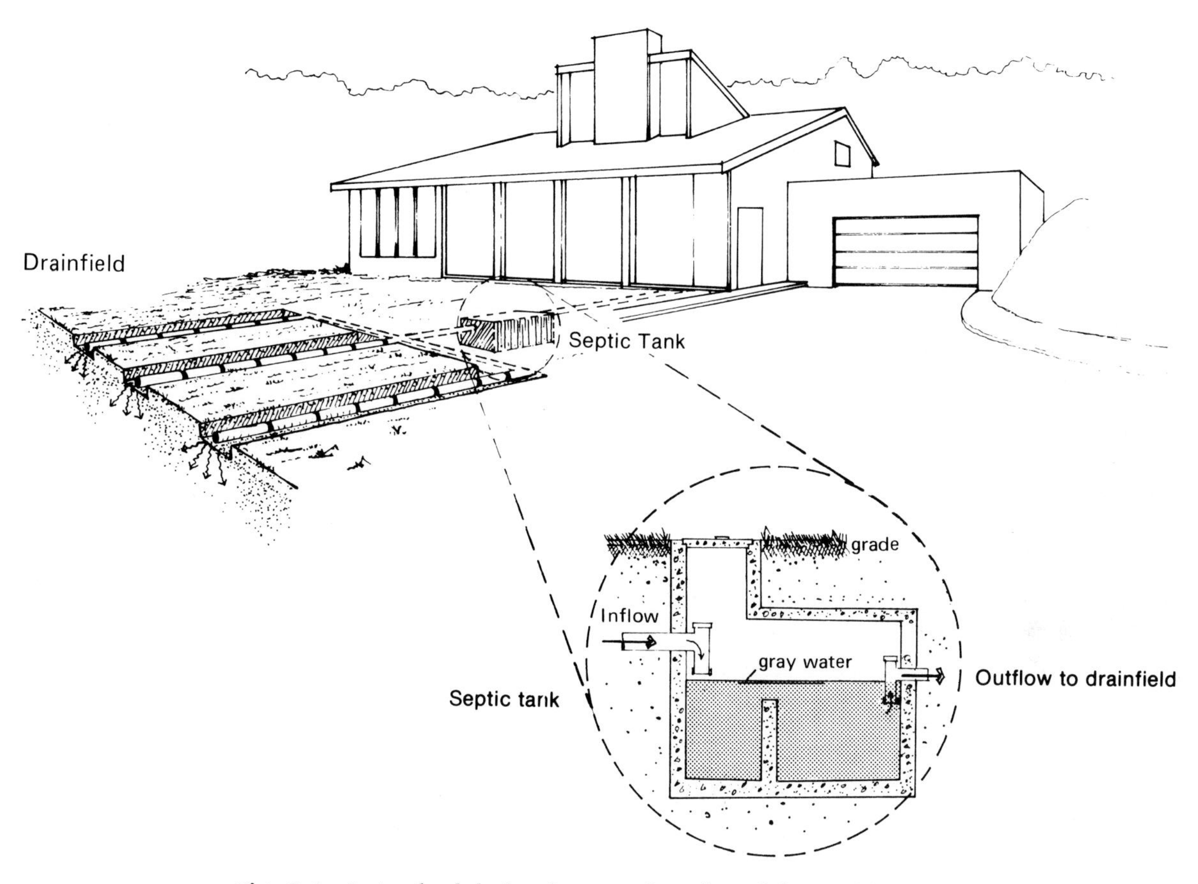

Fig. 6.1 A standard design for a septic tank and drainfield. Solid waste accumulates in the septic tank while the gray water flows into the drainfield where it seeps into the soil.

CONTROLS ON PERCOLATION

The percolation rate of a soil is controlled by three factors: *soil texture*, *water content*, and *slope*. Fine-textured soils generally transmit water more slowly than coarse-textured soils do, and thus have lower capacities for wastewater absorption. On the other hand, fine-textured soils are more effective in filtering chemical and bacterial contaminants from wastewater. The ideal soil, then, is a textural mix of coarse particles (to transmit water) and fine particles (to act as an effective biochemical filter).

Soil moisture tends to reduce permeability in any soil; therefore, soils with high watertables are limited for wastewater disposal. Slope influences percolation inasmuch as it affects the inclination of the drainfield. If the slope of the ground exceeds three to four feet per one hundred feet, it is difficult to arrange the drainfield at an inclination gentle enough (two to four inches per one hundred feet is recommended) to prohibit gray water from flowing rapidly down the drain pipes and concentrating at the lower end of the drainfield. In addition, the soil layer must be deep enough so that bedrock does not retard the entry of percolating water into the subsoil.

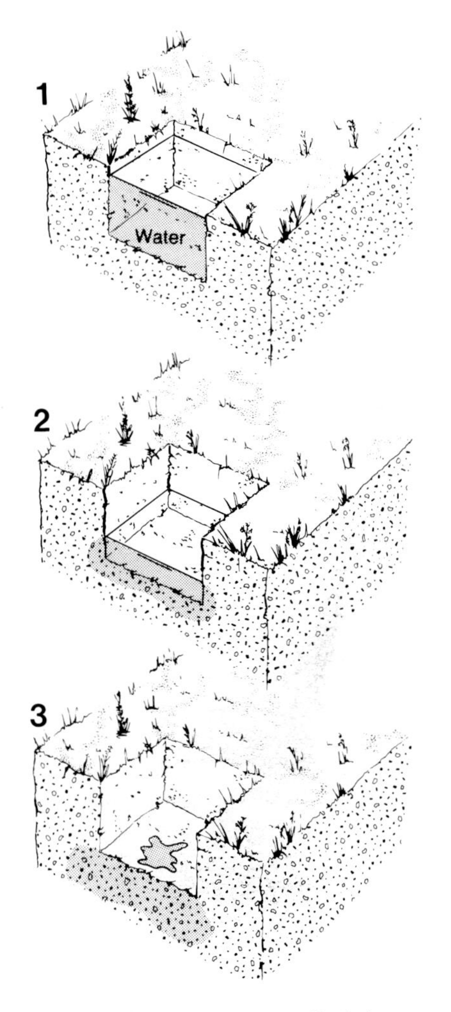

Fig. 6.2 The basic idea of the percolation test. A small pit is excavated, filled with water, allowed to drain, then refilled with water and allowed to drain again. The rate of fall in the water surface is the "perc" rate. (From W. M. Marsh and J. Dozier, *Landscape: Introduction to Physical Geography,* © 1981, Addison-Wesley, Reading, Massachusetts. Page 162. Used by permission.)

ENVIRONMENTAL IMPACT AND SYSTEM DESIGN

Soil-absorption systems are a standard means of sewage disposal throughout the world. In the United States, as much as 25 percent of the population relies on these systems; in developing countries such as Mexico, fully 80 to 90 percent of the population uses some form of soil absorption for waste disposal, including pit-style privies. Not surprisingly, a high percentage of these systems do not function properly, often resulting in serious health problems and environmental damage.

The causes of system failure are usually tied to one or more of the following: improper siting and design of the drainfield, overloading (that is, over-use), inadequate maintenance of septic tank and tile system, and loss of soil percolation capacity due to clogging of interparticle spaces or groundwater saturation of the drainfield bed. Failure often results in the seepage of wastewater into the surface layer of soil and onto the ground. Here humans are apt to come into contact with it, and it may enter lakes and streams, contaminating water supplies and fostering the growth of algae and related organisms (Fig. 6.3).

Fig. 6.3 An algae bloom resulting in part from seepage of nutrients into the lake from drainfields.

Avoidance of these problems begins with site analysis and soil evaluation. Soils with percolation rates of less than one inch per hour are considered unsuitable for soil-absorption systems. For soils with acceptable percolation rates, additional criteria must be applied: slope, soil thickness (depth to bedrock), and seasonal high watertable. Given that these criteria are satisfied, the system can now be designed. The chief design element is drainfield size and it is based on two factors: (1) the actual soil percolation rate, and (2) the rate at which wastewater will be released to the soil, that is, the loading rate. For residential structures, the loading rate is based on the number of bedrooms; each bedroom is proportional to two persons. The higher the loading rate and the lower the percolation rate, the larger the drainfield required (Fig. 6.4).

Successful operation of the system necessitates regular maintenance, especially the removal of sludge from the septic tank, avoidance of overloading, and the lack of interference from high groundwater. The latter may occur during unusually wet years or because of the raising of a nearby reservoir, for example. On the average, the lifetime of a drainfield is fifteen to twenty-five years, depending on local conditions. At this time the soil may be too wet to drain properly, because the watertable has been raised after years of recharge from drainfield water, and/or the soil may have become clogged with minute particles. In addition, the buildup of chemicals such as phosphorus may become excessive, thereby reducing the filtering capacity of the soil. In any event, old drainfields should be abandoned and a new one constructed in a different location.

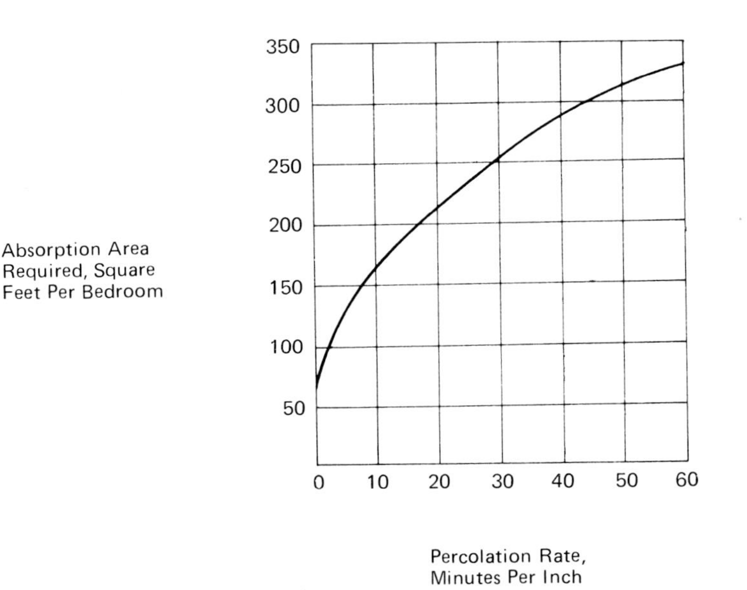

Fig. 6.4 Absorption area requirements based on soil-percolation rates. (From U.S. Public Health Service, Publication No. 526, 1957.)

CASE STUDY

Planning for Wastewater Disposal Using Soil Maps, Nova Scotia

Michael D. Simmons

Although population growth in Nova Scotia has been slow, the 1960s and 1970s saw a relatively large number of housing starts as a result of the break-up of households, the availability of government housing assistance, and a general trend toward rural and suburban settlement. Many of these new homes were built on unserviced lots within commuting distance of urban centers. To help in planning where homes should and should not be built, a task force was appointed in 1972 by the Province of Nova Scotia to determine soil suitability for on-site wastewater disposal.

The basic soil units and maps of the Nova Scotia soil survey served as the geographic base. The criteria for wastewater disposal were based on guidelines established by the Nova Scotia Department of Public Health:

- The percolation rate must be less than thirty minutes per inch; a rate of up to sixty minutes per inch may be acceptable with suitable modifications.
- Depth to bedrock must be four feet or more below the disposal field; that is, about six feet below the surface.
- The groundwater table must also be at least four feet below the disposal field and the drainfield site must not be in a marshy area or an area subject to flooding.
- Topography, or relative elevations within the lot, must be considered.

In addition to the factors covered by the guidelines, three other limitations were considered:

- Prolonged or periodic saturation of the surface soil layers associated with a perched watertable or flooding.
- Excessively rapid percolation, permitting contamination of groundwater supplies.
- Excessive slope of the land surface.

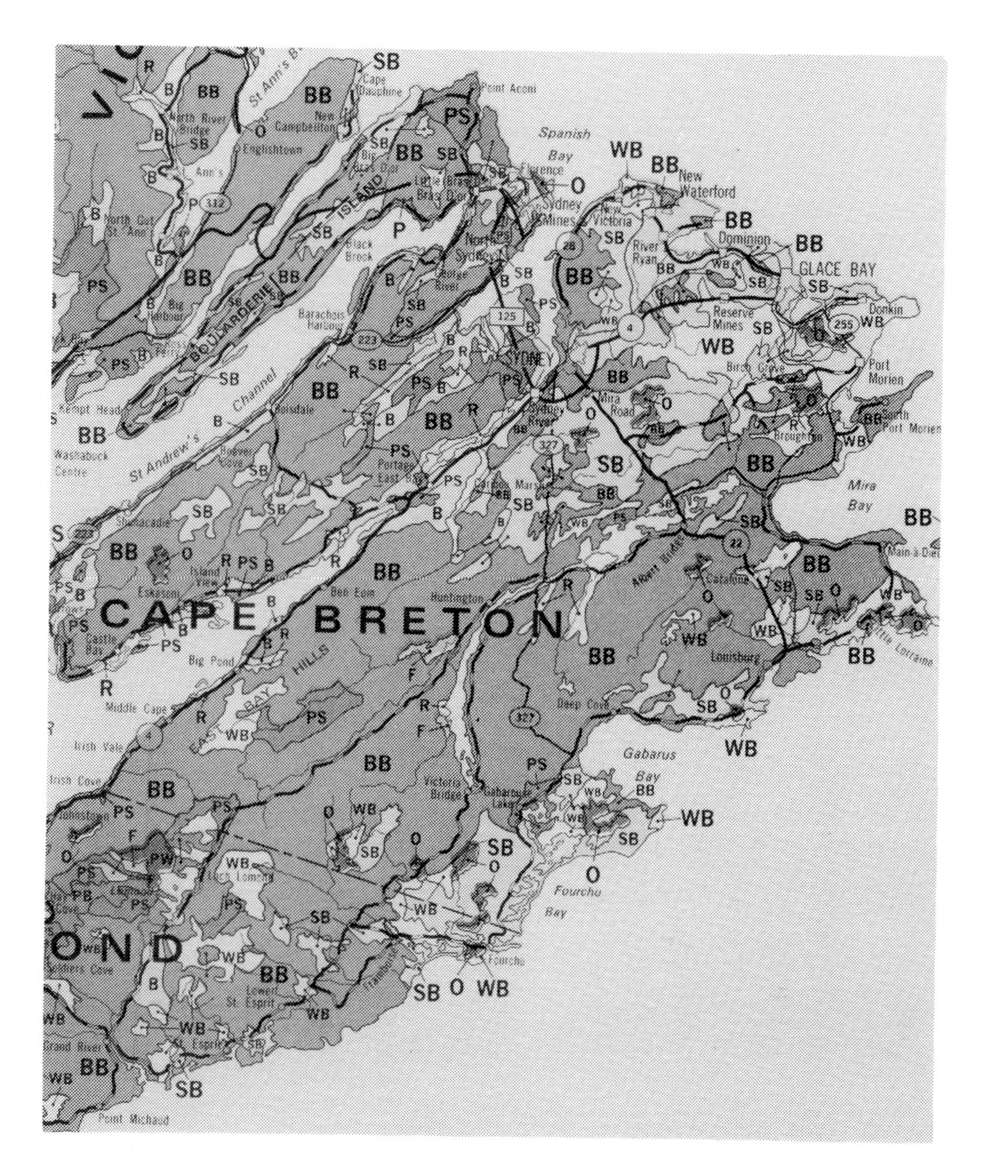

SOILS OF NOVA SCOTIA
LIMITATIONS FOR SEPTIC TANK ABSORPTION FIELDS[1]

Legend

Map Unit Symbol	Primary Limitation	Probability of Occurrence[2]	Secondary Limitation	Probability of Occurrence[2]	Acres ('000)	Hectares ('000)	% of Province
N	NONE TO MODERATE LIMITATIONS	85	NONE	-	709	287	5.5
B	BEDROCK	65	NONE	-	1016	411	7.9
BB	BEDROCK	90	NONE	-	6439	2606	49.4
P	PERCOLATION	90	NONE	-	420	170	3.3
PS	PERCOLATION	90	PERIODIC SATURATION	80	1322	535	10.3
PW	PERCOLATION	90	PROLONGED SATURATION	90	264	107	2.0
PB	PERCOLATION	90	BEDROCK	75	498	201	3.9
SB	PERIODIC SATURATION	90	BEDROCK	80	1073	434	8.3
WB	PROLONGED SATURATION	90	BEDROCK	80	388	157	3.1
R	RAPID PERCOLATION	80	NONE	-	163	66	1.2
WR	PROLONGED SATURATION	90	RAPID PERCOLATION	80	30	12	0.2
F	FLOODING	90	PERCOLATION (RAPID or SLOW)	80	171	69	1.4
(solid black)	OTHER UNSUITABLE AREAS [Swamps, peat bogs, salt marsh, coastal beach, eroded land]	95		-	444	179	3.5

[1]NOTE: Site modifications may be introduced which overcome these limitations. Techniques include the addition of fill, drainage to lower the water table and possibly deep tillage to increase the percolation rate. Detailed investigation of local site conditions are required to determine the feasibility and cost of overcoming the site limitations.

[2]PROBABILITY OF OCCURRENCE

(cont.)

CASE STUDY (cont.)

These criteria were compared with the soil survey data base and six operational criteria were defined: (1) percolation rate too slow, (2) bedrock less than six feet from surface, (3) prolonged saturation, (4) seasonal saturation, (5) susceptibility to flooding, and (6) percolation rate too high. In addition to the soil survey, data from public health inspectors, well-drilling logs, as well as additional field tests were used in the mapping program.

In attempting to synthesize the data, inconsistencies were often uncovered between the soil descriptions included in the soil survey and observations based on field inspection. This was attributed to two factors: the scale of mapping and the time or season of field testing. Soil moisture varies seasonally; therefore, in periods of low moisture, soils may appear to perform satisfactorily if subjected only to a single percolation test. To identify soils with seasonal limitations, it was found that percolation tests should be performed only in May, early June, and November in most years.

The scale of mapping was a more difficult problem. Because the Province is covered by fourteen soils survey reports at small map scales (one inch to one mile, one inch to two miles, and one inch to one-third mile), the detail necessary for site specific investigations was missing. For this reason, it was necessary to use a mapping technique based on the probability of occurrence of a limitation. This consisted of identifying the one or two major limitations of each soil type and then sampling that soil unit to determine the consistency of occurrence of those limitations. The overall consistency of occurrence ranged from 80 to 90 percent, although in some areas with a bedrock limitation, bedrock was only identified in 50 percent of the sites. A probability of 80 percent indicates that a given limitation could be expected in four out of five development sites selected at random within a given soil mapping unit.

Since mapping was based on the Nova Scotia soil survey, the largest scale maps that could be produced by this project were 1:50,000. These maps have been put to a variety of uses: first, as public information vehicles to illustrate the extent and distribution of problem soils; and second, in communities where there has been a reluctance to accept on-site test results, to provide corroborating evidence of problem soils. This has not only strengthened local planning regulatory processes, but has provided guidelines for municipal and regional development plans as well.

Michael D. Simmons is a resource planner with Maritime Resource Management Service, Amherst, Nova Scotia.

ASSESSING SOIL SUITABILITY FOR ON-SITE DISPOSAL

The suitability of an individual site for wastewater disposal is best determined by field inspection and percolation tests. For planning problems involving large areas, however, this is not always possible to carry out in detail, and it is necessary to resort to less expensive methods for an assessment of soil suitability. In most cases, the investigator must rely on existing maps and data, as is illustrated by the Nova Scotia case study. The best sources in the United States are the county soil reports produced by the Soil Conservation Service; in Canada, they are soils maps produced by the provinces or regional agencies. Where SCS studies are not available, the investigator must turn to other sources and synthesize existing maps and data into information meaningful to wastewater disposal problems. The maps in Fig. 6.5 are part of a set of maps produced by the United States Geo-

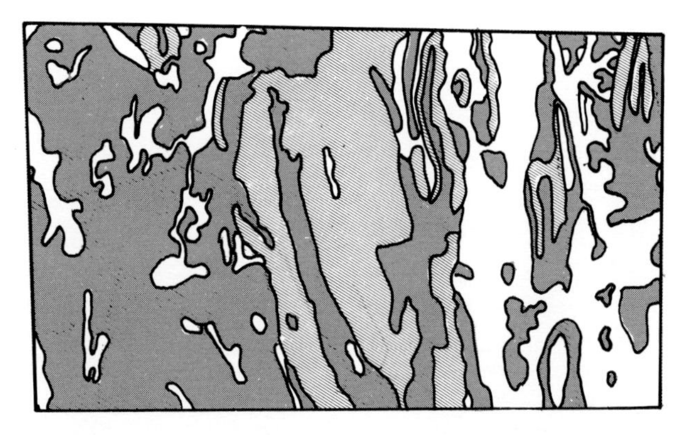

Slope

- 0 to 3 percent
- 3 to 15 percent
- 15 percent and greater

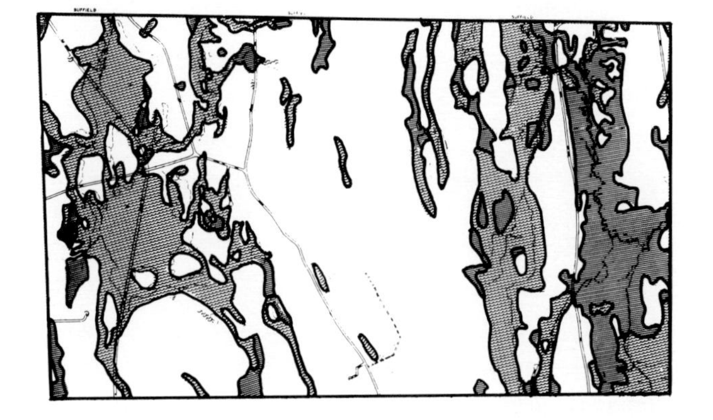

Seasonal High Watertable

- Areas well drained
- High water table 1 to 2 months of the year
- High water table 2 to 12 months of the year

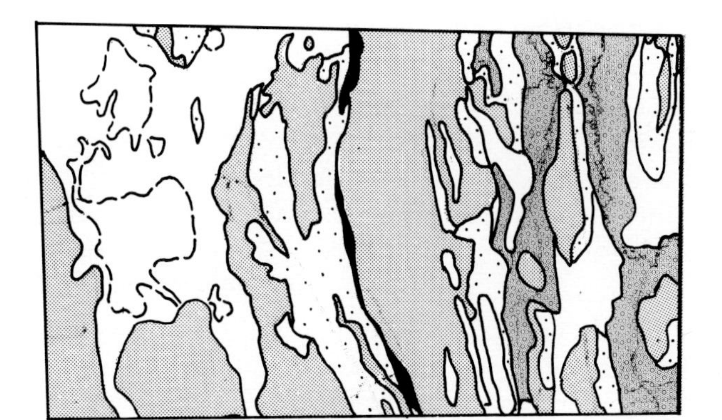

Unconsolidated Materials

- Till
- Compact till
- Clay deposit
- Sliderock deposit
- Swamp
- Sand or sand and gravel deposit

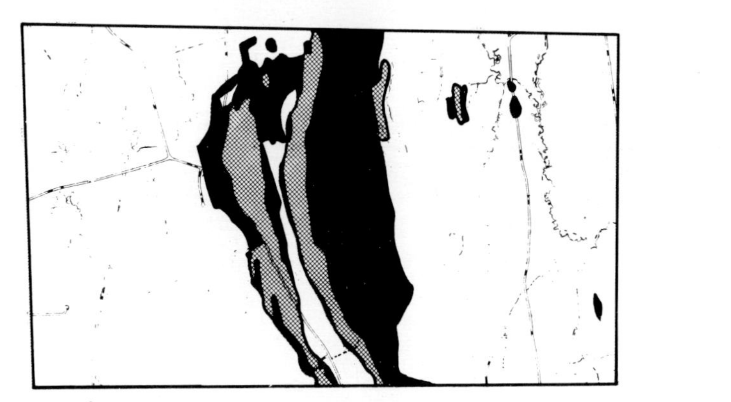

Depth to Bedrock

- Bedrock within 2 feet
- Bedrock within 10 feet
- Bedrock deeper than 10 feet

0 4000 FEET

Fig. 6.5 A series of maps compiled as part of a land capability study that can be used to assess on-site wastewater disposal suitability. (From the U.S. Geological Survey.)

logical Survey for a land use capability study in Connecticut. Three of these maps (slope, seasonal high watertable, and depth to bedrock) are directly applicable to wastewater disposal problems; however, the fourth map, called "unconsolidated materials," must be used in place of soil type or soil texture. The six classes of materials must be translated into their soil counterparts, and the soils, in turn, into their potential for wastewater disposal; for example:

Material	*Approximate Soil Equivalent*	*Potential for Wastewater Disposal*
till	loam	good
compacted till	loam, perhaps clayey, that drains poorly	fair/poor
clay deposit	clay; clayey loam	poor
sliderock deposit	rock rubble	poor
swamp	organic: muck, peat	poor
sand or sand and gravel deposits	sand; sandy loam; gravel	fair

With this in hand, we can proceed to combine the four maps and produce a suitability map for wastewater disposal. One procedure for this sort of task involves assigning a value or rank to each trait according to its relative importance in wastewater absorption. In this case, three criteria are given for each trait and each can be assigned a numerical value (Table 6.1). (Unconsolidated materials are reduced to good, fair, and poor according to the translation above.) For any area or site, the relative suitability for wastewater disposal can be determined by simply summing the four values. The final step is to establish a numerical scale defining the limits of the various suitability classes. The classes should be expressed in qualitative terms, for example, as high, medium, or low, because as numerical values they are meaningless and often misleading in terms of their relative significance.

Modern soil surveys by the Soil Conservation Service present a somewhat different arrangement. First, all data and information are organized and presented according to soil type. Second, in many areas the SCS provides information on soil suitability for wastewater disposal by SAS; in that

Table 6.1 Criteria and Numerical Values for SAS Suitability for Maps in Figure 6.5

	Material	*Depth to Bedrock*	*Seasonal Watertable*	*Slope*
3 (good)	till	>10 ft.	well drained	0–3%
2 (fair)	sand, sand and gravel, compacted till	2–10 ft.	1–2 mos. high watertable	3–15%
1 (poor)	clay, sliderock, swamp	<2 ft.	2–12 mos. high watertable	>15%

case, all one need do is to map the various SCS classes and record the criteria and rationale for each. However, the criteria and rationale used are not always clear, or one may wish to add or delete certain criteria depending on the land use problem and/or local conditions. In such instances, the procedure would be largely the same as that outlined previously, except that the mapping unit is already established (Fig. 6.6) and the data base may be different. County soil reports often include data on permeability, seasonal high watertable, and depth to bedrock as well as texture and slope. Where data are available for these criteria, the steps listed in Table 6.2 can be used to build a soil suitability map for wastewater disposal.

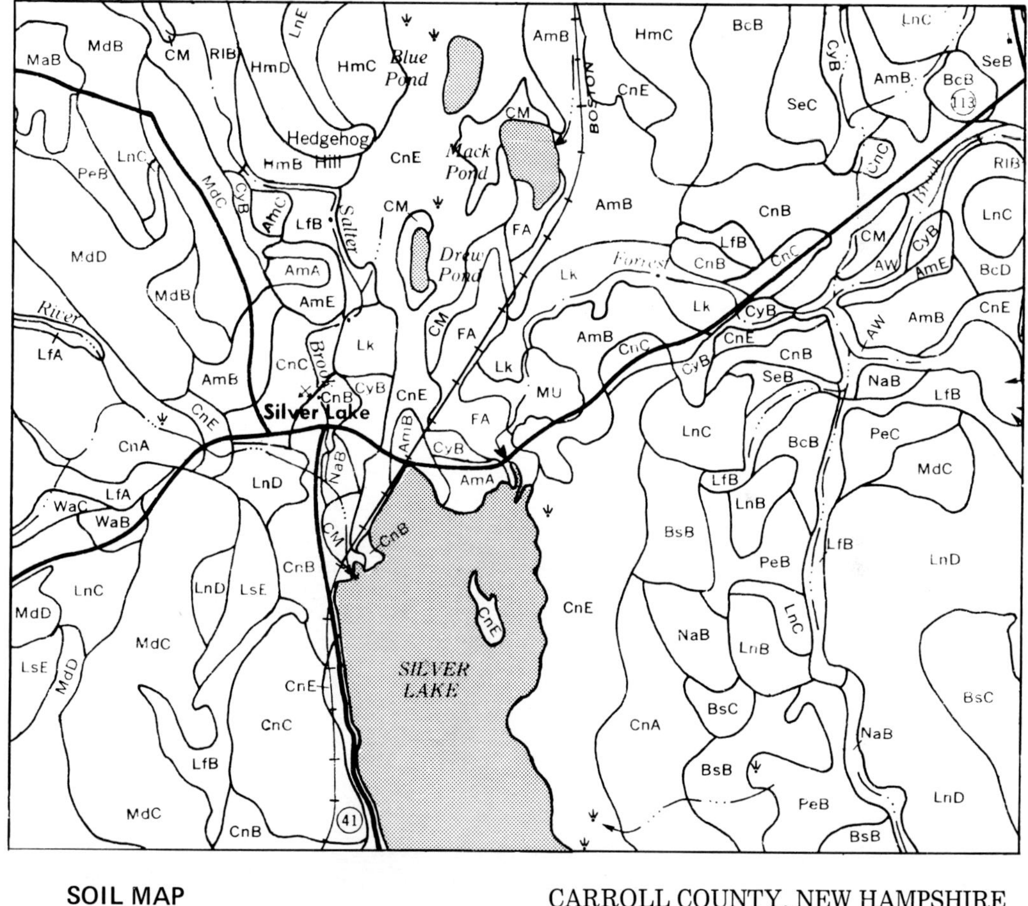

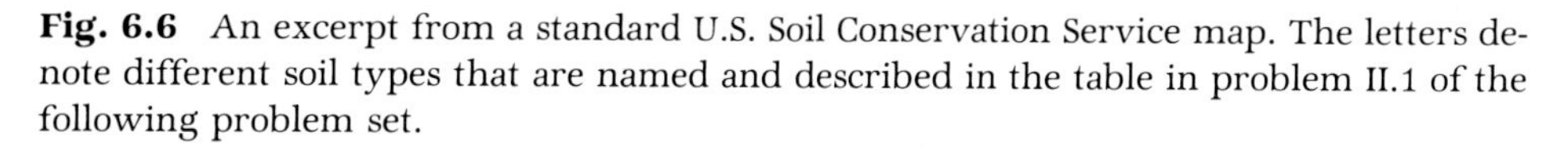

Fig. 6.6 An excerpt from a standard U.S. Soil Conservation Service map. The letters denote different soil types that are named and described in the table in problem II.1 of the following problem set.

Table 6.2 Procedure for Soil Suitability Classification for SAS

Step 1. Select the traits for evaluation and build a matrix listing each soil type and each trait, for example:

			TRAIT		
Soil	*Texture*	*Depth to Bedrock*	*Depth to Seasonal Watertable*	*Permeability*	*Slope*
Becket:					
Berkshire:					
Chocorua:					

Step 2. Set up a ranking system such as that in Table 6.1 and complete the matrix by assigning a numerical value to each soil trait.

Step 3. Total the ranking values for each soil series.

Step 4. Based on these totals, classify the soils as high, medium, or low.

Step 5. Go to the soils map and color the areas that correspond to each suitability class. For this you will need to devise a three-part color scheme and map legend.

PROBLEM SET

I. The following map shows the land use in the area covered by the maps in Figure 6.5:

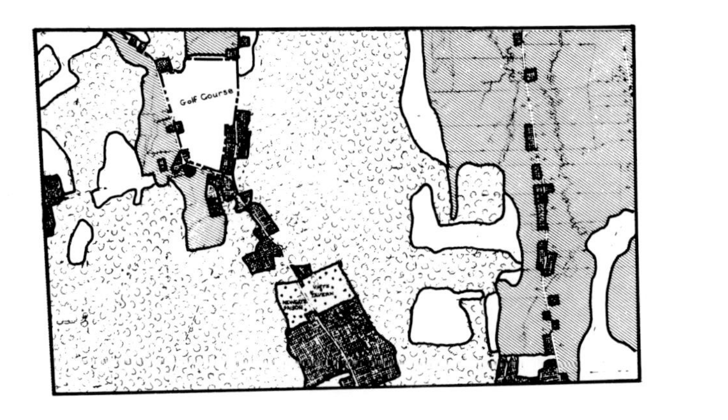

Existing Land Use

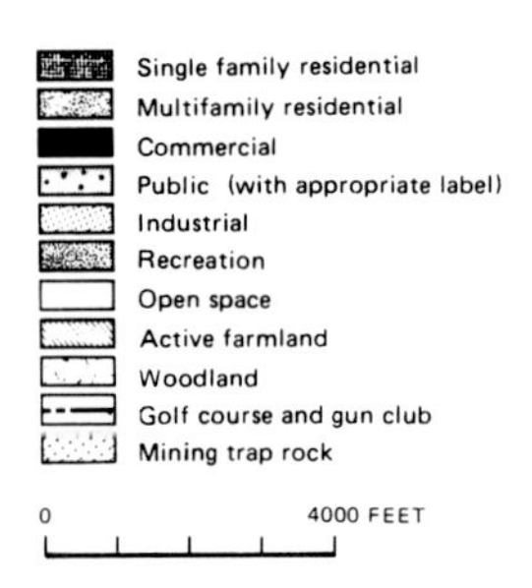

1. Delimit the areas of residential land use on a sheet of tracing paper and, placing it over each map, define the residential areas that fall into the "poor" category in Table 6.1. Shade in those areas that have two or more poor rankings.
2. Provide a brief explanation of the nature of the limitations in each shaded area in a form that would be suitable to a nontechnical audience.

II. The following soils data pertain to the area shown in the United States Soil Conservation Service map in Fig. 6.6.

1. Using these data, classify the suitability of soils for wastewater disposal based on a soil-absorption system. This will require devising a numerical evaluation scheme along the lines of the one in Table 6.1 and following the procedure outlined in Table 6.2.

2. Based on the above results, devise a map showing the distribution of three suitability classes.

3. Suppose you have to explain to an unhappy person who holds land in the area classed "low suitability" just what is meant by this description. Emphasize how the system works, and what the consequences of malfunction could be. By the way, alternatives to the standard drainfield SAS do exist and have been found to be quite suitable for certain settings. These include terraced drainfield systems, earth-mound systems, composting systems of various types, and neighborhood-scale pipe collection systems using large drainfields, lagoons, and/or portable treatment systems.

Soil Data

Soil Series and Map Symbols	*Depth to Bedrock (feet)*	*Depth to Seasonal High Water Table (feet)*	*Texture, at Depths (inches)*	*Permeability (inches/hour)*	*Slope*[1] *(percent)*
Adams: AmA, AmB, AmC, AmE	>8	>5	Loamy sand (0–18) Sand (18–32) Coarse sand (32–50)	>6.0 >6.0 >6.0	A = 0–3% B = 3–8% C = 8–15%
Alluvial land: AW		0 (Subject to frequent flooding)	Variable	No valid estimates can be made	D = 15–25% E = >25%
Becket: BcB, BcD	>4	>2½	Fine sand loam (0–18) Fine sandy loam (18–24) Gravelly loamy sand (24–42)	0.6–2.0 0.6–2.0 0.2–0.6	
Berkshire: BsB, BsC	>4	>4	Fine sandy loam (0–13) Fine sandy loam (13–24) Gravelly fine sandy loam (924–48)	0.6–2.0 0.6–2.0 0.6–2.0	
Chocorua: CM	>8	0	Organic material (0–31) Coarse sand (31–42)	0.6–2.0 >6.0	
Colton: CnA, CnB, CnC, CnE	>8	>5	Gravelly loamy fine sand (0–8) Gravelly loamy coarse sand (8–16)	>6.0 >6.0	

(cont.)

Soil Data (cont.)

Soil Series and Map Symbols	*Depth to Bedrock (feet)*	*Depth to Seasonal High Water Table (feet)*	*Texture, at Depths (inches)*	*Permeability (inches/hour)*	*Slope[1] (percent)*
Croghan: CyA, CyB	>8	1½–2	Loamy fine sand (0–28)	>6.0	
			Loamy fine sand (28–50)	>6.0	
Hermon: HmC, HmD, HnD, HmE, HnE	>4	>4	Fine sandy loam (0–4)	2.0–6.0	
			Fine sandy loam and gravelly sandy loam (4–19)	2.0–6.0	
			Gravelly loamy sand (19–50)	>6.0	
Limerick: Lk	>8	0–1 (periodically flooded)	Silt loam (0–36)	0.6–2.0	
			Sand and fine gravel (36–50)	>6.0	
Lyman: LnC, LnE, LsD, LsE, LfB, LnB	1–2	<0 (above bedrock)	Fine sandy loam (0–7)	0.6–2.0	
			Fine sandy loam (7–18)	0.6–2.0	
Marlow: MaB, MdB, MdC, MdD	>4	>2½	Fine sandy loam (0–6)	0.6–2.0	
			Gravelly fine sandy loam (6–20)	0.6–2.0	
			Fine sandy loam (20–42)	0.2–0.6	
Muck and Peat: MU, FA		0 (Subject to flooding or ponding)	Variable	No valid estimates can be made	
Peru: PeB, PeC	>4	1–2½	Fine sandy loam (0–18)	0.6–2.0	
			Sandy loam (18–24)	0.6–2.0	
Naumbury: NaB			Gravelly fine sandy loam (24–50)	0.2–0.6	
Ridgebury: RIB	>4	0–1	Fine sandy loam (0–6)	0.6–2.0	
			Fine sandy loam and sandy loam (6–20)	0.6–2.0	
			Fine sandy loam and gravelly fine sandy loam (20–50)	0.2–0.6	
Skerry: SeB, SeC	>4	1–2½	Fine sandy loam (0–19)	0.6–2.0	
			Gravelly fine sandy loam (19–23)	0.6–2.0	
			Gravelly loamy sand (23–45)	0.2–0.6	
Waumbek: WaB, WaC	>4	1½–2½	Fine sandy loam (0–7)	2.0–6.0	
			Fine sandy loam and loamy fine sand (7–19)	2.0–6.0	
			Gravelly loamy fine sand and gravelly loamy sand (19–26)	>6.0	

[1]See suffix letter (A, B, C, D, E) of map symbol.

SELECTED REFERENCES FOR FURTHER READING

Clark, John W., et al. "Individual Household Septic-Tank Systems." In *Water Supply and Pollution Control.* 3rd ed. New York: IEP/Dun-Donnelley, 1977, pp. 611–621.

Cotteral, J. A., and Norris, D. P. "Septic-Tank Systems." *Proceedings American Society of Civil Engineers,* Journal of Sanitary Engineering Division 95, no. SA4, 1969, pp. 715–746.

Environmental Protection Agency. "Alternatives For Small Wastewater Treatment Systems." *EPA Technology Transfer Seminar Publication,* EPA-625/4-77-011, 1977.

Huddleston, J. H., and Olson, G. W. "Soil Survey Interpretation for Subsurface Sewage Disposal." *Soil Science,* vol. 104, 1967, pp. 401–409.

Public Health Service. "Manual of Septic Tank Practice." United States Public Health Service Publication No. 526, Washington, D.C.: Government Printing Office, 1957.

7

STORMWATER DISCHARGE AND LANDSCAPE CHANGE

INTRODUCTION

One of the most serious problems associated with land development is the change in the rate and amount of runoff reaching streams and rivers. Both urbanization and agricultural development effect an increase in overland flow, resulting in greater magnitude and frequencies of peak flows on streams. The impacts of this change are serious, both financially and environmentally: damage from flooding is increased, water quality is reduced, and channel erosion is accelerated.

Responsible planning and management of the landscape depend on accurate assessment of the changes in runoff brought on by land development. In the United States and parts of Canada this problem has reached epidemic proportions, and in most communities developers are now required to provide analytical forecasts of the changes in overland flow and stream discharge expected from a proposed development. Such forecasts are used not only as the basis for recommending alternatives to traditional stormwater systems in order to reduce environmental impact, but also for evaluating the performance of an entire watershed that is subject to many development proposals.

OVERLAND FLOW

Most precipitation reaching the ground is disposed of in two ways. Some is absorbed directly by the soil in a process known as *infiltration.* The remainder, called *overland flow,* runs off the surface, eventually joining streams and rivers. The ratio of infiltration to overland flow varies drastically in the modern landscape. In heavily vegetated terrain, infiltration appears to be so high that overland flow is practically negligible; streams gain their discharge instead from groundwater, interflow, and channel precipitation. In dry areas and areas where forest and ground cover have been cleared and replaced by agriculture, settlements, and related land uses, overland flow is, by contrast, substantial.

If we examine the ground to determine what factors control overland flow, we will find that land cover (vegetation and land use), soil type, and surface inclination (slope) are the chief controls. Figure 7.1 illustrates the relationship among overland flow, soil, and vegetative cover on a hillslope. As a general rule, overland flow increases with slope, decreases with soil organic content and particle size, increases with ground coverage by hard surface material such as concrete and asphalt, and decreases with vegetative cover.

For a particular combination of these factors, a *coefficient of runoff* can be assigned to a surface. This is a dimensionless number between 0 and 1.0 that represents the proportion of a rainfall remaining on the ground after infiltration has occurred. A coefficient of 0.60, for example, means that 60 percent of rainfall (or snowmelt) is available for overland flow, whereas 40 percent was lost to infiltration.[1] Table 7.1 lists some standard coefficients of

1. Actually several factors contribute to the uptake of rain by the landscape, including interception by vegetation, detention by surface objects, and evaporation.

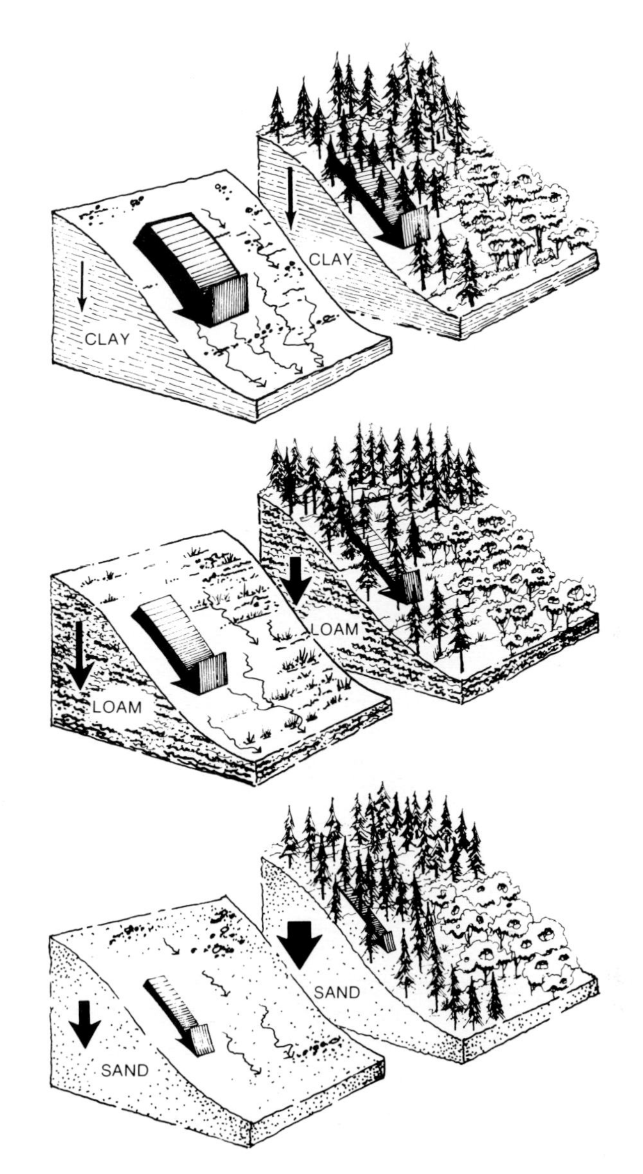

Fig. 7.1 Schematic illustration showing the relative changes in overland flow with soil type and vegetation. (From W. M. Marsh and J. Dozier, *Landscape: An Introduction to Physical Geography*, © 1981, Addison-Wesley, Reading, Massachusetts. Fig. 11.4. Reprinted with permission.)

runoff for rural areas based on slope, vegetation, and soil texture. For urban areas, coefficients are mainly a function of the hard surface cover (Table 7.2).

COMPUTING RUNOFF FROM A SMALL WATERSHED

The runoff generated from a small watershed (less than two to three square kilometers in area) can be computed by means of the *rational method.* This method is based on a formula that combines the coefficient of

Table 7.1 Coefficients of Runoff for Rural Areas

Topography and Vegetation	*Open Sandy Loam*	*Clay and Silt Loam*	*Tight Clay*
Woodland			
Flat (0–5% slope)	0.10	0.30	0.40
Rolling (5–10% slope)	0.25	0.35	0.50
Hilly (10–30% slope)	0.30	0.50	0.60
Pasture			
Flat	0.10	0.30	0.40
Rolling	0.16	0.36	0.55
Hilly	0.22	0.42	0.60
Cultivated			
Flat	0.30	0.50	0.60
Rolling	0.40	0.60	0.70
Hilly	0.52	0.72	0.82

Table 7.2 Coefficients of Runoff for Selected Urban Areas

Area	Coefficient
Commercial:	
Downtown	0.70–0.95
Shopping centers	0.70–0.95
Residential:	
Single family (5–7 houses/ac)	0.35–0.50
Attached, multifamily	0.60–0.75
Suburban (1–4 houses/ac)	0.20–0.40
Industrial:	
Light	0.50–0.80
Heavy	0.60–0.90
Railroad yard	0.20–0.80
Parks, Cemetery	0.10–0.25
Playgrounds	0.20–0.40

runoff with the intensity of rainfall and the area of the watershed. The outcome gives the peak streamflow (discharge) for one rainstorm:

$$Q = A \cdot C \cdot I$$

where

Q = discharge in cubic feet per second
A = area in acres
C = coefficient of runoff
I = intensity of rainfall in feet/hour

Besides the obvious need to measure the area of the watershed and determine the correct coefficient of runoff, it is necessary to select the appropriate rainfall intensity value. This begins with the identification of the desired size and duration of the rainstorm, for example, 30-minute 5-year,

60-minute 10-year, or the 2-hour 100-year.[2] The quantity of rain produced by this storm for any area can be read from a rainfall intensity map (Fig. 7.2). The second step is based on two facts: (1) within the period of a rainstorm, say, 60 minutes, the actual intensity of rainfall initially rises, hits a peak, and then tapers off; and (2) the time taken for runoff to move from the perimeter to the mouth of the watershed, called the *concentration time*, varies with the size and conditions of the watershed. Combining these two facts, you can see that in a small watershed the concentration time may be less than the duration of the storm. Therefore, in order to accurately compute the peak discharge for the storm in question, you must use the value of rainfall intensity that corresponds to the time of concentration. The graph in Fig. 7.3 gives representative curves for various storms and shows how to use the graph based on an example storm intensity of 1.75 inches per hour and a concentration time of 16.5 minutes.

ESTIMATING THE CONCENTRATION TIME

Reliable estimates of the concentration time are clearly important in computing the discharges from small watersheds. The approach to this problem generally involves making separate estimates for: (1) the time of overland flow; and (2) the time of channel flow, and then summing the two. The graph in Fig. 7.4 can be used to estimate the time of overland flow if three things are known: (1) the length of the path (slope) from the outer edge of the watershed to the head of channel flow; (2) the predominant ground cover; and (3) the average slope of the ground to the head of channel flow. If these are not known, one must resort to an approximation based on representative overland flow velocities (Table 7.3).

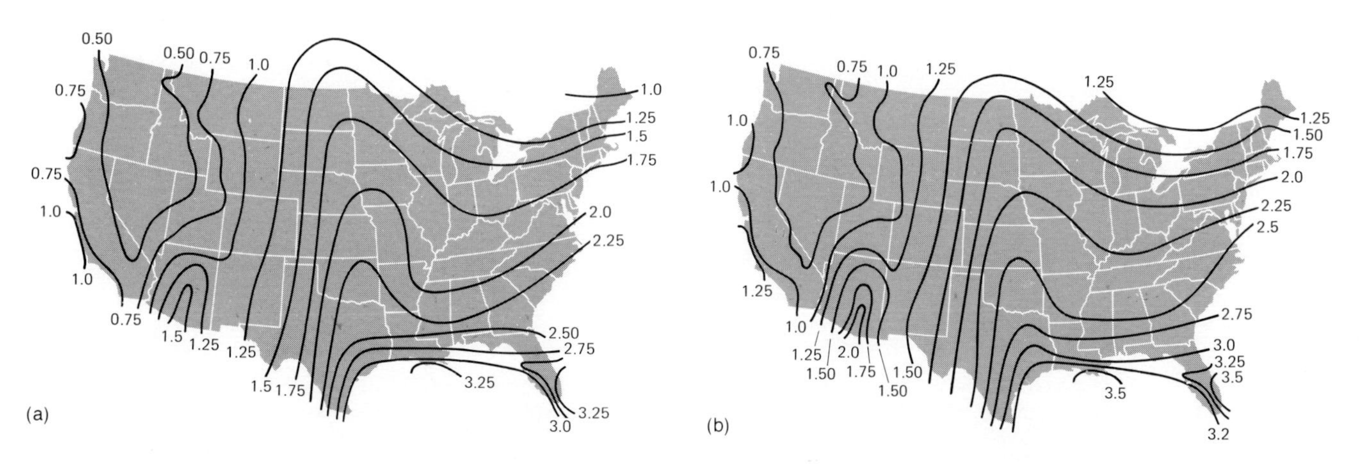

Fig. 7.2 The amounts of rainfall that can be expected in one hour from the 5-year and 10-year storms. (Based on U.S. Department of Agriculture data.)

2. The size and duration selected are often recommended by local agencies responsible for stormwater management. The 10-year and 100-year storms of 60 minutes duration are currently the most common "design" storms for stormwater computations in the United States.

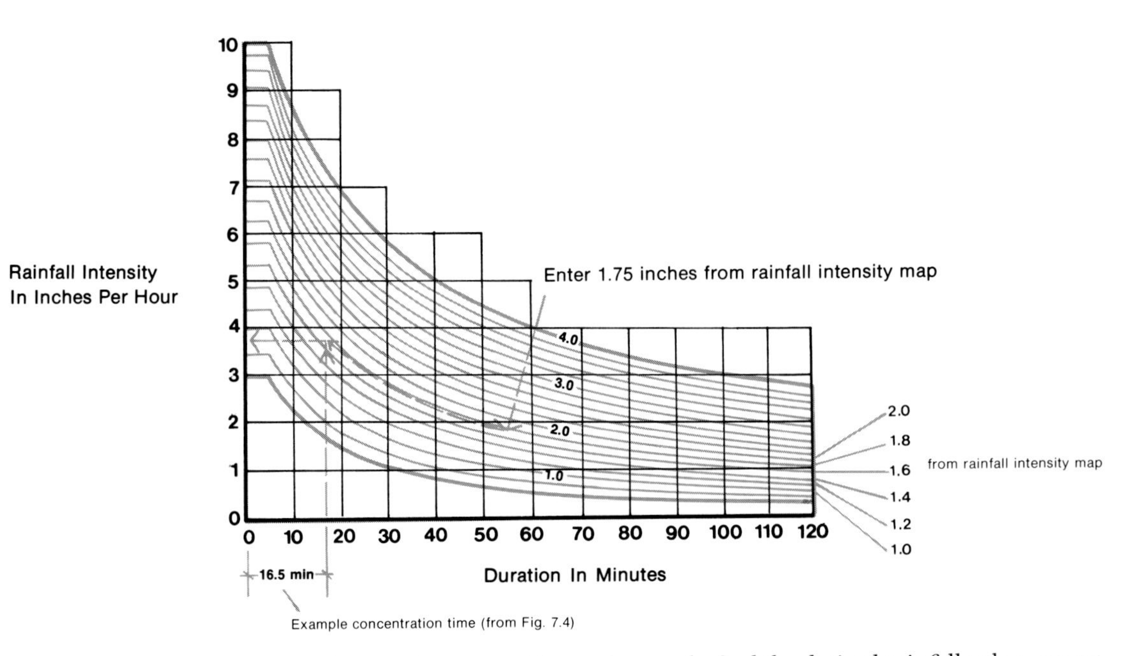

Fig. 7.3 Rainfall intensity curves. To use the graph, find the desired rainfall value among the curves and read the concentration time on the base of the graph as shown by the example. These particular curves do not work for all locations; curves can be obtained locally for most areas. (From U.S. Department of Agriculture, Miscellaneous Publication 204.)

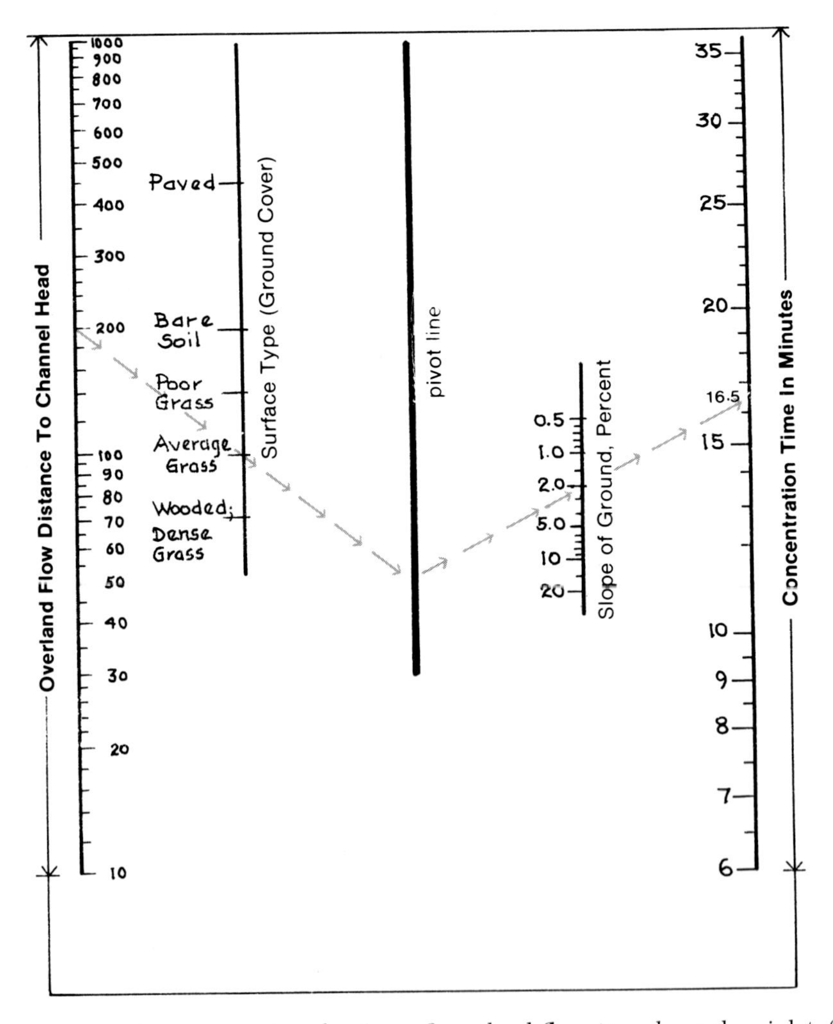

Fig. 7.4 A graph for estimating the time of overland flow to a channel or inlet. (From U.S. Department of Agriculture, Miscellaneous Publication 204.)

Table 7.3 Overland Flow Velocities

Distance to Channel Head	*Pavement*	*Turf*	*Barren*	*Residential*
Around 100 ft.	0.33 ft/sec	0.20 ft/sec	0.26 ft/sec	0.25–0.33 ft/sec
Around 500 ft.	0.82 ft/sec	0.25 ft/sec	0.54 ft/sec	0.50–0.65 ft/sec

The velocity of channel flow is generally much greater than that of overland flow, and if the slope, roughness, and geometry of the channel are known, velocity can be computed using the Manning formula:

$$v = 1.49\frac{R^{2/3}s^{1/2}}{n}$$

where

v = velocity, ft or m/sec.
R = hydraulic radius (wetted perimeter of channel divided by cross-sectional area of channel)
s = slope or grade of channel
n = roughness coefficient

Roughness Coefficients (n)	
Natural stream channel	
gravel bottom with cobbles and boulders	0.040–0.050
cobble bottom with boulders	0.050–0.070
Unlined open channel, soil bottom	0.020–0.025
Grassy swale, 12-inch grass	0.090–0.180
Overland flow surface	
compact clay	0.030
dense sod or shrubs	0.035–0.040

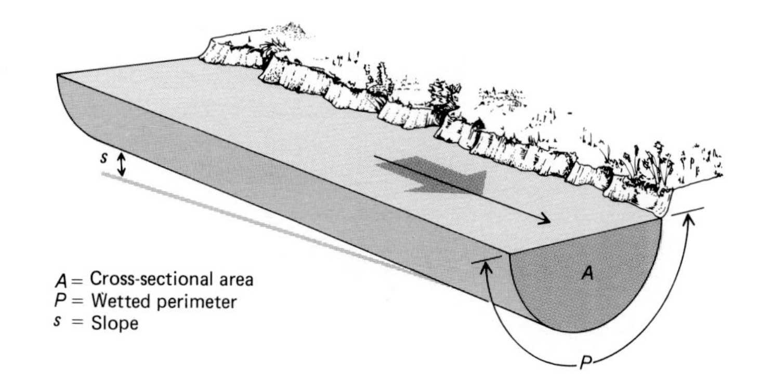

USING THE RATIONAL METHOD

As a forecasting device, the rational method is suited best not only to watersheds that are small but also to those that are partially or fully developed. Beyond that, the reliability of the method depends on the accuracy of the values used for the coefficient of runoff, the concentration time, the drainage area, and the rainstorm intensity. With the exception of the latter, the necessary data may be generated through field observations and surveys or from secondary sources, in particular, topographic maps, soils maps, and aerial photographs.

Given that the pertinent data are in hand, the following procedure may be used to compute the discharge resulting from a specified rainstorm. Steps 1 through 7 apply to peak discharge and step 8 to total discharge.

1. Define the perimeter of the watershed and measure the watershed area.
2. Subdivide the watershed according to cover types, soils, and slopes. Assign a coefficient of runoff to each, and measure each subarea.
3. Determine the percentage of the watershed represented by each subarea and multiply this figure by the coefficient of runoff of each. This will give you a coefficient adjusted according to the size of subarea.
4. Sum the adjusted coefficients to determine a coefficient of runoff for the watershed as a whole.
5. Determine the concentration time using the graph in Figure 7.4 and the Manning formula.
6. Select a rainfall value for the location and rainstorm desired. Using this value and the concentration time, identify the appropriate rainfall intensity from the curves in Fig. 7.3.
7. Multiply the watershed area times the coefficient of runoff times the rainfall intensity value to obtain peak discharge in cubic feet per second. (Actually, the answer is in acre inches per hour; however, these units are so close to cubic feet per second that the two are interchangeable.)
8. The total amount of discharge produced as a result of a storm can also be computed with the rational formula except in this case the one-hour rainfall value is used, and one would solve for acre feet or cubic feet per hour.

TRENDS IN STORMWATER DISCHARGE

The clearing of land and the establishment of farms and settlements has been the dominant geographic change in the North American landscape in the past 200 years. Almost invariably this has led to an increase in both the amount and rate of overland flow, producing larger and more frequent peak flows in streams. With the massive urbanization of the twentieth century, this trend has become even stronger and has led to increased flooding and flood hazard, to say nothing of the damage to aquatic environments.

From a hydrologic standpoint, the source of the problem can be narrowed down to changes in two parameters: (1) the drastic increase in the coefficient of runoff in response to land clearing, deforestation, and the addition of impervious materials to the landscape; and (2) the corresponding decrease in the concentration time (Fig. 7.5). In agricultural areas, the latter is brought about by the construction of field drains and ditches as well as the straightening and deepening of stream channels. With urbanization, ditches are replaced with stormsewers, small streams are piped underground, and gutters are added to streets, all of which may reduce concentration times by more than tenfold. Changes in a third parameter, rainstorm intensity, may also contribute to discharge increases; storm magnitudes appear to be on the rise in large metropolitan areas in the United States. At the present, however, it is impossible to generalize about this trend because it is apparent only in certain metropolitan areas.

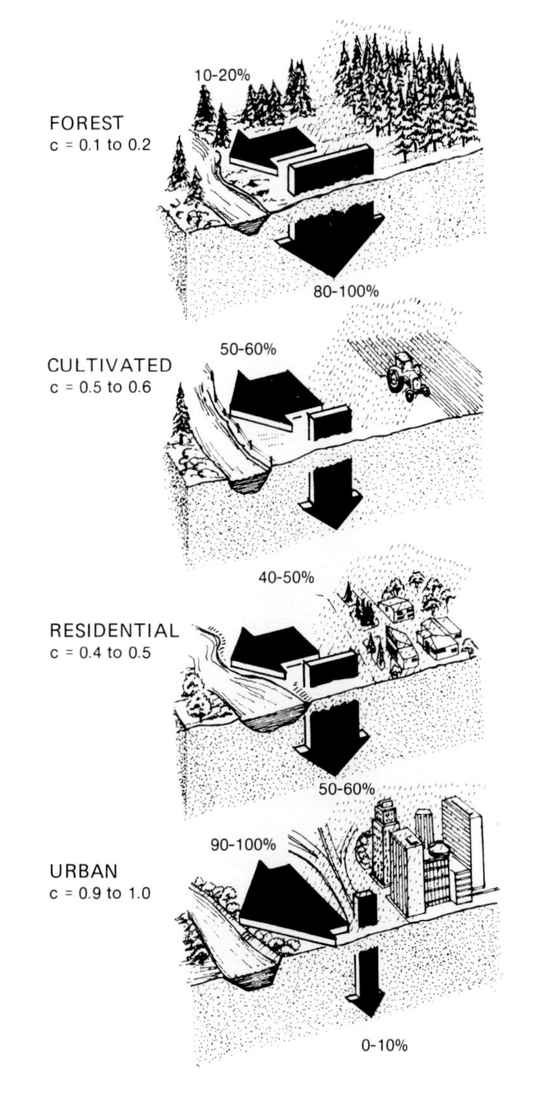

Changes in the coefficient of runoff with land use and land cover. (From W. M. Marsh and J. Dozier, *Landscape: An Introduction to Physical Geography*, © 1981, Addison-Wesley, Reading, Massachusetts. Fig. 13.20. Reprinted with permission.)

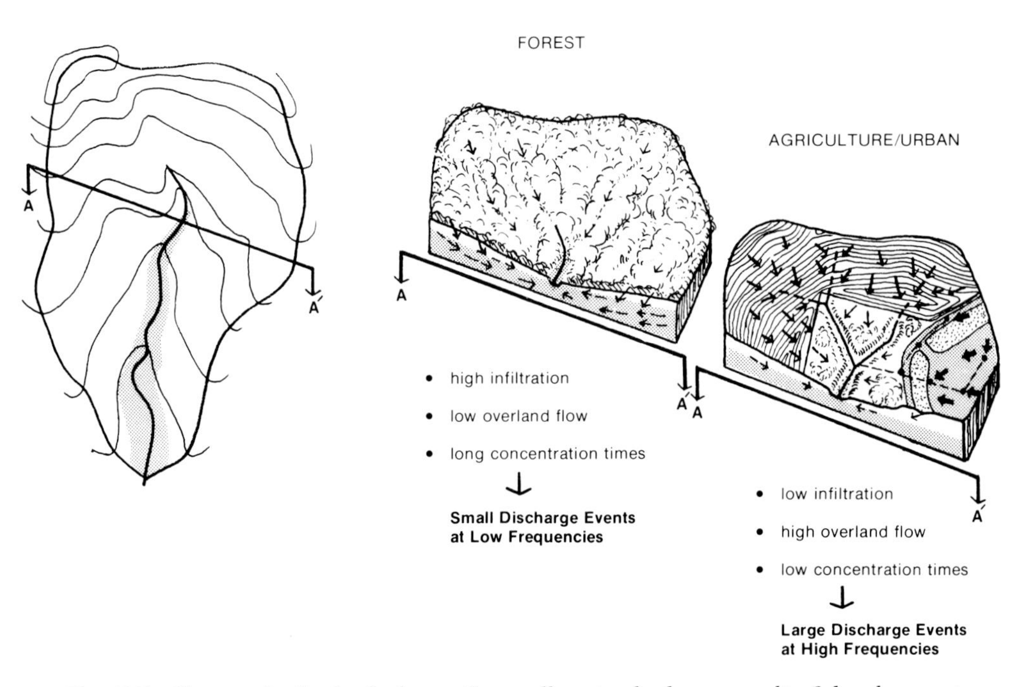

Fig. 7.5 Changes in the hydrology of a small watershed as a result of development.

STORMWATER PLANNING AND MANAGEMENT

Community ordinances increasingly place stringent performance standards on developers, often calling for zero net increase in stormwater discharge from a site as a result of development. This means that the rate of release across the border of a site can be no greater after development than before. A basic dilemma thus arises because most forms of development affect an increase in runoff, but at the same time themselves require relief from stormwater.

Three strategies may be used to achieve onsite management of stormwater: (1) store the excess water on site, releasing it slowly over a long time; (2) return the excess water to the ground, where it would have gone before development; (3) plan the development such that runoff is not significantly increased. The first is the most common strategy and it involves the construction of detention basins. These are ponds sized to store the design storm and then allow it to discharge at a specified rate (labeled 'A' in Fig. 7.6). For instance, if a design storm generates a total runoff of 4.2 acre feet (182,952 cubic feet) in one hour, and the allowable release rate is 12 cubic feet per second (43,200 cubic feet per hour), then about 3.2 acre feet of gross storage are needed.

The second strategy utilizes soil infiltration and is usually accomplished by the construction of a trough or pit, which is backfilled with crushed rock or gravel and sodded over. Stormwater is directed into the pit, from which it percolates into the ground. Since the void spaces constitute only 30 to 35 percent of the gravel, a large number of such facilities may be needed to store excess storm water on most sites (labeled 'B' in Fig. 7.6). On the other hand, in dry regions, this method may prove to be an effective means of recharging local soil moisture supplies.

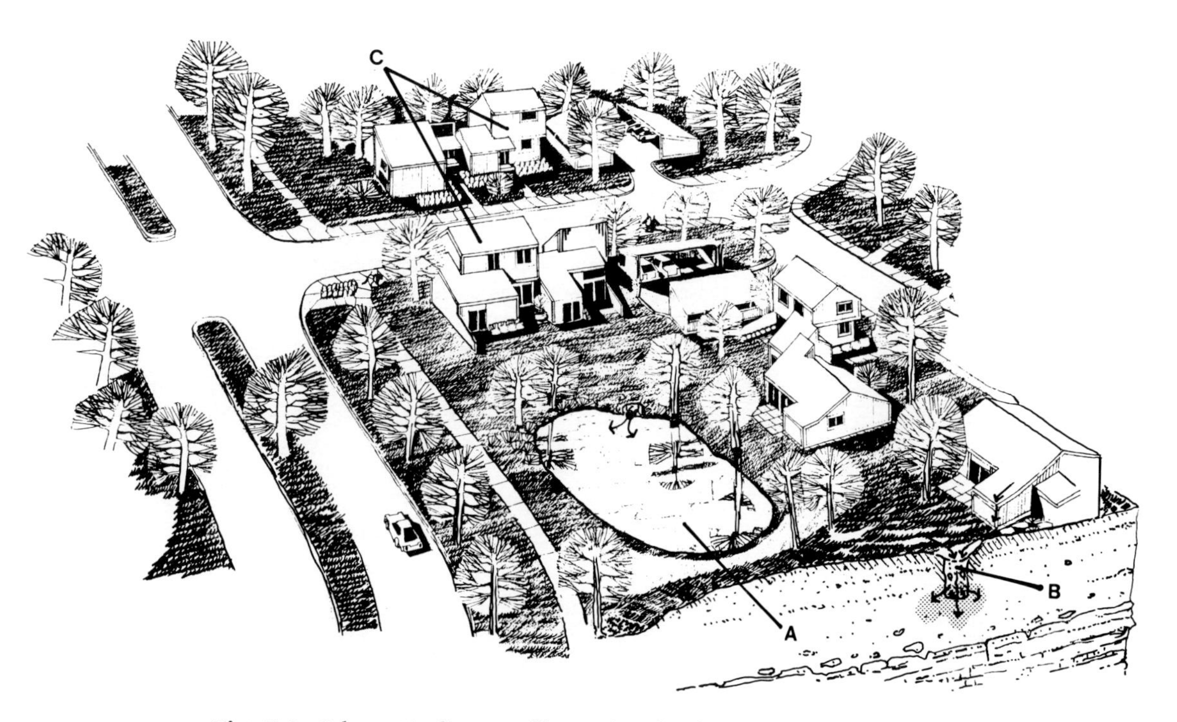

Fig. 7.6 Schematic diagram illustrating the three strategies that may be used to manage stormwater onsite: (A) onsite detention; (B) dry well disposal; (C) clustered development. (Adaptation on an original drawing by Howard Deardorff. Used by permission.)

The third strategy eliminates the need for corrective measures and relies on prudent site planning to resolve the problem. Strict attention is paid to surface materials, avoiding impervious materials wherever possible, and to density ratios of developed to open space. Planned unit developments (PUD) are one means of reducing impervious area by clustering buildings and related facilities (labeled 'C' in Fig. 7.6). Although it is possible to achieve zero net increase in runoff solely through careful site planning, especially when former farmland is involved, most successful stormwater management requires a combination of strategies.

THE CONCEPT OF PERFORMANCE

Any effort to plan and manage the environment rests on a concept about how the environment should perform. When plans are formulated, the objective is either to guide and structure future change in order to avoid undesirable performance and/or to improve performance in an environment, setting, or system whose existing performance is judged to be inadequate. The phrase *judged to be inadequate* is important because any judgment on performance is based on human values. "That stream floods too often and it poses a danger to local inhabitants," is a value judgment by someone about environmental performance. Likewise, an ordinance restricting development from wetlands reflects a societal value about either the performance of wetlands, the performance of land use, or both.

CASE STUDY

Watershed Management Planning in an Arid Environment

Richard A. Meganck

Northern Mexico, like the American Southwest, is a region of rapid economic growth with limited water resources. Surface water supplies are scarce, and in most areas development is heavily dependent on groundwater. Locally groundwater reserves are often linked to basins that are fed by precipitation on surrounding mountain slopes. From the flanks of the basins, water is transmitted through loose deposits into aquafers on the valley floors, and from these aquafers, water is extracted by cities, farms, and industry. Although the mountains receive more precipitation than the valleys and have better moisture balances, the rate of recharge of valley aquafers is generally low. To maintain a water supply over the long run, extraction rates should not exceed recharge rates, and sources of recharge should be well managed.

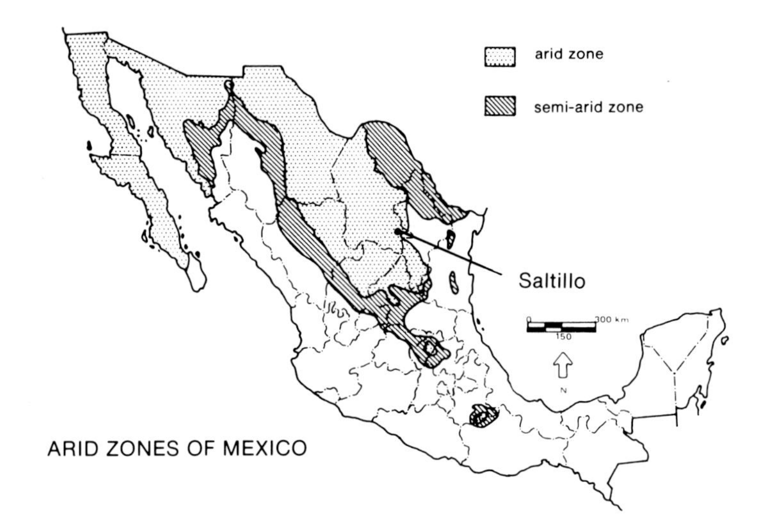

Saltillo, Mexico, with a population of 300,000, is situated in a desert basin of the Sierra Madre Oriental. Stimulated by a variety of tax incentives, automobile and steel plants have recently been constructed, and the population of Saltillo is expected to double within five years. As a result, land prices have quadrupled, a present shortage of 10,000 homes and 5,000 laborers exists, and demand for basic services has increased greatly. Most notable among these is the demand for water, 50 percent of which is pumped from a mountain basin just 7 km south of the city, the San Lorenzo Canyon. The canyon's primary land use is agriculture (mostly grazing) with 64 percent of the land held under *ejidos* (government subsidized agricultural villages) and 36 percent under private ownership.

Over the past several decades, poor land use practices have led to such degradation of the canyon that both its hydrologic and agricultural resources have declined seriously. Virtually all the original forest cover has been removed; overgrazing has promoted serious soil erosion and stream sedimentation; moreover, large amounts of soil have been mined to supply topsoil for gardens in the villages and cities. Together these activities have reduced the canyon's capacity to capture water and recharge the aquafers that are so essential to Saltillo. At the same time, Saltillo is mining water at increasing rates each year.

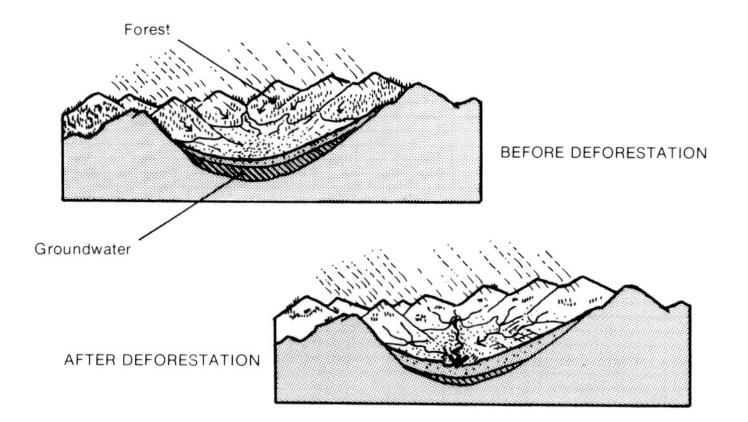

Unless remedial measures are taken in the San Lorenzo Canyon (and similar canyons), Saltillo will soon be faced with severe water shortages. Accordingly, the Mexican government and the Regional Development Program of the Organization of American States have agreed to develop a management plan for the San Lorenzo Canyon that could serve as a model for similar areas. Because of the land ownership situation and the long established land use practices of the ejidos, the plan must address both the social and the physical elements of the problem.

An interdisciplinary planning team was formed, management objectives determined, field data collected and analyzed, and a planning framework defined. Finally, a management program was elaborated aimed at the following goals:

- To insure long-term water production and protection of water recharge zones.
- To provide a more diversified and stable economic base for local rural inhabitants (a political reality for acceptance of the plan by local and state officials).
- To stabilize and protect ecosystem processes in the long term.

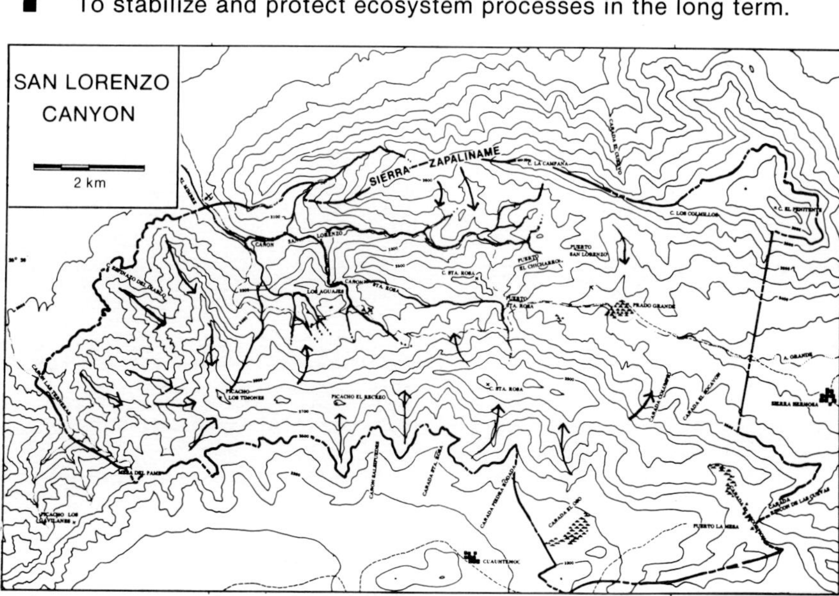

Maps and Photographs by Janet O. Meganck

(cont.)

CASE STUDY (cont.)

Soil mining by traditional methods. Soil is sold for gardens and landscaping in cities.

- To help meet the increasing demand for outdoor recreation and education.
- To resolve potential conflicts with private property owners.

In order to meet these goals and slow the rate of degradation of the canyon, the following actions have been recommended:

1. Enforcement of interim management measures, including an immediate cessation of forestry, grazing, hunting, trapping, and soil mining activities.

Soil mining by modern methods. Loss of soil mass reduces moisture retention and ground recharge potential of watersheds.

2. A request for international funds to help underwrite the change for local rural inhabitants from a forestry/grazing economy to one of improving established crop production and diversification to a fruit and Christmas-tree operations. (Local private enterprise has already proven the economic success of apple and pine tree production.)
3. Reforestation of approximately 6,000 hectares with 6 to 8 million native pines and junipers.
4. The implementation of soil conservation and fire management measures, elimination of undesirable vegetation, protection of vital and endangered habitat, and the eventual reintroduction and management of certain wildlife species.
5. Acquisition of resource development rights on private lands within the proposed multiple use management area through market value purchase, property exchange, life-lease agreements, or other arrangements.

Long-term management of the San Lorenzo Canyon according to this plan should provide a suitable framework to guide ecodevelopment, not only because it addresses the water system as a physical entity, but because it recognizes the complex social and cultural forces at work in the Mexican landscape and the fact that changes in land use practices must be based on sound economic and social incentives.

Richard A. Meganck, a resource planner, is a project director for the Regional Development Program, Organization of American States.

In watershed planning and management, *performance goals* must be set early in the program to determine the direction, quality, and appropriate level of detail of the planning process. Generally, the larger the watershed, the more difficult the task is because large watersheds usually involve many communities and interest groups with different views and policies on the matter. The task is usually less complex for a small watershed, not only because there are fewer players involved, but also because the watershed itself is less complex (Fig. 7.7).

The process of formulating performance goals usually begins with the definition of local (watershed land users) values, attitudes, and policies. Next, regional factors are considered; in particular, policies pertaining to development intensity, stormwater retention, wetlands, open space, and the like. In addition, the values and needs, both articulated and apparent, of downstream riparians must be given serious consideration. Where, for instance, upstream users depend on a watershed and its streams to remove stormwater, and downstream riparians have set a goal around the maintenance of habitat and quality residential land near water courses, the two goals are in conflict, and unless modified, the downstream group stands little chance of success.

A third consideration in setting performance goals is the carrying capacity of the watershed based on its biophysical character. The watershed must be evaluated as an environmental entity and weighed against the proposed goals based on local and regional values and policies. A watershed and streams that could not support a trout population under predevelopment conditions certainly cannot support one after development; therefore, a performance goal of habitat quality suitable to sustain trout would be physically unrealistic. It is usually the role of the environmental specialist in hydrology, soils, ecology, or forestry, for example, to examine the

Fig. 7.7 The performance of this watershed was improved by stormwater retention ponds and reforestation of waterways. The performance goals were to reduce soil loss and downstream flooding. (Photographs by the U.S. Soil Conservation Service.)

watershed, determine its condition and potentialities, and recommend appropriate modifications in proposed performance goals.

Once performance goals have been formulated, *performance standards* and controls have to be defined. Performance standards are the specific levels of performance that must be met if goals are to be achieved; for example, a stormwater discharge might call for zero net increase in peak discharges on first-order streams after a particular date or development density has been reached. This means that in planning new development it is necessary to take into account all changes in land use activities in the watershed at this time and determine from the balance of both positive and negative changes whether special measures or strategies are needed. If, for instance, cropland is being converted to woodlands at the same time woodland is being cleared for new residential development, analysis may show that one change offsets the other, thereby maintaining the performance standard of zero net change in stormwater discharge.

Performance controls are the rules and regulations used to enforce the standards and goals. These may be specific ordinances limiting the percentage of impervious surface, requirements for site plan review and approval, or incentives such as tax breaks for restoring open space. Controls are necessary because without them the plan has no real "teeth" and thus no regulatory strength.

PROBLEM SET

I. Examine the watershed maps in Fig. 7.8. Each map shows the same watershed at three different dates in its development. The watershed is located in central Kentucky.

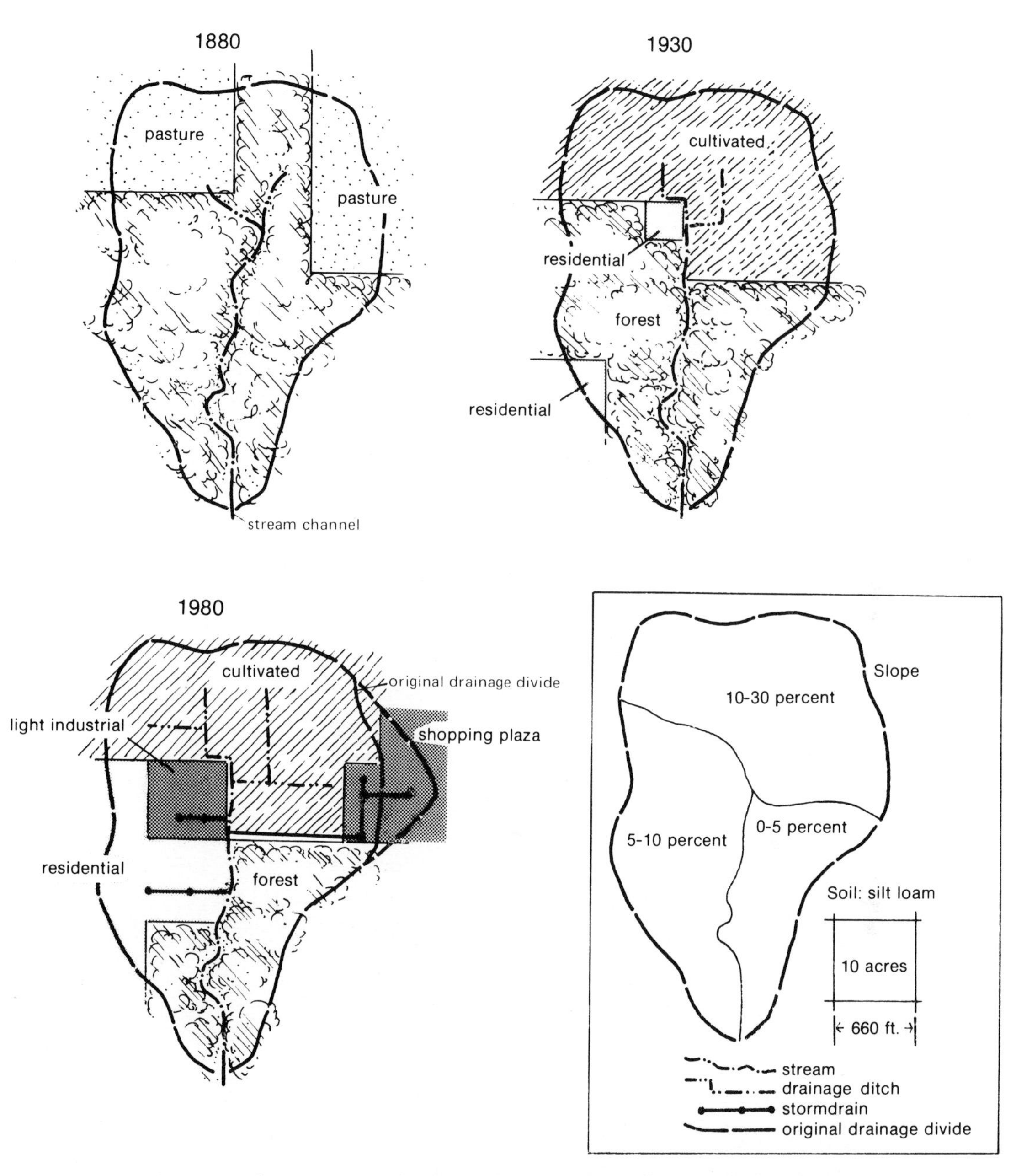

Fig. 7.8 Three stages of land use change in a small watershed, 1880, 1930, and 1980.

1. Using just the average coefficient of runoff, the basin area, and the 10-year 1-hour rainfall for this area, determine the total volume of runoff from the basin for 1880, 1930, and 1980.
2. What is the *peak discharge* for the 1880, the 1930, and the 1980 situation?
3. Construct a graph or graphs showing the change in: (a) peak discharge; (b) concentration time; and (c) the watershed coefficient of runoff between 1880 and 1980.
4. Forecast a peak discharge on the assumption that by the year 2005 the remaining forested area will be converted to residential use.

II. Describe several strategies or measures that could be used to improve the performance of the 1980 watershed in Fig. 7.8. Identify which of the land uses in the watershed at this time would be appropriate for which measure or strategy.

1. Describe what factors would have to be taken into consideration to determine the appropriate *level* of performance for this watershed.

SELECTED REFERENCES FOR FURTHER READING

Dunne, Thomas, and Black, R. D. "Partial-Area Contributions to Storm Runoff in a New England Watershed." *Water Resources Research*, vol. 6, pp. 1296–1311.

Ferguson, Bruce K. *Controlling Stormwater Impact.* American Society of Landscape Architects, Landscape Architecture Technical Information Services, 1981, 31 pp.

Hewlett, J. D., and Hibbert, A. R. "Factors Affecting the Response of Small Watersheds to Precipitation in Humid Regions." In *Forest Hydrology.* Oxford: Pergamon Press, 1967, pp. 275–290.

Horton, Robert E. "The Role of Infiltration in the Hydrologic Cycle." *American Geophysical Union Transactions*, vol. 14, 1933, pp. 446–460.

Marsh, William M., and Dozier, Jeff. "Runoff and Streamflow." In *Landscape: An Introduction to Physical Geography.* Reading, Mass.: Addison-Wesley, 1981, pp. 177–199.

Meganck, Richard. *Multiple Use Management Plan, San Lorenzo Canyon* (Summary). Organization of American States and Universidad Autonoma Agraria, 1981, 72 pp.

Poertner, Herbert G. *Practices in Detention of Urban Stormwater Runoff.* Chicago: American Public Works Association, 1974, 231 pp.

Seaburn, G. E. "Effects of Urban Development on Direct Runoff to East Meadow Brook, Nassau County, Long Island, New York." *U.S. Geological Survey Professional Paper 627–B*, 1969.

U.S. Soil Conservation Service. *Urban Hydrology For Small Watersheds.* Technical Release No. 55, Washington, D.C., Department of Agriculture, 1975.

8

STREAMFLOW AND FLOOD HAZARD

INTRODUCTION

The variability of river flow has been a serious issue for thousands of years in human settlement and land use. Until the past century or so, relatively little was understood about the sources of streamflow, especially as they relate to runoff. This is not surprising, though, because runoff is influenced by so many factors. Different combinations of rainfall, snowmelt, groundwater, soil moisture, and infiltration, for example, can produce a wide range in runoff conditions that result in widely different levels of river flow. In addition, alterations of drainage basins and river channels by humans can have marked influences on flow rates. As land use changes, runoff rates also change, producing corresponding changes in streamflow; accordingly, effective land use planning in river valleys is a source of constant challenge throughout the world.

In the annals of land use studies, river valleys hold a special status because in virtually all agrarian and industrial societies they are among the most attractive and yet hazardous environments for settlement. This, of course, is dramatically illustrated in history by the classical accounts of the great civilizations of the Nile, Indus, and Tigris-Euphrates valleys, which were both nurtured and plagued by their rivers. The dilemma persists in the modern world, and despite efforts to control riverflow, loss of life and property has increased in the past century. In America, river valleys were generally the first geographic settings to receive formal attention in land use planning. This came after 1930 for the most part and has since given rise to local and state ordinances in most areas on the use of floodplains. Nevertheless, the problem grows each year as development slips ahead of floodplain mapping programs and as more or less passive streams are changed to active ones by development in their watersheds.

The quantity of water carried by a stream is termed *discharge.* It is measured in volumetric units of water passing a point on a stream, such as a bridge, over time. The conventional units in English-speaking countries are cubic feet per second, abbreviated "cfs." In recent years, cubic meters have begun to replace cubic feet, but the latter are still widely used in science, engineering, and land planning.

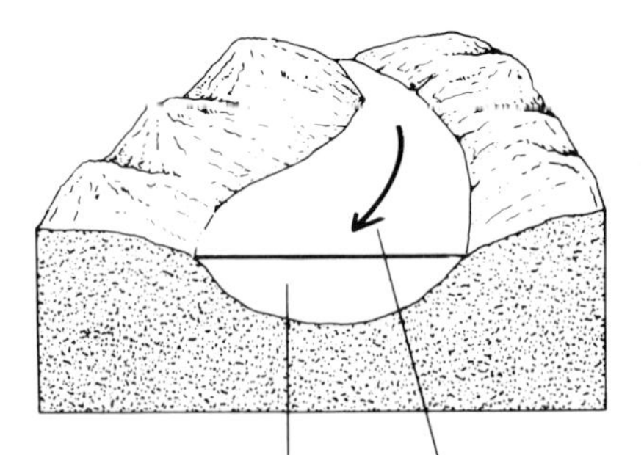

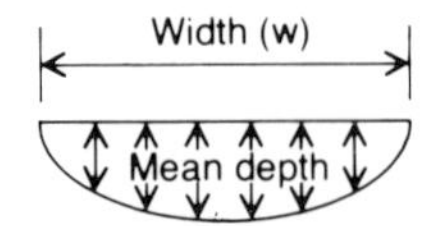

SOURCES OF STREAMFLOW

Streamflow is derived from four sources: channel precipitation, overland flow, interflow, and groundwater. *Channel precipitation* is the moisture falling directly on the water surface, and in most streams, it adds very little to flow. Groundwater, on the other hand, is a major source of discharge. Groundwater enters the stream bed where the channel intersects the watertable and provides a steady supply of water, termed *baseflow*, during both dry and rainy periods. Owing to the large supply of groundwater available to streams and the slowness of the response of groundwater to precipitation events, baseflow changes only gradually over time.

Interflow is water that infiltrates the soil and then moves laterally to the stream channel in the zone above the watertable. Much of this water is transmitted within the soil itself, some of it moving along the soil horizons. Until recently, the role of interflow in streamflow was not understood, but field research has begun to show that next to baseflow, it is the most important source of discharge for streams in forested lands. This research also revealed that overland flow in heavily forested areas makes negligible contributions to streamflow (Fig. 8.1).

Overland flow, however, is a major source of discharge in dry regions, cultivated lands, and urbanized areas. Overland flow is water that moves directly from the surface of the land into a stream. Since it is not filtered through the soil or the groundwater system, overland flow is released to the stream very rapidly, causing sudden increases in discharge. The quickest response times between rainfall and streamflow occur in urbanized areas where gutter and stormsewer systems are used to route overland flow to streams.

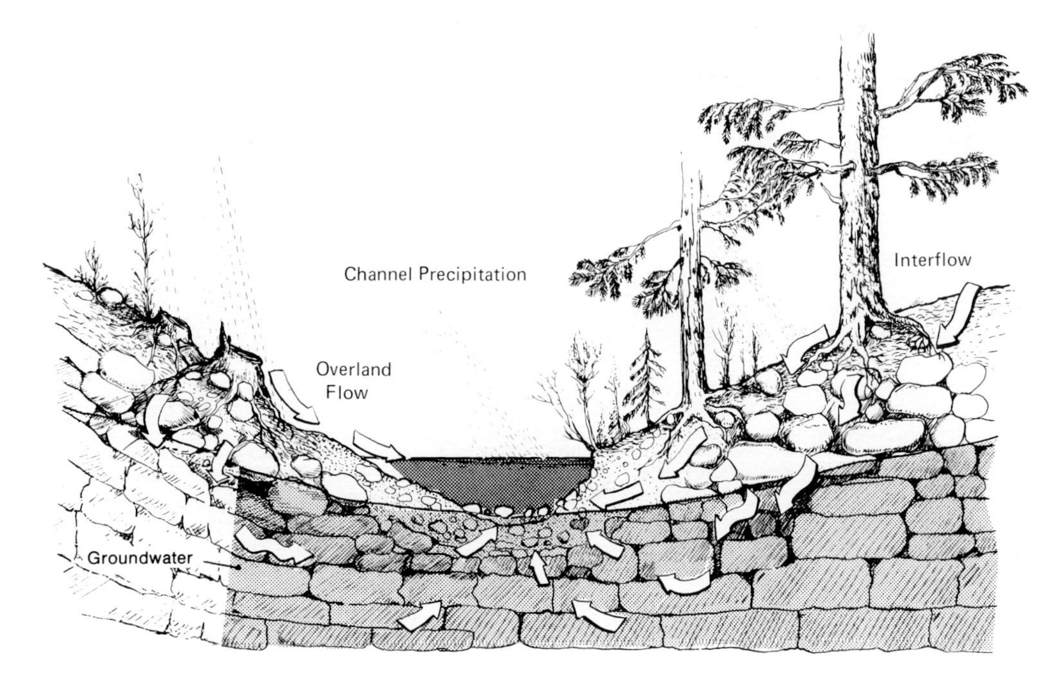

Fig. 8.1 Sources of streamflow: channel precipitation, overland flow, interflow, and groundwater.

METHODS OF FORECASTING STREAMFLOW

If the drainage area of a river is relatively small and no discharge records are available, one must make discharge estimates using the rational method; however, if chronological records of discharge are available, a short-term forecast of discharge can be made for a given rainstorm using a hydrograph. This method involves plotting the discharge from a watershed over time as the watershed and its streams respond to the rainstorm. The method is called the *unit hydrograph method* because it addresses only the runoff produced in a specified period of time, the time taken for a river to rise, peak, and fall in response to a particular storm. Using rainfall data, then, it is possible to forecast streamflow for selected storms, called standard storms. A *standard rainstorm* is a high-intensity storm of some recurrence (average return) period, for example, two years, five years or fifty years, that produces a known amount of rise in the river. One method of unit hydrograph analysis involves expressing the hour-by-hour or day-by-day increase in streamflow as a percentage of total runoff. Plotted on a graph, these data form the unit hydrograph for that storm, which represents the runoff added to the prestorm baseflow (Fig. 8.2).

Flow period	Total flow	Storm-flow
0-12 hrs	63 m^3/s	6 m^3/s
12-24	192	127
24-36	1065	991
36-48	1101	1019
48-60	714	623
60-72	453	354
72-84	275	170
84-96	194	85
96-108	144	28

Peak = 1253; 1175 at hour 36

Flow period	Percentage of total flow
0-12 hrs	0.2%
12-24	3.6
24-36	29.0
36-48	29.8
48-60	18.3
60-72	10.3
72-84	5.0
84-96	2.4
96-100	0.8

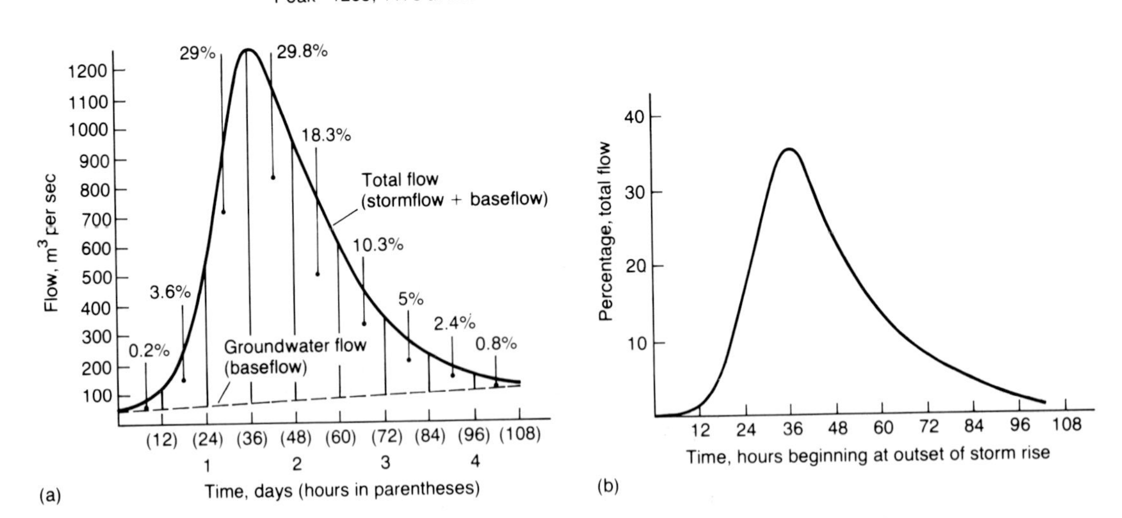

Fig. 8.2 Unit hydrograph for the uppermost part of the Youghiogheny River. Streamflow is expressed as a percentage of total flow. (From W. M. Marsh and Jeff Dozier, *Landscape: Introduction to Physical Geography*, © 1981, Addison-Wesley, Reading, Massachusetts. Fig. 13.23. Reprinted with permission. Data from C. O. Wisler and E. F. Brater, *Hydrology*, 2nd ed., New York: John Wiley, 1959.)

MAGNITUDE AND FREQUENCY METHOD

To forecast the flows in a large drainage basin using the unit hydrograph method would prove to be difficult because in a large basin conditions may vary significantly from one part of the basin to another. This is especially so with the distribution of rainfall, because the basin is rarely covered evenly by an individual rainstorm. As a result, the basin does not respond as a unit to a given storm, making it difficult to construct a reliable hydrograph. Instead, we turn to the *magnitude and frequency method* and calculate the probability of the recurrence of large flows based on records of past years' flows. In the United States, these records are maintained by the Hydrological Division of the U.S. Geological Survey for most rivers and large streams. For a basin with an area of 10,000 square km or more, it is not uncommon for the river system to be gauged at five to ten places. The data from each gauging station apply to the part of the basin upstream of that location.

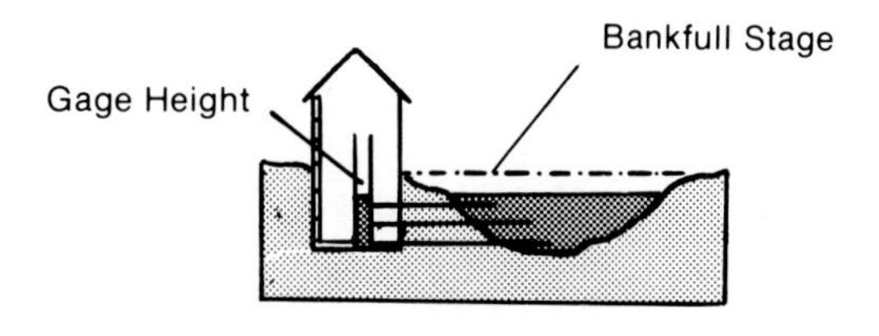

Gauging (Gaging) Station

Since we are mainly concerned with the largest flows, only *peak annual flows* are recorded for most rivers. With the aid of some statistical techniques, we can use these flow data to determine the probability of recurrence of a given flow on a river. Key data available for each gauging station include:

- *Peak Annual Discharge:* the single largest flow in cubic feet per second (cfs).
- *Date of Peak Annual Discharge:* the month and year in which this flow occurred.
- *Gauge Height:* the height of water above channel bottom.
- *Bankfull Stage:* gauge height when channel is filled to bank level.

The following steps outline a procedure for calculating the recurrence intervals and probabilities of various peak annual discharges:

- First, *rank* the flows in order from highest to lowest; that is, list them in order from biggest to smallest values. For two of the same size, list the oldest first.
- Next, determine the *recurrence interval* of each flow:

$$t_r = \frac{n + 1}{m}$$

where:

t_r = the recurrence interval in years
n = the total number of flows
m = the rank of flow in question

- Third, determine the probability (p) of any flow:

$$p = \frac{1}{t_r}$$

This will yield a decimal, say 0.5, which can be converted to a percent (50 percent) by multiplying by 100.

- In addition, it is also useful to know *when* peak flows can be expected; accordingly, the monthly frequency of peak flows can be determined based on the dates of the peak annual discharges. Further, the percentage of peak annual flows that actually produce floods can be determined by comparing the stage (elevation) represented by each discharge with the bankfull elevation of the gauging station, that is, any flow that exceeds bankfull elevation is a flood.

FORECASTING RIVER FLOW BASED ON DISCHARGE RECORDS

Given several decades of peak annual discharges for a river, it is possible to make limited projections to estimate the size of some large flow that has not been experienced during the period of record. The technique most commonly used involves projecting the curve (graph line) formed when peak annual discharges are plotted against their respective recurrence intervals. In most cases, however, the curve bends rather strongly, making it difficult to plot a projection accurately (Fig. 8.3). This problem can be overcome by plotting the discharge and/or the recurrence interval data on logarithmic graph paper or logarithmic/probability paper. Once the plot is

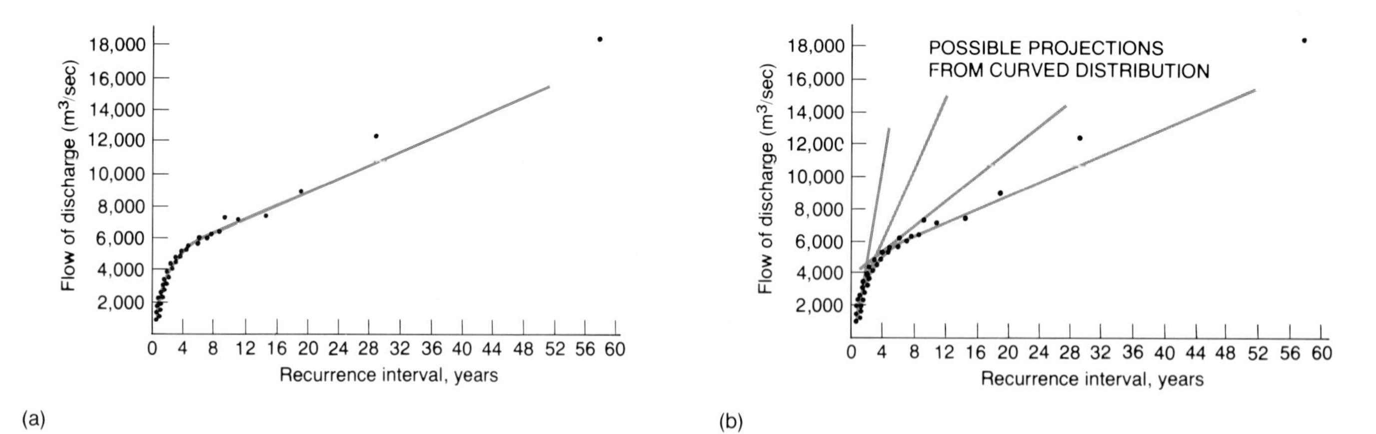

Fig. 8.3 (a) Graph based on recurrence interval and discharge of peak annual flows, Eel River, California, for the period 1911–1969. (b) This graph shows why it is impossible to plot one straight graph line from a curved distribution. (From W. M. Marsh and Jeff Dozier, *Landscape: Introduction to Physical Geography*, © 1981, Addison-Wesley, Reading, Massachusetts. Figs. 31.3 and 31.4. Reprinted with permission.)

straightened, a line can be ruled through the points. The process of making a projection can be accomplished by merely extending the line beyond the points and then reading the appropriate discharge for the recurrence interval in question. In Fig. 8.4 the curve was straightened by using a logarithmic scale on the discharge (vertical) axis and a probability scale on the recurrence interval (horizontal) axis. The projected 500-year flow is estimated to be 20,000 m^3/sec (706,000 cfs) based on a curve representing fifty-eight years of data.

Making forecasts of river flows based on discharge records is extremely helpful in planning land use and engineering bridges, buildings, and highways in and around river valleys. On the other hand, this sort of forecasting technique has several distinct limitations. First, the period of record is very short compared to total time the river has been flowing; therefore, forecasts are based only on a glimpse of the river's behavior. Second, throughout much of North America, watersheds have undergone such extensive changes because of urban and agricultural development, forestry, and mining that discharge records for many streams today have a different meaning than those several decades ago. Third, climatic change in some areas, for example, near major metropolitan regions, may be great enough over fifty or one hundred years to produce measurable changes in runoff. Together, these factors suggest that forecasts based on projections of past flows should be taken as approximations of future flows, and for greatest reliability, they should be limited to events not too far beyond the years of record, say, 50-year, 100-year, and 200-year flows.

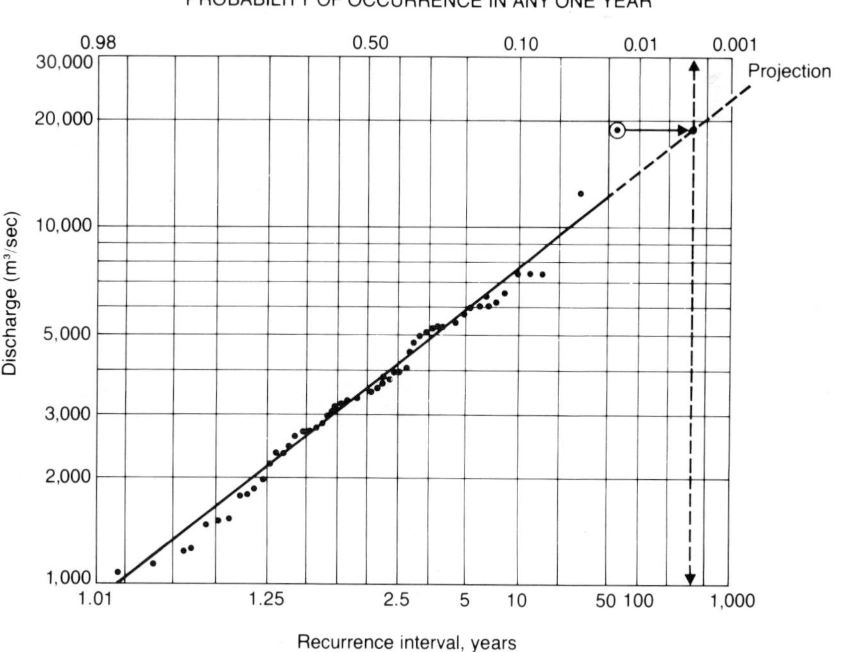

Fig. 8.4 Straightened curve of the distribution shown in Fig. 8.3. An estimate of the magnitude of flows at intervals beyond 58 years is now possible as shown for the 500-year flow which would have a discharge of approximately 20,000 m^3/sec (706,000 cfs). (From W. M. Marsh and Jeff Dozier, *Landscape: Introduction to Physical Geography*, © 1981, Addison-Wesley, Reading, Massachusetts. Fig. 31.5. Reprinted with permission.)

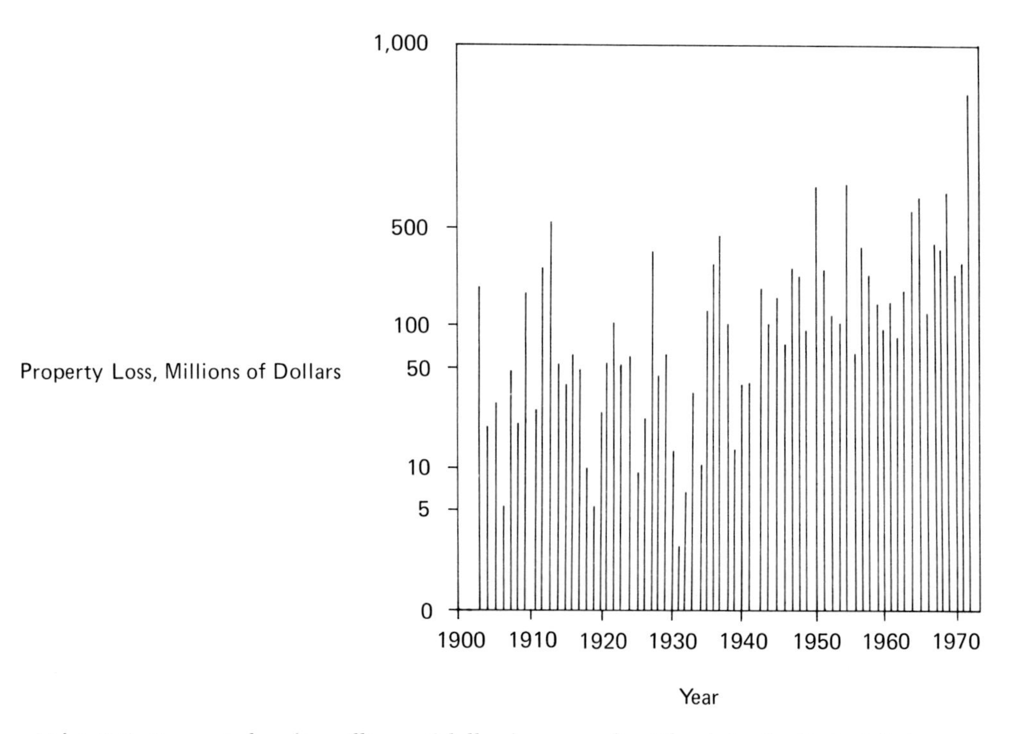

Fig. 8.5 Property loss (in millions of dollars) as a result of floods in the United States over seventy years. (Data from U.S. National Weather Service.)

APPLICATIONS TO LAND PLANNING

The principal use of discharge data in land planning is in the assessment of flood hazard. Property damage from floods in the United States and Canada has increased appreciably in this century, and in order to curb this trend, improved understanding, among other things, of the hydrologic behavior of streams and rivers is necessary (Fig. 8.5).

Detailed analysis of a river's flood potential is a difficult and expensive task. Basically, it involves translating a particular discharge into a level (elevation) of flow and then relating that flow level to the topography in the river valley. In this manner one can determine what land is inundated and what is not. This procedure requires accurate discharge and topographic data as well as a knowledge of channel conditions. The latter is necessary in order to compute flow velocity, which is the basis for determining the water elevation in different reaches of the channel, that is, just how high a given discharge will reach.

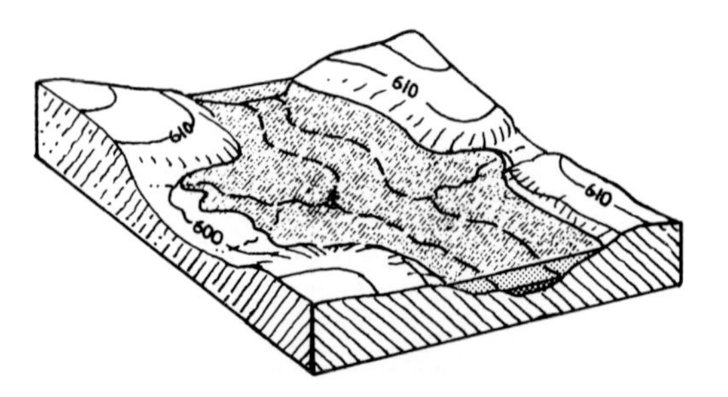

Flood Backwaters and Valley Contours

CASE STUDY

Increased Flooding in an Urban Watercourse, Metro Toronto

J.C. Mather

Highland Creek drains an area of 42.6 square miles in the borough of Scarborough, part of the Toronto metropolitan region. On August 27 and 28, 1976, two severe thunderstorms, producing 1.69 inches and 2.24 inches of rain respectively, caused significant flooding on the creek with damage reaching nearly $2 million. The August 27 storm had an estimated return period of twenty years, yet produced a flood equivalent to that of a one-in-thirty-seven-year event. The storm of the following day, though more severe in terms of rainfall, resulted in less damage, partially because its peak flow was routed through a number of pools and small reservoirs cleared by the earlier stormflow.

The source of the storms was two cold fronts associated with a pair of mid-latitude cyclones, which passed through southern Canada in quick succession. Strong squall lines developed along the fronts, and weather analysis showed that the heaviest rainfall was actually generated by a number of thunderstorm centers in the squall lines. During the periods of peak rainfall, it appears that most of the watershed was covered by the storms.

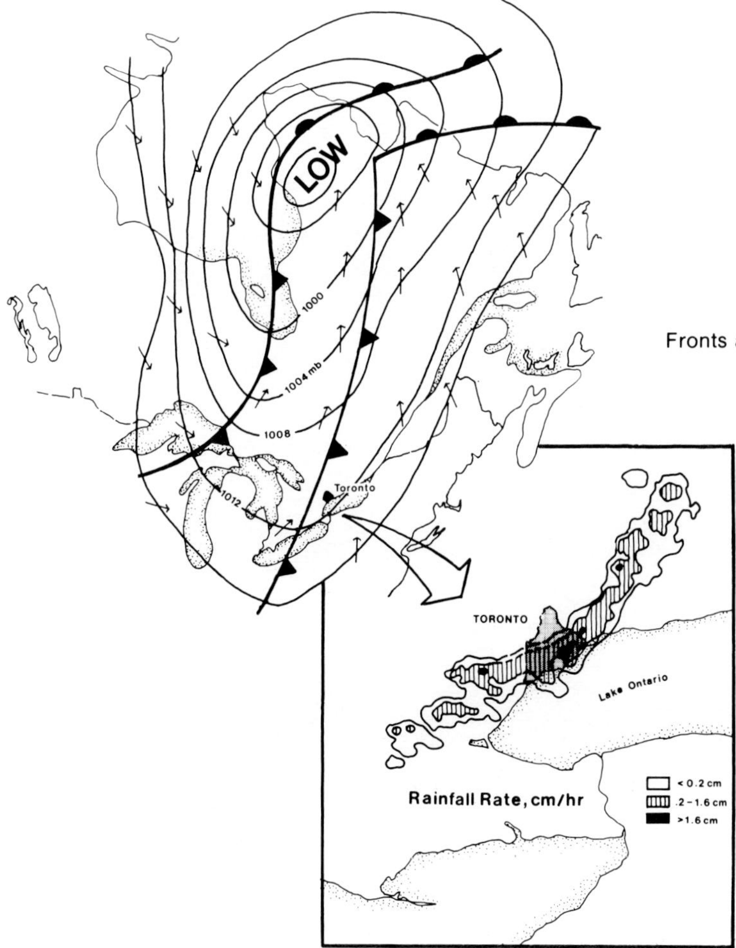

Fronts and Associated Precipitation, August 28

(cont.)

CASE STUDY (cont.)

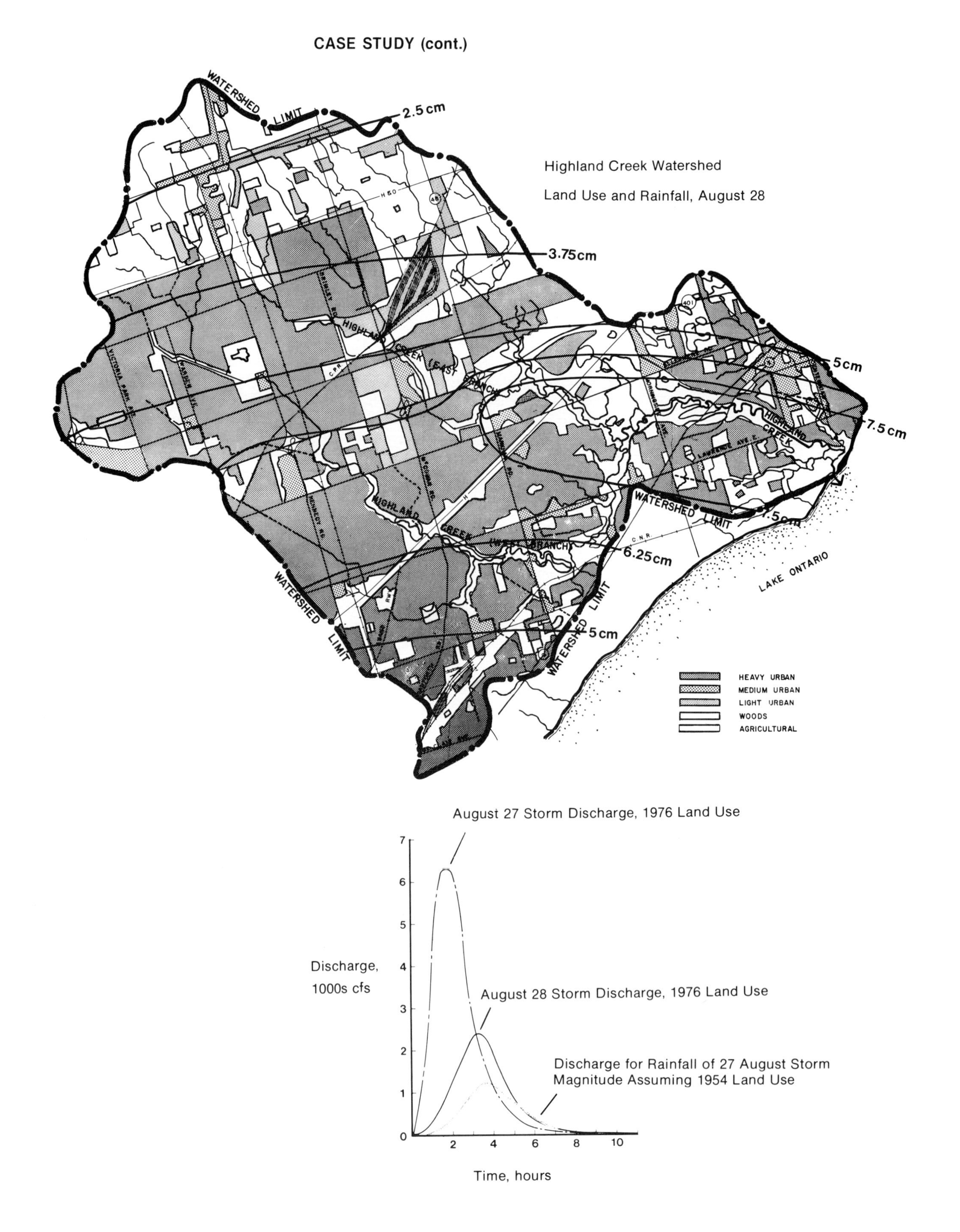

A comparison of hydrographs for Highland Creek based on 1954 and 1976 conditions reveals that for the same storm the 1976 peak discharge was more than six times greater than its 1954 counterpart. This dramatic increase is attributed to urbanization of the watershed, channeling of tributaries, massive loadings from stormsewer discharge, and the removal of forest from the floodplain. These factors have combined to produce larger amounts and higher rates of runoff, thereby increasing both the magnitude and frequency of flood flows on Highland Creek.

The policy of the Metro Toronto and Region Conservation Authority to restrict the encroachment of private development on floodplains helped minimize the severity of damage to land use by the two floods. The creek valley now serves two main functions: as open space and as a stormwater collection area. Continued performance in these capacities is possible, provided that action is taken to keep the two functions compatible. This will mean restricting structures from the floodplain and stream channel, locating open space facilities away from high flow zones, protecting banks from erosion, and investigating alternative approaches to stormwater management that do not rely exclusively on stormsewers and stream channeling.

J. C. Mather is head of the Flood Control Section, Water Resource Division, The Metropolitan Toronto and Region Conservation Authority.

For planning purposes, however, neither the time nor money is usually available for such detailed analysis. We must, instead, turn to existing topographic maps and existing discharge data and make a best estimate of the extent of various frequencies of flow. The accuracy of such estimates not only depends on the reliability of the data, but also on the shape of the river valley. Where the walls of a river valley rise sharply from the floodplain, the limits of flooding are often easy to define; however, where a river flows through broad lowlands the problem can be very difficult (Fig. 8.6).

The U.S. National Flood Insurance Program is based on a definition of the 100-year "floodplain," and in most areas, the boundaries of this zone are delineated using the method previously described (see Fig. 1.4). According to this program, two zones are actually defined: (1) the regulatory *floodway*, the lowest part of the floodplain where the deepest and most frequent floodflows are conducted; and (2) the *floodway fringe*, on the margin of the regulatory floodway, an area that would be lightly inundated by the 100-year flood. Buildings located in the regulatory floodway are not eligible for flood insurance, whereas those in the flood fringe are eligible provided that a certain amount of floodproofing is established (Fig. 8.7).

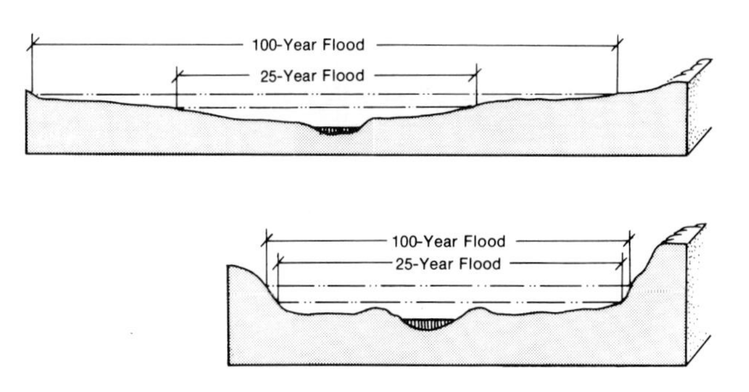

Fig. 8.6 Diagrams showing the differences in the extent of the 25-year and 100-year floods related to valley shape.

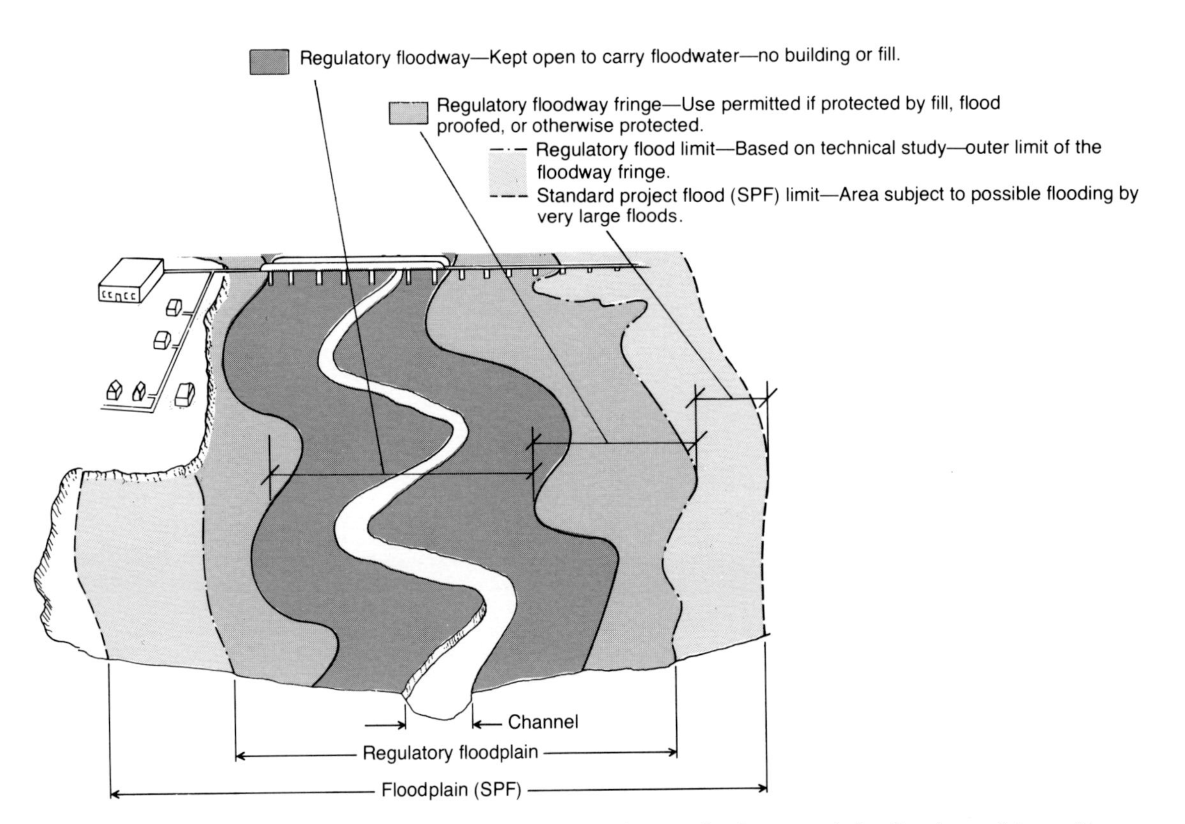

Fig. 8.7 Illustrations defining the regulatory floodway and the floodway fringe. (From U.S. Water Resources Council, *Regulation of Flood Hazard Areas*, vol. 1, 1971.)

The flood insurance program is intended to serve as one of the controls to limit development in floodplains. Other controls include zoning restrictions against vulnerable land uses and educational programs to inform prospective settlers of the hazards posed by river valleys.

In the case of river valleys where development is already substantial, the only means of reducing damage is to relocate flood-prone land uses or reduce the size of hazardous river flows. Because of the high costs and the sociological problems associated with relocation, planners often turn to a flow reduction option. This calls for structural (engineering) changes such as building reservoirs, dredging channels, diverting flows, and/or constructing embankments to confine flow, all of which are expensive and often deleterious to the environment.

OTHER METHODS FOR DEFINING FLOOD-PRONE AREAS

Many streams are too small to qualify for an official gauging station, yet they merit serious consideration in local land use planning. Communities bordering on such streams are often pressed for information not only on flood hazards, but on many other features of the stream valley that are important to planning decisions. Typical concerns include soil suitability for foundations and basements, siting of sanitary landfills and wastewater disposal facilities, the role of wetlands in the floodplain environment, and the distribution of forests and wildlife habitats. Much of this information can be generated from published sources, in particular, topographic contour maps, aerial photographs, and soil maps.

Of all the features of the river valley, the *floodplain* is the most important from a planning standpoint. Defined according to geomorphic criteria, the floodplain is the low-lying land along the stream, the outer limits of which may be marked by steep slopes, the valley walls. The floodplain is important for several reasons: first, excluding the stream channel itself, the floodplain is generally the lowest part of the stream valley and thus is most prone to flooding; second, floodplain soils are often poorly drained because of the nearness of the watertable to the surface and saturation by floodwaters; and third, floodplains are formed by incremental erosion and deposition associated with the lateral migration of streams in their valleys; therefore, the borders of the floodplain can be taken as a good indicator of the extent of alluvial soil. These features are the key indicators for floodplain mapping.

MAPPING FLOODPLAINS

Floodplains can be delineated in four ways, according to: (1) topography, (2) vegetation, (3) soils, and (4) the extent of past flood flows.

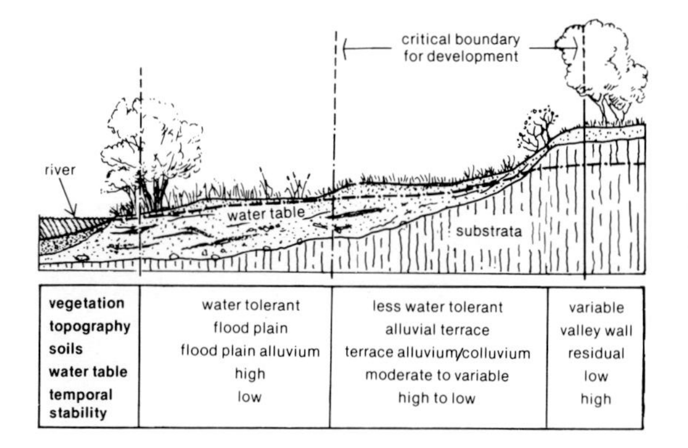

Topographic contour maps can be used to delineate floodplains if the valley walls are high enough to be marked by two or more contour lines; however, those that carry only a single contour, or fall within the contour interval, cannot be delineated even though they may be topographically distinct to the field observer. In the latter case, aerial photographs can be helpful in mapping this feature. Viewed with a standard stereoscope, topographic features appear exaggerated, thereby making identification of valley features a relatively easy task in many instances (Fig. 8.8).

Aerial photographs also enable one to examine vegetation and land use patterns for evidence of floodplains and wet areas. In the prairies and the major farming regions, such as the Great Plains and the Corn Belt, the floodplains of streams and rivers are often demarcated by belts of forest (labeled 'a' in Fig. 8.9). Similarly, in forested areas the floodplain tree covers are sometimes different in composition, and once these differences are known, they can be used to help identify floodplains (labeled 'b' in Fig. 8.9).

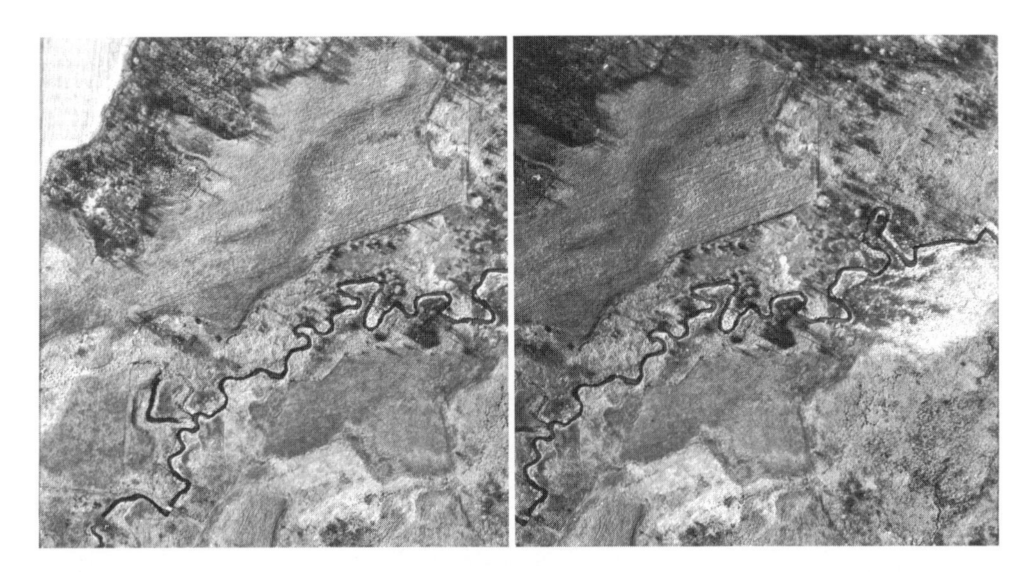

Fig. 8.8 Stereopair of aerial photographs showing a small floodplain. Though not detectable on standard topographic contour maps, the floodplain stands out vividly on the aerial photographs.

Because river floodplains are built from river deposits, the soils of floodplains often possess characteristics distinctively different from those of the neighboring uplands (see Fig. 5.5). They are generally described in the U.S. Soil Conservation Service reports as soils having high watertables, seasonal flooding, and diverse composition. Therefore, soil maps prepared by the SCS can also be used in the definition of floodplains. A note of caution is called for in using SCS maps, however, because the SCS bases

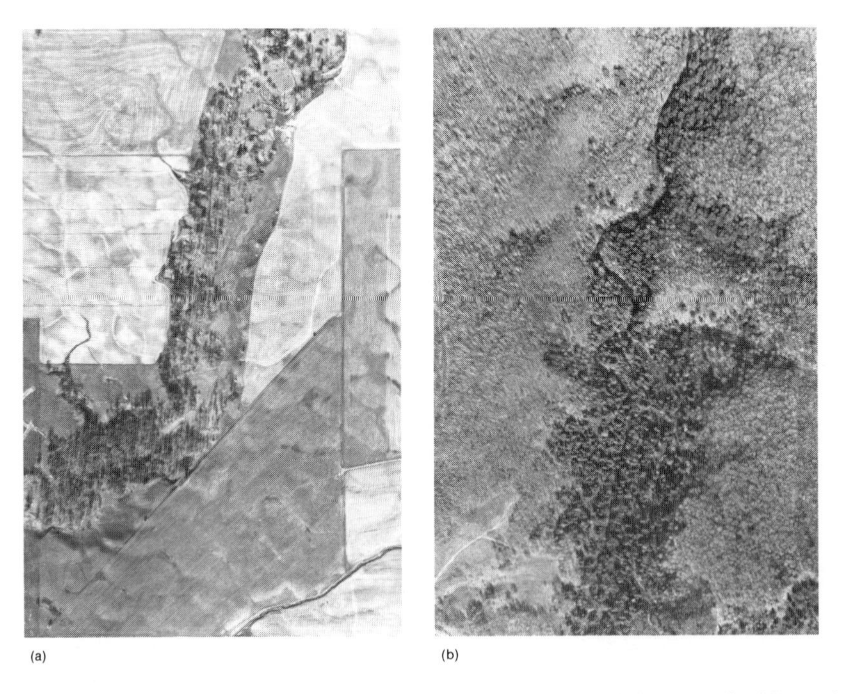

Fig. 8.9 Floodplain demarcated by a belt of trees (a), and floodplain marked by a difference in forest composition (b).

much of its own decisions for drawing the boundaries of different soil types on topography, vegetation, and land use patterns. Therefore, one must be aware that a good correlation between floodplains based on SCS soil boundaries and floodplains based on observable vegetation, topography, and land use may constitute circular reasoning.

Finally, floodplains may be defined according to the extent of past floodflows. Evidence of past flows may be gained from firsthand observers who are able to pinpoint the position of the water surface in the landscape and from features such as organic debris on fences and deposits on roads that can be tied to the flood (Fig. 8.10). This method usually demands extensive fieldwork, including interviews with local residents.

The map overlay technique is most commonly employed in synthesizing all these data. Although it has been used by geographers for decades, this technique was popularized among planners, landscape architects, and related professionals by Ian L. McHarg, a planner who has gained recognition through his book, *Design with Nature* (1969), and his strong defense of balanced human-environment approaches to planning. As noted earlier, one overlay scheme involves assigning numerical values to the various classes of each component of the landscape (vegetation, drainage, soils, and the like) according to their land use suitability. The values are then summed and the results used to delineate areas or zones of different development potentials (Fig. 8.11). One should be cautioned against attaching too much meaning to the numerical results, however, because the values are not intrinsic quantities but rather arbitrary numbers assigned to qualitative classes.

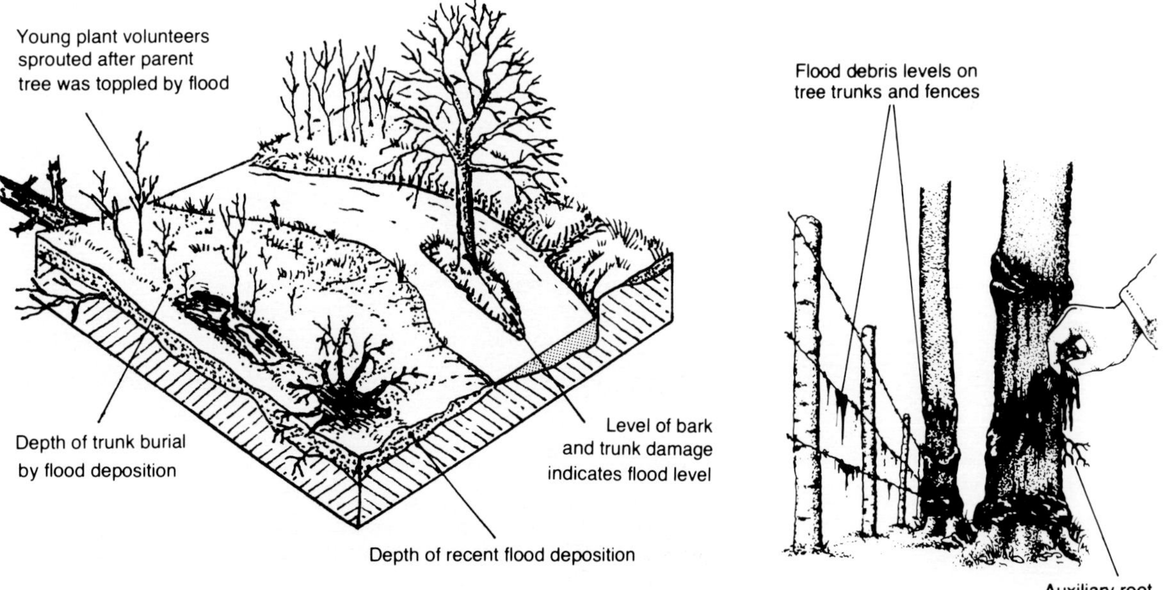

Fig. 8.10 In the field, evidence of the levels of recent floods can be gained from debris stranded on fences and marks on tree trunks. Past floods are sometimes evidenced in downed trees and buried vegetation. (From *Environmental Analysis for Land Use and Site Planning*, by W. M. Marsh. Copyright © 1978, McGraw-Hill, New York. Used with the permission of McGraw-Hill Book Company.)

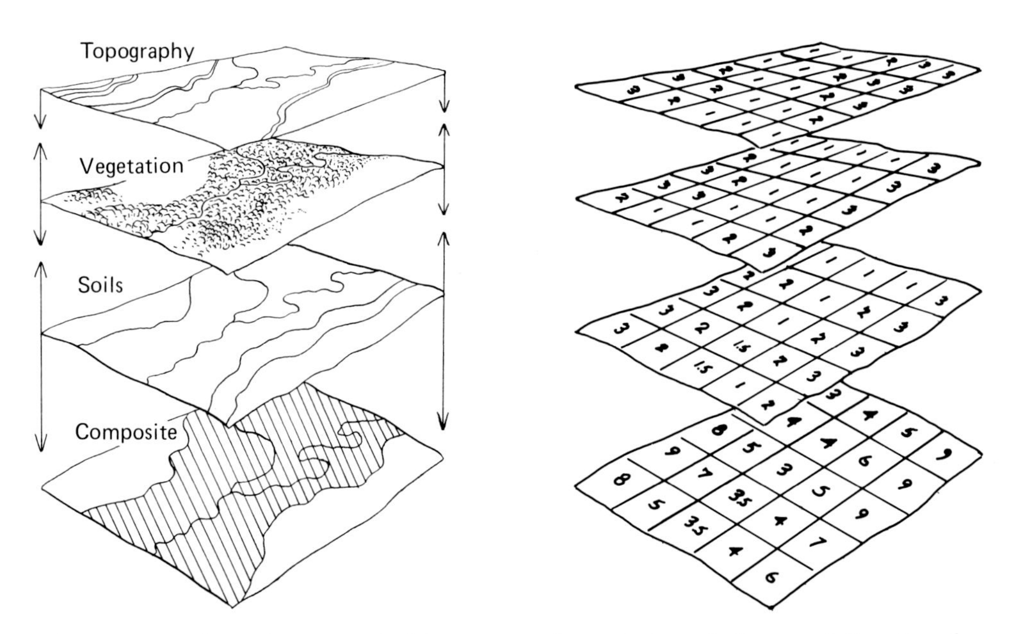

Fig. 8.11 A schematic portrayal of the map overlay technique. Left, maps delineating the pertinent features; Right, these features codified into a numerical scheme based on land use suitability.

PROBLEM SET

I. Select a set of peak annual discharge data representing thirty or more years of river flow and respond to the following:

1. Determine the probability of recurrence of (a) the bankfull discharge (or the discharge nearest to it), and (b) one or two flood flows representing years of your choice. This will require ranking the flows and computing the recurrence intervals and probabilities for the appropriate discharges.
2. Plot the discharge data against the respective recurrence intervals and draw a graph line that represents the best fit among the dots. Describe the relationship (shape of the curve) and indicate its limitations for projecting flows of magnitudes beyond those shown by the data.
3. Construct a graph showing the monthly frequency of (a) peak annual discharges, and (b) flood flows.

II. Using a stereoscope, examine the aerial photographs in Fig. 8.8.

1. Delineate the floodplain. Are there parts of the floodplain that appear more susceptible to flooding than others? Where?
2. Which of the following land uses would be inappropriate for this area (floodplain) and why: sanitary landfill, parking lots, ballfields, single-family residences?

SELECTED REFERENCES FOR FURTHER READING

Burby, Raymond J., and French, S. P. "Coping With Floods: The Land Use Management Paradox." *Journal of the American Planning Association*, vol. 47, no. 3, pp. 289–300.

Dunne, Thomas and Leopold, Luna B. "Calculation of Flood Hazard." In *Water In Environmental Planning.* San Francisco: W. A. Freeman, 1978, pp. 279–391.

Erikson, Kai T. *Everything In Its Path: Destruction of Community In the Buffalo Creek Flood.* New York: Simon and Schuster, 1976, 284 pp.

Kochel, R. Craig, and Baker, Victor R. "Paleoflood Hydrology." *Science*, vol. 215, no. 4531, 1982, pp. 353–361.

Marsh, William M., and Dozier, Jeff. "The Magnitude and Frequency of Landscape Processes." In *Landscape: An Introduction to Physical Geography.* Reading, Mass.: Addison-Wesley, 1981, 636 pp.

Rahn, Perry H. "Lessons Learned From the June 9, 1972, Flood in Rapid City, South Dakota." *Bulletin of the Association of Engineering Geologists*, vol. 12, no. 2, 1975, pp. 83–97.

Schneider, William J., and Goddard, J. E. "Extent of Development of Urban Flood Plains." *U.S. Geological Survey Circular 601-J*, 1974, 14 pp.

White, Gilbert F. *Flood Hazard in the United States: A Research Reassessment.* Boulder: The University of Colorado, Institute of Behavior Science, 1975.

Wolman, M. Gordon, "Evaluating Alternative Techniques For Floodplain Mapping." *Water Resources Research*, vol. 7, 1971, pp. 1383–1392.

9

WATERSHEDS, DRAINAGE NETS, AND LAND USE

INTRODUCTION

Overland flow moves only a short distance over the ground before it gathers into minute threads of water. These threads merge with one another, forming rivulets capable of eroding soil and shaping a small channel. The rivulets in turn join to form streams and the streams join to form rivers, and so on. This system of channels, characterized by streams linked together like the branches of a tree, is called a *drainage network*, and it represents nature's most effective means of getting liquid water off the land. The area feeding water to the drainage network is the drainage basin, or watershed, and for a given set of geographic conditions, the size of the main channel and its flows increase with the size of the drainage basin.

Early in the history of civilization, humans learned about the advantages of channel networks, for both distributing water and removing it from the land. The earliest sewers were actually designed to carry stormwater, and they were constructed in networks similar to those of natural streams. Whether they knew it or not, the ancients followed the *principle of stream orders* in the construction of both stormsewer and field irrigation systems. This principle describes the relative position, called the *order*, of a stream in a drainage network and helps us to understand the relationships among streams in a complex flow system.

Modern land development often alters drainage networks by obliterating natural channels, adding man-made channels, or by changing the size of drainage basins. Such alterations can have serious environmental consequences, including increased flooding, loss of aquatic habitats, reduced water supplies during low flow periods, and lowered water quality. In land use planning generally little attention is paid to drainage networks as geographic entities; however, they are taken seriously by civil engineers who seek to maintain or improve their performance in stormwater removal when basins are developed.

THE ORGANIZATION OF NETWORKS AND BASINS

The principle of stream orders is built on a classification system based on the rank of streams within the drainage network. First-order streams are channelized flows with no tributaries. Second-order streams are those with at least two first-order tributaries. Third-order streams are formed when at least two second-order streams join together, and so on (Fig. 9.1).

The number of streams of a given order that combine to form the next higher order generally averages around 3.0, and is called the *bifurcation ratio*. From a functional standpoint, this ratio tells us that the size of the receiving channel must be at least three times the average size of the tributaries. For drainage nets in general, a comparison of the total number of streams in each order to the order itself reveals a remarkably consistent relationship, called the *law of stream orders*, in which stream numbers decline with increasing order.

Given a classification by order of the streams in a drainage net, we can examine the relationship between orders and other hydrologic characteristics of the river system such as drainage area, stream discharge, and stream

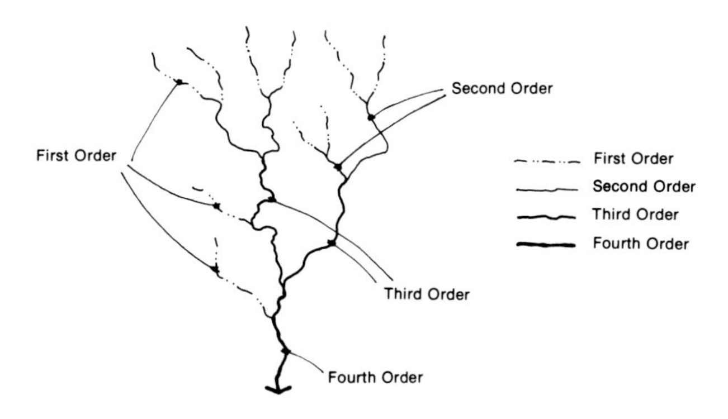

Fig. 9.1 Stream order classification according to rank in the drainage network. This follows the scheme originally defined by Robert Horton.

lengths. This provides a basis for comparing drainage nets under different climatic, geologic, and land use conditions; for analyzing selected aspects of river flow, such as changes due to urbanization; and for defining zones with different land use potentials. Table 9.1 lists and defines several factors involved in drainage network and basin analysis.

Drainage basins or areas can also be ranked according to stream order principle. First-order drainage basins are those emptied via first-order streams; second-order basins are those in which the main channel is of the second order; a fifth-order basin would be one in which the trunk stream is of the fifth order. Just as all high-order streams are products of a complete series of lower streams, high-order (large) basins are comprised of a complete series of lower-order basins, each set inside the other. This is sometimes referred to as a *nested hierarchy*, as is illustrated in Fig. 9.2.

In accounting for the combined areas of the basins that make up a larger basin, however, not all the land area is taken up by the lower-order

Table 9.1 Factors Important to the Analysis of Drainage Networks

- *Number of Streams*—the sum total of streams in each order.
- *Bifurcation Ratio*—(branching ratio) Ratio of the number of streams in one order to the number in the next higher order.

$$BR = \frac{N}{N_u}$$

where:

BR = bifurcation ratio
N = the number of streams of a given order
N_u = the number of streams in the next highest order

- *Drainage Basin Order*—Designated by highest order (trunk) stream draining a basin.
- *Drainage Area*—Total number of square miles or square kilometers within the perimeter (divide) of a basin.
- *Drainage Density*—Total length of streams per square mile or square kilometers of drainage area.

$$\text{Density} = \frac{\text{sum total length of streams (mi. or km)}}{\text{drainage area (mi}^2\text{ or km}^2\text{)}}$$

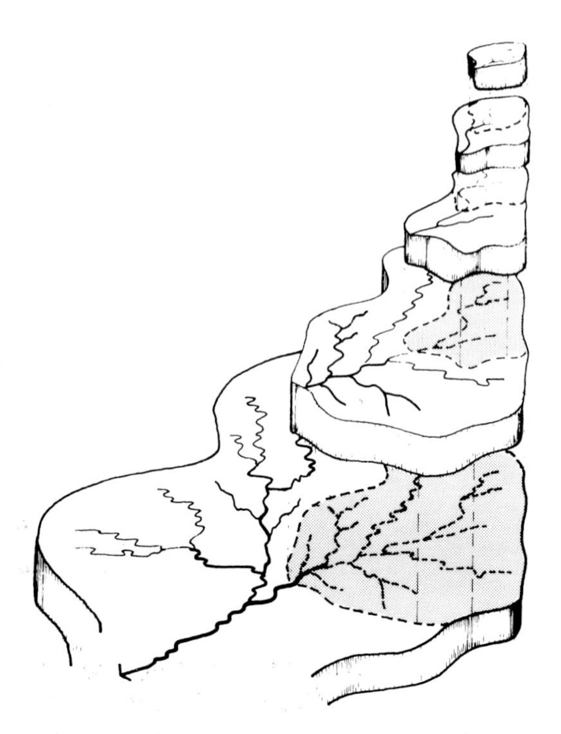

Fig. 9.2 Illustration of the nested hierarchy of lower-order basins within a large drainage basin.

A drainage basin in arid mountainous terrain. (Photograph by Richard and Janet Meganck.)

basins. Invariably a small percentage of the land drains directly into higher-order streams without passing through the numerical progression of lower-order basins (Fig. 9.3). This is called *nonbasin drainage area*, and it generally constitutes 15 to 20 percent of the total drainage area in basins of second order or larger.

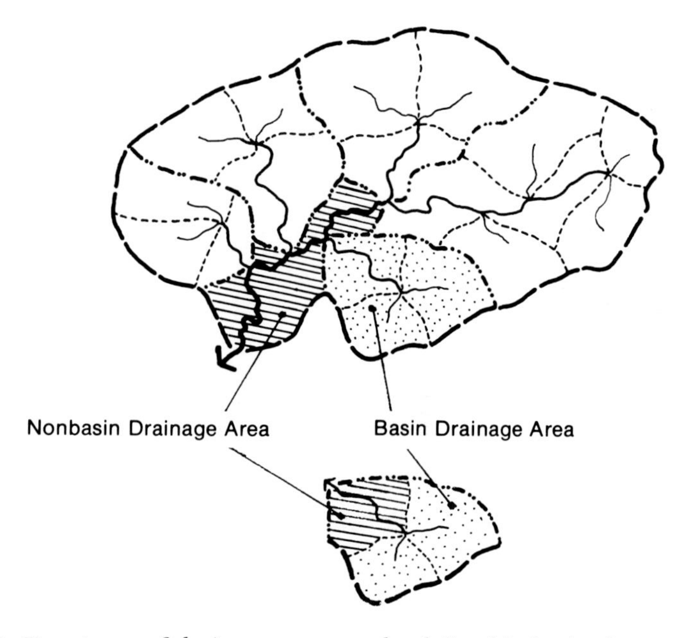

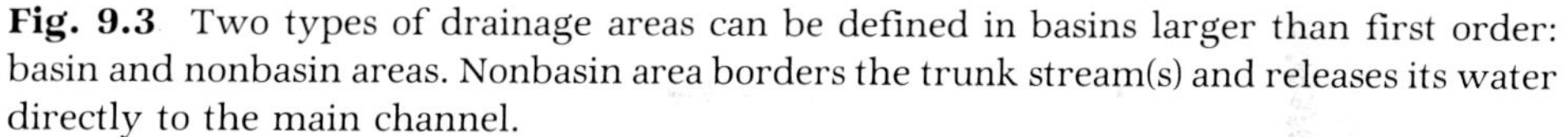
Fig. 9.3 Two types of drainage areas can be defined in basins larger than first order: basin and nonbasin areas. Nonbasin area borders the trunk stream(s) and releases its water directly to the main channel.

MAPPING THE DRAINAGE BASIN

The process of mapping the individual drainage basins within a large drainage net requires finding the drainage divides between channels of a particular order. This is best accomplished with the use of a topographic contour map; however, field inspection may be needed in developed areas in order to find culverts, diversions, and other drainage alterations. As a first step, channels should be traced and ranked, taking care to note the scale and the level of hydrographic detail provided by the base map. This is important because maps of large scales will show more first-order streams, owing to the fact that they are usually drawn at a finer level of resolution than smaller scale maps.

CASE STUDY

Drainage Area and Highway Location: Impact of Highway Salting on Water Quality

Robert L. Melvin

The application of salt to highways to reduce ice and snow buildup is a widespread practice in the United States and Canada. The amounts of salt applied vary geographically, depending on snowfall, winter temperatures,

(cont.)

CASE STUDY (cont.)

traffic levels, and state and local highway maintenance policies. In the Midwest and northeastern United States, annual salt (sodium chloride) applications typically amount to ten tons per lane mile; that is, twenty tons per mile of a two-lane highway. Salt is highly soluble in a humid environment, therefore such doses can pose a threat to surface and groundwater quality as well as to other components of the environment such as vegetation, soils, and animal habitats. In this century, salt concentrations in the Great Lakes have more than tripled, and 30 to 60 percent of this increase is attributed to road salting.

An initial assessment of the effects of highway salting can be made on the basis of drainage area, highway density, and salt application data. Assuming rain and snow are, over many years, evenly distributed throughout small drainage areas, then differences in the average annual flow of streams can be attributed principally to differences in the size of the drainage area. The larger the drainage area, the greater the dilution of highway salt. Therefore, the ratio of salt application to drainage area can be used as an index of the relative impact of road salt on water quality: the higher the ratio, the lower the relative dilution. The ratio index can also be weighted where necessary for variations in average annual streamflow owing to factors other than drainage area. The following example uses the ratio between salt applied and drainage area to compare the relative effect of the salting of a segment of a two-lane highway in the Rutty Creek drainage basin with the salting of two segments of two similar highways in the

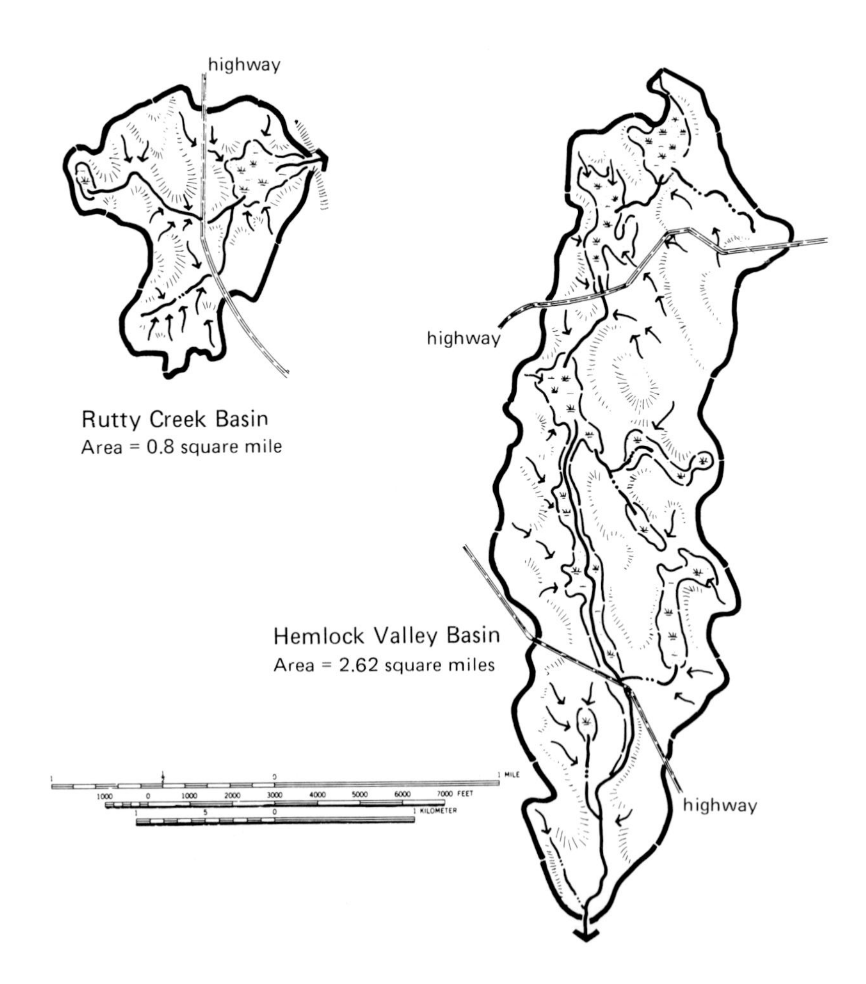

Hemlock Valley Brook drainage basin, both in Connecticut.The average annual salt application per lane mile of highway is approximately 8.8 tons.

Rutty Creek Drainage Basin

Area = 0.80 square mile
Lane-miles of highway = 1.66 miles
Ratio of salt application to drainage area

$$= \frac{8.8 \text{ tons} \times 1.66 \text{ lane-miles}}{0.8 \text{ square mile}}$$

= 18.3 tons per square mile

Hemlock Valley Brook Drainage Basin

Area = 2.62 square mile
Lane-miles of highway = 3.5 miles
Ratio of salt application to drainage area

$$= \frac{8.8 \text{ tons} \times 3.5 \text{ lane-miles}}{2.62 \text{ square miles}}$$

= 11.8 tons per square mile

A situation that is conducive to the direct release of road salt into a waterbody.

This appraisal indicates significantly less dilution and more environmental impact from highway salting in the Rutty Creek basin. A more precise evaluation could be made using information on streamflow variability at the sites and salt applications per storm event. Relatively simple evaluations of the type described can identify potential problem areas, leading to detailed field studies and to subsequent alteration of the pattern of deicing agents, the kind of agents used, or the total quantity of agents applied.

Robert L. Melvin is a hydrologist with the U.S. Geological Survey

In areas of complex terrain, the task of locating basin perimeters can be tedious, but it can be streamlined somewhat if the pattern of surface runoff (overland flow) is first demarcated. This can be done by mapping the direction of runoff using short arrows drawn perpendicular to the contours over the entire drainage area. Two basic patterns will appear: divergent and convergent. Where the pattern is divergent, a drainage divide (basin perimeter) is located (Fig. 9.4).

By connecting all the drainage divides between first-order streams, a watershed can be partitioned into first-order basins. This will account for 80 to 85 percent of the total drainage area, the remainder being nonbasin drainage. From this pattern, second-order basins can be traced, and they are comprised of two or more first-order basins plus a fraction of nonbasin area.

Since perennial streams are fed by both surface water and groundwater, it is also important to be aware that the drainage basins for these two sources may be different. In most instances, however, it is impossible to identify such an arrangement without the aid of an area-wide groundwater study. A clue to existence of a significant difference in the two can sometimes be found in first- or second-order streams that have exceptionally large or small base flows for their surface drainage area relative to similar streams in the same region.

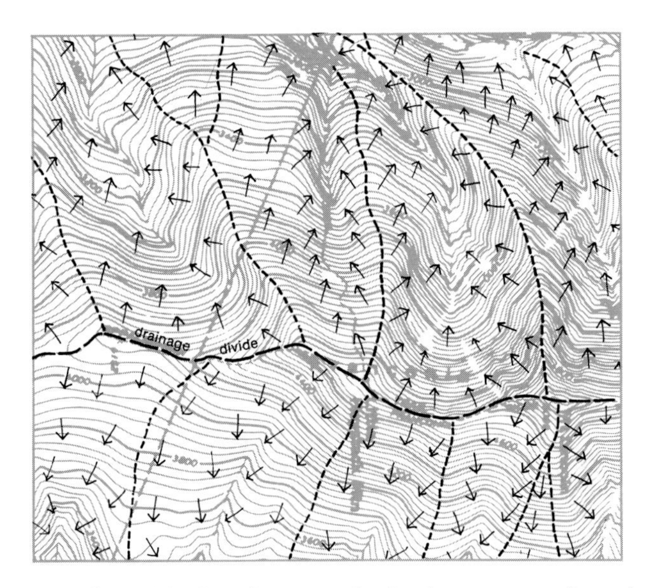

Fig. 9.4 Mapping and partitioning the watershed using vectors of overland flow.

TRENDS IN THE DEVELOPMENT OF SMALL DRAINAGE BASINS

Clearing and development of land often has a pronounced influence on drainage networks and basins. Deforestation and agriculture may initiate soil erosion and gully formation. As gullies advance, they expand the drainage network, thereby increasing the number of first-order streams and the drainage density. The main hydrologic consequence of this is shortened concentration times because the distance that water must travel as overland flow or interflow is reduced (Fig. 9.5). Discharges are in turn larger, large flows occur with higher frequencies, erosion is greater, and water quality can be expected to decline.

Urbanization Urbanization also leads to considerable change in the shape and density of a drainage network. One of the first changes that takes place is a "pruning" of natural channels; i.e., removal of parts of the network. These channels are often replaced by ditches and underground channels in the form of storm and sanitary sewers. Although the *natural network* may be pruned, the *net effect* of urbanization is usually an increase in total channels and in turn an increase in the overall drainage density (Fig. 9.6). Coupled with the lower infiltration rates of urbanized areas, this leads to increased amounts of runoff and shorter concentration times for the drainage basin, both of which produce larger peak discharges. As a result, both the magnitude and frequency of peak discharges are increased for receiving streams and rivers (Fig. 9.5).

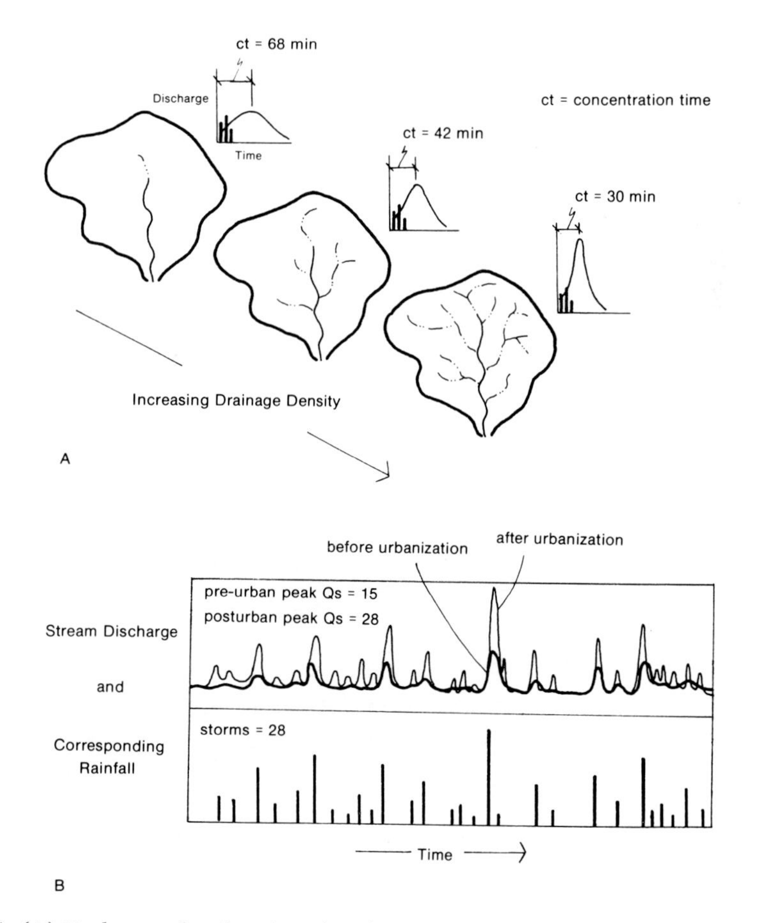

Fig. 9.5 (A) Hydrographs showing the changes brought about by increased drainage density: concentration time is reduced, causing greater peak discharges for a given rainstorm. (B) A schematic portrayal of the magnitude and frequency record of stream discharges to rainfall before and after urbanization.

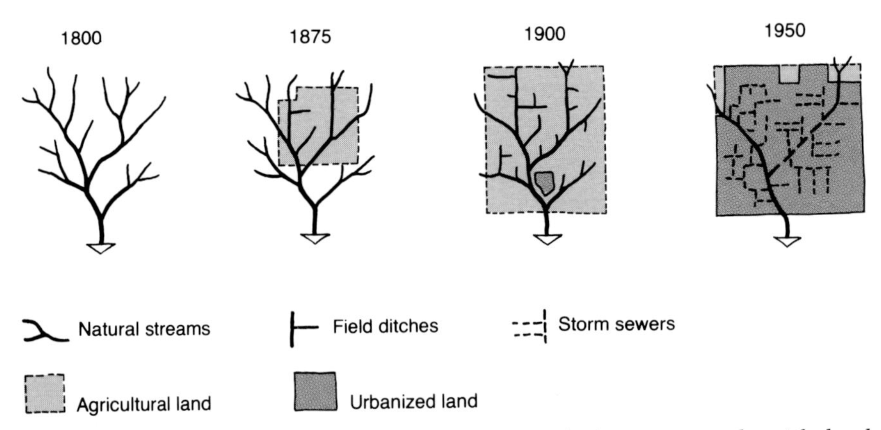

Fig. 9.6 Pruning, grafting, and intensification of a drainage network with land use change. (Adapted from Peter Van Dusen, "Spatial Organization of Watershed Systems," Ph.D diss., Ann Arbor: University of Michigan, 1971.)

Stormsewers are underground pipes that conduct surface water by gravity flow from streets, buildings, parking lots, and related facilities to streams and rivers. The pipes are usually made of concrete, sized to the area they serve, and capable of transmitting stormwater at a very rapid rate. Studies show that stormsewers and associated impervious surfaces increase the frequency of flood flows on streams in fully urbanized areas by as much as sixfold (Fig. 9.7).

The graph in Fig. 9.7 assumes that the size of the drainage basin has remained unchanged with sewering. In many instances, however, the size of the basin is also increased as sewering takes place, as is illustrated in the case of the Reeds Lake watershed (Fig. 9.8). The additional drainage area is completely urbanized and laced with stormsewers and produces peak discharges perhaps four to five times larger than drainage areas of comparable size without stormsewers and such heavy development.

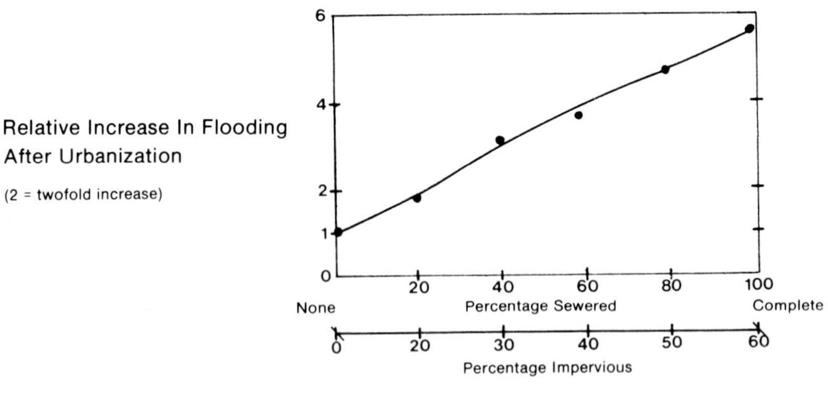

Fig. 9.7 Increased frequency of flood flows related to stormsewering and impervious surface. (From Luna B. Leopold, "Hydrology for Urban Land Planning—A Guidebook on the Effects of Urban Land Use," *U.S. Geological Survey Circular 554*, 1968.)

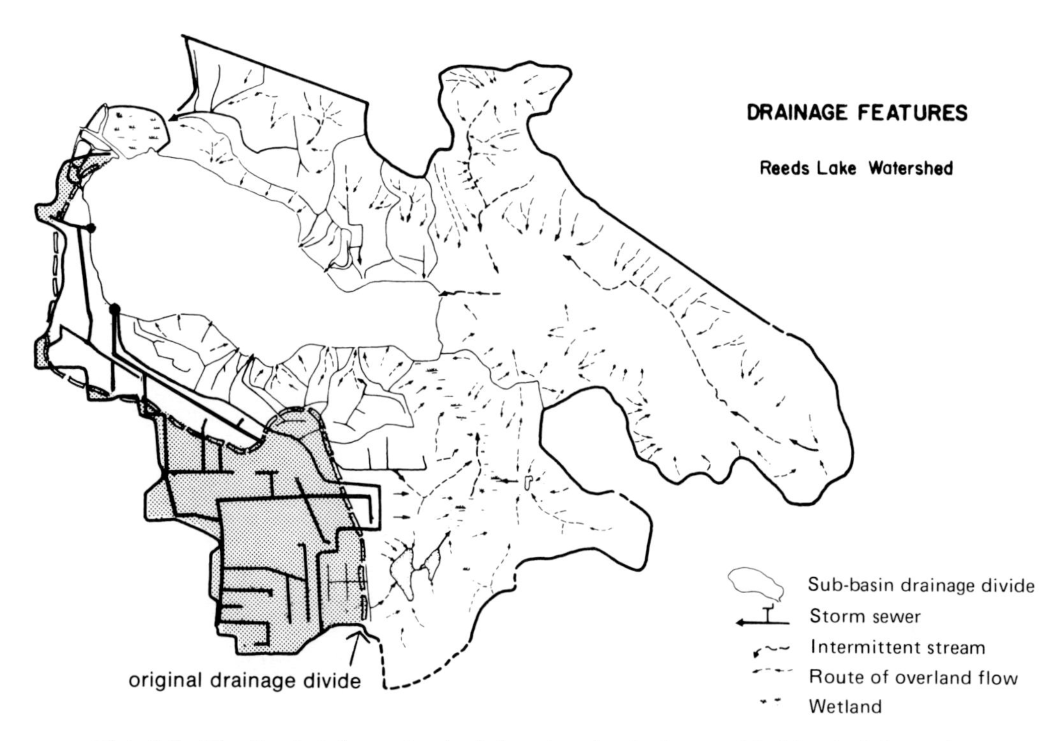

Fig. 9.8 The Reeds Lake watershed showing the drainage added (in tint) through stormsewering.

Agriculture Modern agriculture is also responsible for altering drainage basins and stream networks. In humid regions, farmers often find it necessary to improve field drainage to facilitate early spring plowing and planting. In addition to cutting ditches through and around fields and deepening and straightening small streams, networks of drain tiles are often installed in fields. Drain tiles are small perforated pipes (originally ceramic but now plastic) buried just below the plow layer. They collect water that has infiltrated the soil and conduct it rapidly to an open channel. Though the effects of tile systems on streamflow have not been documented, these systems undoubtedly serve to increase the magnitude and frequency of peak flows locally in much the same way as yard drains and stormsewers do in cities.

Lakes and Reservoirs The watersheds that serve both natural and manmade impoundments follow the same organizational principles as river watersheds, with the impoundment itself representing a segment of some order in the drainage network or a node linking streams of lower orders. On the other hand, the land use patterns are different in one important respect: The heaviest development is usually found in the nonbasin drainage area (see Fig. 10.6 in Chapter 10). This is a discontinuous belt of shoreland that encircles the waterbody and is highly attractive to residential, recreational, and commercial development. As the development takes place, the need for storm drainage usually arises and yard drains, ditches, and stormlines are constructed where, under natural conditions, channels never existed. In addition, the shoreland zone is often expanded inland with stormsewer construction, thereby capturing additional runoff and rerouting it directly to the waterbody.

PLANNING AND MANAGEMENT CONSIDERATIONS

Since small drainage basins are the building blocks of large drainage systems, it is essential that watershed planning and management programs address the small basin. Most small basins (primarily first, second, and third orders) are comprised of three interrelated parts: (1) an outer, upland zone that generates overland flow and ephemeral channel flows; (2) a low area or collection zone in the upper basin where runoff from the upland zone accumulates; and (3) a central conveyance zone represented by a valley and stream channel through which water is transferred from the collection zone to higher order channels (Fig. 9.9). The hydrologic behavior of each zone is different and each in turn calls for different planning and management strategies.

The *upland zone* is generally least susceptible to drainage problems. It provides the greatest opportunity for site-scale stormwater management because most sites in this zone have little upslope drainage to contend with and because surface flows are generally small and diffused. By contrast, the *collection zone* in the upper basin is subject to serious drainage problems. Seepage is common along the perimeter, and groundwater saturation can be expected in the lower central areas during much of the year. During periods of runoff, this zone is prone to inflooding, caused by massive stormwater loading from upland surfaces. Inflooding is very common in rural areas of modest local relief; in fact, it is probably the most common source of local flood damage today in much of the Midwest and southern Ontario.

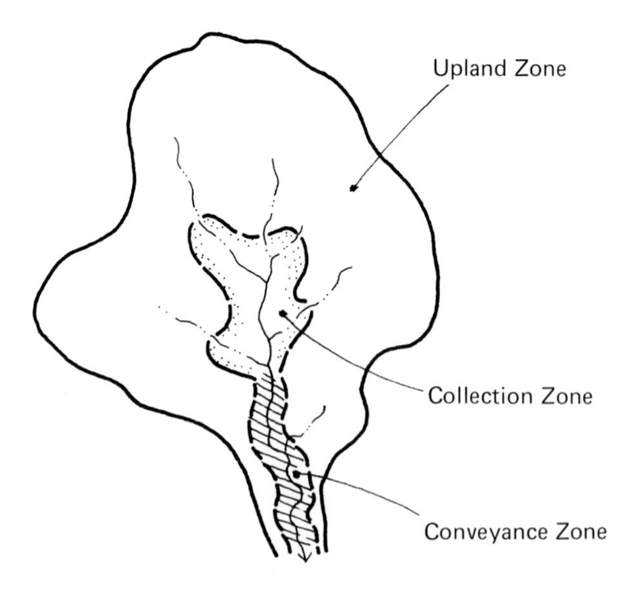

Fig. 9.9 The three main hydrologic zones of a small drainage. The lower two zones are the least suited to development.

The central *conveyance zone* contains the main stream channel and valley, including a small floodplain. The flows in this zone are derived from the upper two zones as well as from groundwater inflow directly to the channel. Groundwater contributions provide the stream baseflow and constitute the vast majority of the stream discharge over the year. Stormwater, on the other hand, is derived mainly from the upper zones, and though small in total, it constitutes the bulk of the largest peak flows in developed or partially developed basins. The central conveyance zone is also subject to flooding, but in this case it is outflooding caused by the stream overtopping its banks. Because this zone is prone to comparatively large floods, especially if the upper basin is developed, it is generally least suited to development and most difficult to manage hydrologically. In hilly and mountainous terrain, such as coastal California, where slopes tend to be unstable during wet periods, it is this zone and the upper collection zone that receive the landslide and mudflow debris from slope failures.

Recognition of the constraints and opportunities associated with each of these drainage zones is an important step toward the formation of development guidelines for small basins. In particular, this provides a rationale for defining the spatial patterns of land units, including buildable land, open space, and special use areas on a basin-by-basin basis. What it does not provide, however, is a rationale for establishing density guidelines. *Density* is a measure of the intensity of development; defined, for example, as percentage impervious surface, total building floor space per acre, or population density. For water and land management, percentage impervious surface is commonly used to define density.

Establishing the appropriate density for a drainage basin should be based on performance goals and standards (see Chapter 7 for a discussion of performance concepts) and basin carrying capacity. The carrying capacity of a drainage basin is a measure of the amount and type of development it is able to sustain without suffering degradation of water features, water quality, biota, soils, and land use. While this concept is easy to envision, determination of a basin's carry capacity may be difficult to derive. As a general rule, 30 percent development may be the advisable maximum for most basins, but this can vary with the style of development and the character of the basin.

Basins in hilly or mountainous terrain often have the lowest carrying capacity because of the abundance of steep, unbuildable slopes and the rapid rate of stormwater transfer. In addition, the soil mantle may be thin, which not only minimizes the basin's capacity to filter and retain infiltration water, but often increases the tendency for slope failure during wet periods. For basins with less relief, lower gradients, deep soil mantles, and good soil drainage, the carrying capacity may be considerably higher. The acceptable level, however, should take into account factors in addition to the physical character of the basin itself; for example, the nature of the development and drainage conditions downstream, and the character of existing and proposed development within the basin. With respect to the latter, higher densities (that is, resident, worker, dwelling unit, or economic activity per unit area of land) may be allowable where effective stormwater control such as on-site retention or dry wells can be employed effectively (see Fig. 7.6 in Chapter 7). Lower densities would be recommended where standard stormsewers are widely used or where stormwater is apt to become contaminated from manufacturing area runoff, accidental spills, or industrial storage area runoff.

LAND USE PLANNING IN THE SMALL DRAINAGE BASIN

The procedure for building land use plans for small drainage basins begins with a definition of the stream system, patterns of runoff, and the three hydrologic zones. Each zone should be analyzed for soils, slopes, vegetation, and existing land use to determine its limitations for development. Steep slopes, runoff collection areas, unstable and poorly drained soils, forested areas in critical runoff zones, and areas prone to flooding and seepage should be designated as nonbuildable. The remaining area is more or less the developable land, and the bulk of it usually lies in the upland zone of the basin (labeled 'a' in Fig. 9.10).

The next step is to define land use units. These are physical entities of land, with the fewest constraints to development in general, that set the spatial scale and general configuration of development (labeled 'b' in Fig. 9.10). Within this framework, development schemes may be evaluated and either discarded or assigned to the appropriate land unit. To achieve the desired performance, options for the siting of different activities should be exercised; for instance, high-impact activities may be assigned to land units that are buffered by forest and permeable soils from steep slopes and runoff collection areas. In general, the recommended guidelines for site selection and site planning in small drainage basins are: (1) maximize the distance of stormwater travel from the site to a collection area or stream; (2) maximize the concentration time by slowing the rate of stormwater runoff; (3) minimize the volume of overland flow per unit area of land; (4) utilize or provide buffers such as forests and wetlands to protect collection areas and streams from development zones; and (5) divert stormwater away from or around critical features such as steep slopes, unstable soils, or valued habitats (labeled 'c' in Fig. 9.10).

PROBLEM SET

I. Examine the topographic contour map on page 180 and note drainage nets A and B upstream from the points marked by bullets.

1. Trace the stream patterns in each basin and identify the order of each designated stream. Using small arrows, show the pattern of surface runoff that feeds these streams.
2. Delineate the perimeters of both basins and mark them with a solid line. Where the perimeter is uncertain near highways and residential areas, mark the area with an asterisk (or number) and provide a footnote explaining the uncertainty. Based on the streams actually shown, define the first-order basins within each and demarcate them with a broken line.

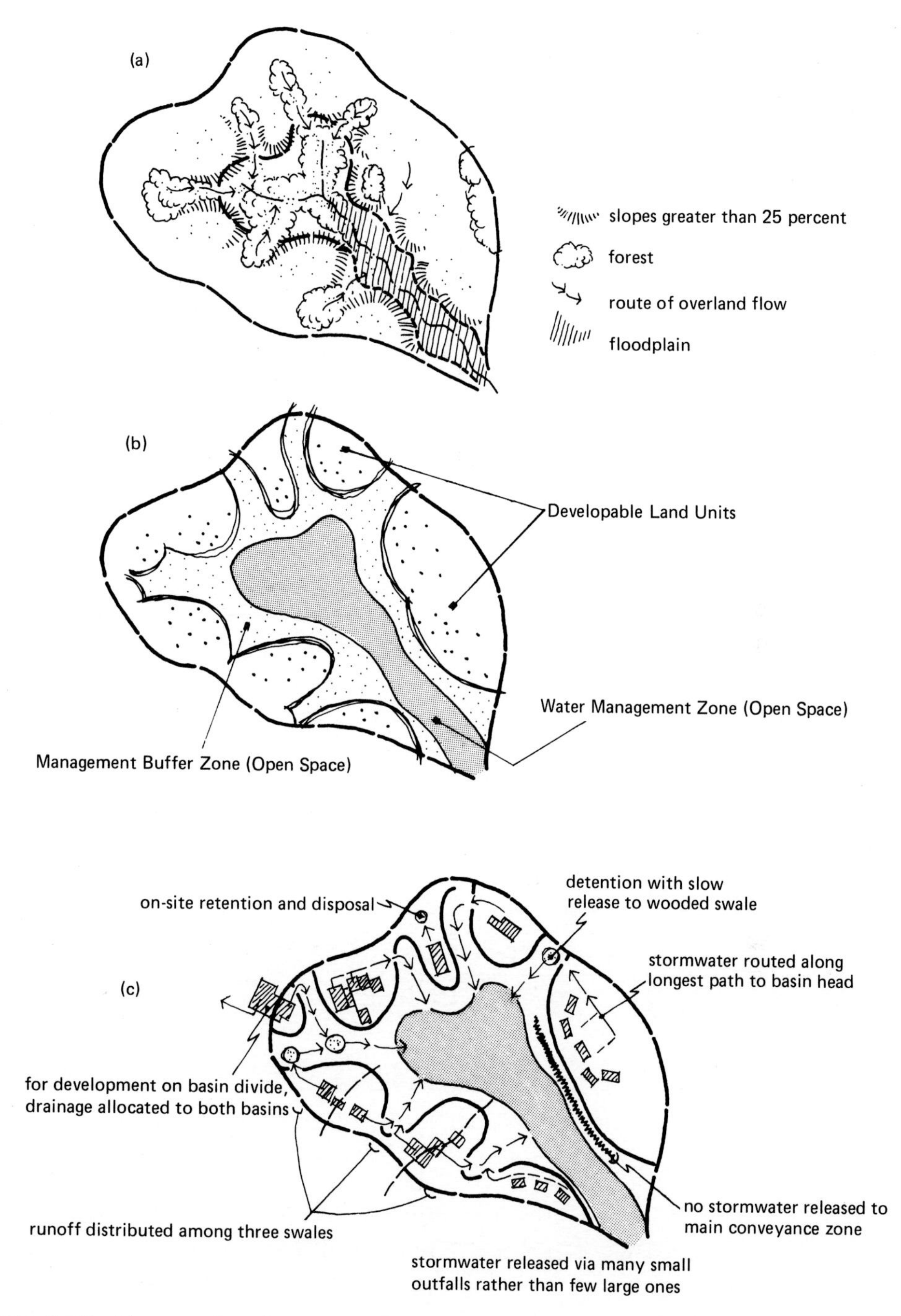

Fig. 9.10 Maps produced as a part of a procedure for land use planning in a small drainage basin: (a) the three hydrologic zones and the significant features of each; (b) the land use units; and (c) some strategies and guidelines in site selection and site planning.

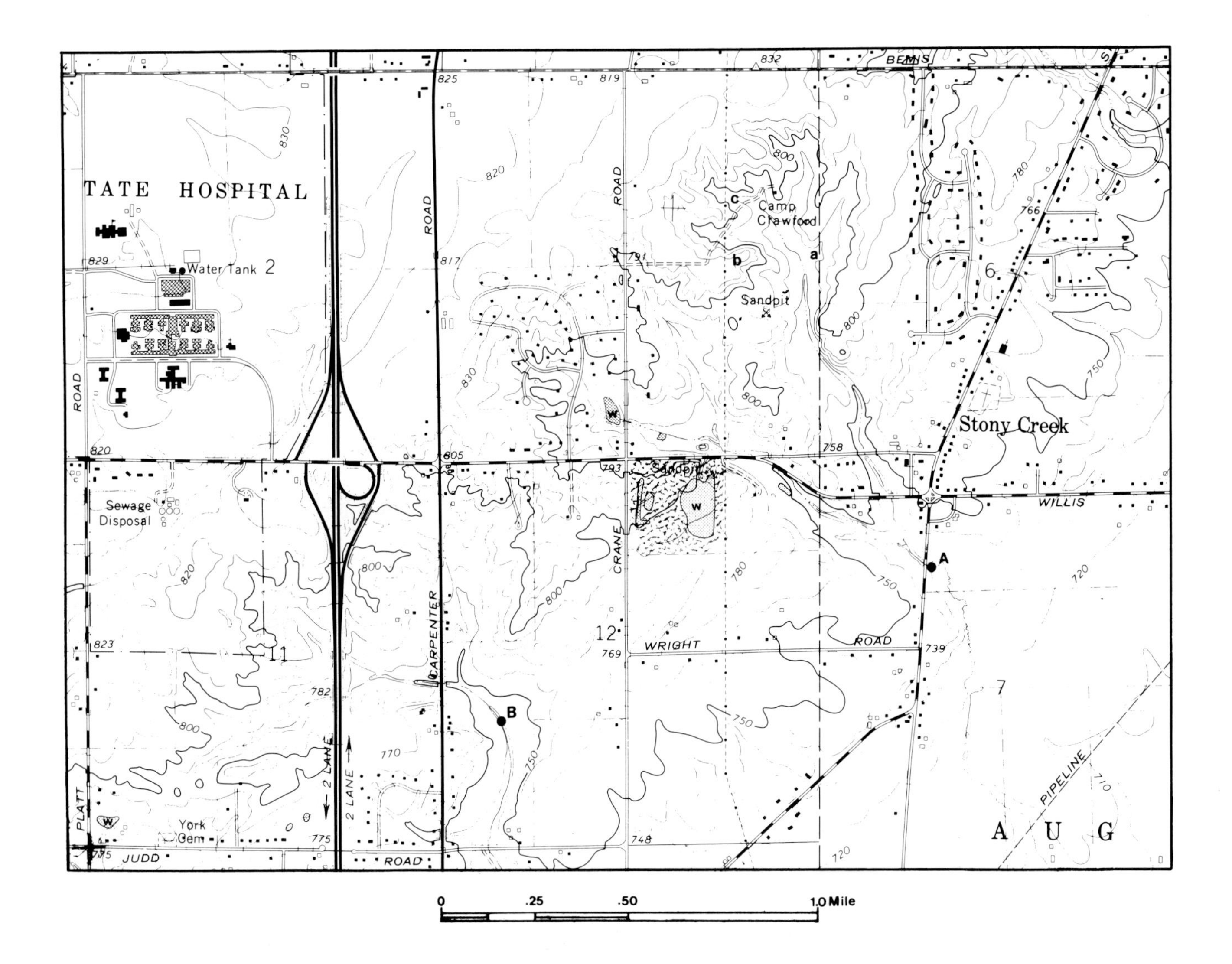

3. Based on the patterns of road, highways, and buildings, which first-order basin is most heavily developed? Shade this basin. Assuming that each building shown in this basin represents two acres of development (based on the combined acreage of buildings, walks, drives, and public roads), what is development density (as a percent of basin area)?

4. In each first-order basin delineate: (1) the zone of overland flow, (2) the upper collection zone, and (3) the central conveyance zone. (Use a separate map for this.) For the uppermost basin (the one containing Camp Crawford), delineate the land units that are suitable for residential development. For the unsuitable area, identify the major constraints and mark them on the map.

II. A developer has options on three sites in the Camp Crawford basin (at points a, b, and c) and you are asked to recommend one for residential development. Based on the results of I.4 above, and the earlier discussion on basin planning and management, define the apparent opportunities and constraints of each site and rank them according to suitability for residential development.

III. Assuming each paved road (represented by solid and broken lines) receives ten tons of salt per lane mile per year, determine the total annual salt loading of basins A and B. Besides reducing salt application rates per lane mile, suggest some other methods to reduce salt loading in basin A.

SELECTED REFERENCES FOR FURTHER READING

Copeland, O. L. "Land Use and Ecological Factors in Relation to Sediment Yield." *U.S. Department of Argiculture, Misc. Publication 970.* Washington, D.C., 1965, pp. 72–84.

Dunne, Thomas, and Leopold, L. B. "Drainage Basins." In *Water in Environmental Planning.* San Francisco: W. H. Freeman, 1978, pp. 493–505.

Dunne, Thomas; Moore, T. R.; and Taylor, C. H. "Recognition and Prediction of Runoff-Producing Zones in Humid Regions." *Hydrological Sciences Bulletin,* vol. 20, no. 3, 1975, pp. 305–327.

Gregory, K. J., and Welling, D. E. *Drainage Basin Form and Process.* New York: Halsted Press, 1973, 456 pp.

Horton, R. E. "Erosional Development of Streams and Their Drainage Basins: Hydrophysical Approach to Quantitative Morphology." *Geological Society of America Bulletin,* vol. 56, 1945, pp. 275–370.

Leopold, L. B. "Hydrology for Urban Land Planning—A Guidebook on the Effects of Urban Land Use," *U.S. Geological Survey Circular 554,* 1968, 18 pp.

Leopold, L. B., and Miller, J. P. "Ephemeral Streams: Hydraulic Factors and Their Relation to the Drainage Net." *U.S. Geological Survey Professional Paper 282–A,* 1956.

Strahler, A. N. "Quantitative Geomorphology of Drainage Basins and Channel Networks." In *Handbook of Applied Hydrology* (ed. Ven te Chow). New York: McGraw-Hill, 1964.

10

WATER QUALITY, RUNOFF, AND LAND USE

- Introduction
- Nutrient Loading and Eutrophication
- Nutrient Mass Budget Concept
- Land Use and Nutrient Loading of Runoff
- Estimating Nutrient Loading
- Planning for Water Quality Management in Small Watersheds
- The Spatial Framework for Planning
- Problem Set
- Selected References for Further Reading

INTRODUCTION

The quality of water in lakes and streams has become a national issue in the United States and Canada. In the past two decades, both countries have enacted complex bodies of law calling for nation-wide pollution control programs. In the United States, abatement programs have relied overwhelmingly on mechanical (engineering) approaches, illustrated, for example, by the modern sewage treatment plant.

The success of these abatement programs has generally been good for pollution of the point source variety. Point sources of water pollution are those characterized by concentrated outfalls from high-intensity land uses. Most examples fall under the general headings of industrial process water and municipal sewage. The particular types of pollutants associated with point sources are listed in Table 10.1. The success of the industrial and municipal pollution abatement programs, when weighed against the U.S. national water quality goals, to the surprise of many, has been substantially less than program proponents originally forecast.

The main explanation for this is that in the modern landscape the magnitude of contributions from nonpoint sources were seriously underestimated. These are spatially diffused sources, which emanate from relatively large areas, and enter streams and lakes via overland flow, precipitation, atmospheric fallout, interflow, and groundwater (Fig. 10.1). The prime contributors include residential area stormwater, septic system seepage, acid rain, and agricultural runoff. Because of the size of the source areas, the numerous outfalls involved, and the sporadic nature of the flows, nonpoint pollution does not lend itself to abatement using conventional mechanical systems. Instead, it is generally agreed that abatement must be approached as a land management problem, focusing on the sites, activities, and conditions that produce the pollutants.

Table 10.1 Pollutants Associated with Point Sources

Pollutant	*Source*
Oxygen-Demanding Elements (oxidizable nitrogen, carbonaceous material, and certain reducing compounds)	municipal sewage, industrial waste
Disease-Bearing Agents (mainly bacteria)	raw or partially treated sewage
Nutrients (mainly phosphorus and nitrogen)	raw or partially treated sewage
Inorganic and Synthetic Chemicals (e.g., PBB and PCB)	industrial waste, accidental spills
Heavy Metals (e.g., arsenic, mercury, and lead)	industrial waste
Radioactive Materials (e.g., thorium-230, radium-226, and strontium-90)	leakage from disposal and manufacturing sites
Heat (e.g., in cooling water)	power plants, institutions, industry

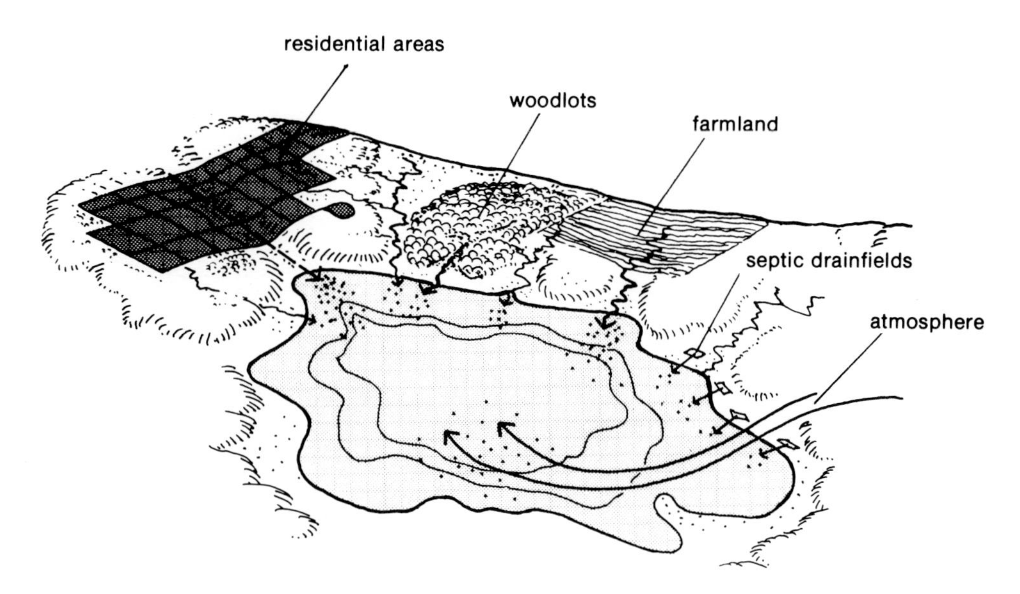

Fig. 10.1 Examples of nonpoint sources of water pollution.

NUTRIENT LOADING AND EUTROPHICATION

Among the many problems caused by water pollution, nutrient loading is one of the most serious and widespread in North America. Nutrients are dissolved minerals that nurture growth in aquatic plants such as algae and bacteria. Among the many nutrients found in natural waters, nitrogen and phosphorus are usually recognized as the most critical ones, because when both are present in large quantities they can induce accelerated rates of biological activity. Massive growths of aquatic plants in a lake or reservoir will lead to: (1) a change in the balance of dissolved oxygen, carbon dioxide, and micro-organisms; and (2) an increase in the production of total organic matter. These changes lead to further alterations in the aquatic environment, most of which are decidedly undesirable from a human use standpoint:

- Increased rate of basin in-filling by dead organic matter.
- Decreased water clarity.
- Shift in fish species to rougher types such as carp.
- Decline in aesthetic quality; e.g., increase in unpleasant odor.
- Increased cost of water treatment by municipalities and industry.
- Decline in recreational value.

Together, the processes of nutrient loading, accelerated biological activity, and the build up of organic deposits are known as *eutrophication*. Often described as the process of aging a water body, eutrophication is a natural biochemical process that works hand-in-hand with geomorphic processes to close out water bodies. Driven by natural forces alone, an inland lake in the mid-latitudes may be consumed by eutrophication within several thousand years, but the rate varies widely with the size, depth, and bioclimatic conditions of the lake. In practically every instance, however, land development accelerates the rate by adding a surcharge of nutrients and sediment to the lake. So pronounced is this increase that scientists refer

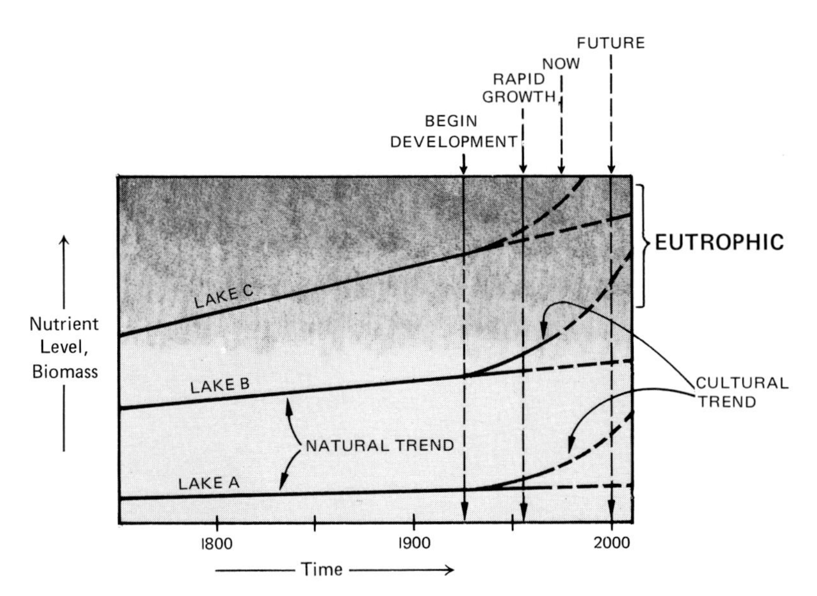

Fig. 10.2 Graph illustrating the concepts of natural and cultural eutrophication. The three sets of curves represent water bodies at three stages of natural eutrophication as they enter the cultural stage of their development.

to two eutrophication rates for water bodies in developed areas: natural and cultural (Fig. 10.2).

For plants to achieve a high rate of productivity, the environment must supply them with large and dependable quantities of five basic resources: heat, light, carbon dioxide, water, and nutrients. According to the biological principle of limiting factors, plant productivity can be controlled by limiting the supply of any one of these resources. In inland and coastal waters, heat, light, carbon dioxide, and water are abundant on either a year-round or seasonal basis. On the other hand, nutrients often tend to be the most limited resource, not just in terms of quantities but also in terms of types. If dissolved nitrogen and phosphorus are both available in ample quantities, productivity is usually high; however, if either one is scarce, productivity may be retarded.

When introduced to the landscape in the dissolved form, phosphorus and nitrogen show different responses to runoff processes. Nitrogen, which is generally more abundant, tends to be highly mobile, moving with the flow of soil water and groundwater to receiving water bodies. If introduced to a field as fertilizer, for example, most of it may pass through the soil in the time it takes infiltration water to percolate through the soil column, as little as weeks in humid climates. In contrast, phosphorus tends to be retained in the soil, being released to the groundwater very slowly. As a result, under natural conditions most waters tend to be phosphorus-limited, and when a surcharge of phosphorus is directly introduced to a water body, accelerated rates of productivity can be triggered. Accordingly, in water management programs aimed at limiting eutrophication, phosphorus control is often the primary goal. In the study of inland lakes, a classification scheme has been devised based on the total phosphorus (both organic and inorganic forms) content of lake water (Table 10.2).

Table 10.2 Levels of Eutrophication Based on Dissolved Phosphorus

Level	*Total Phosphorus, mg/l*	*Water Characteristics*
Oligotrophic (pre-eutrophic)	less than 0.025	no algal blooms or nuisance weeds; clear water; abundant dissolved oxygen
Early eutrophic	.025–.045	↓
Middle eutrophic	.045–.065	
Eutrophic	.065–.085	
Advanced eutrophic	greater than .085	algal blooms and nuisance aquatic weeds throughout growing season; poor light penetration; limited dissolved oxygen

Representative mean annual values of phosphorus in phosphorus-limited water bodies.

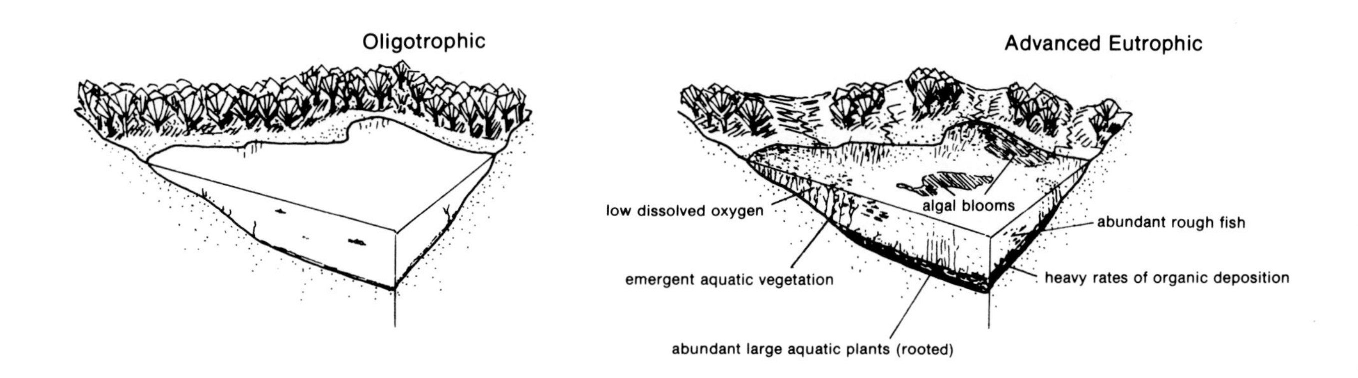

NUTRIENT MASS BUDGET CONCEPT

For any body of water it is possible to compute the nutrient budget by tabulating inputs, outputs, and storage of phosphorus and nitrogen over some time period. Inputs come from four main sources: point sources, surface runoff (streamflow, stormwater, and the like), subsurface runoff (chiefly groundwater), and the atmosphere. Outputs take mainly three forms: streamflow, seepage into the groundwater system, and burial of organic sediments containing nutrients (Fig. 10.3). Storage of nutrients is represented by plants and animals, both living and dead, which release synthesized nutrients upon decomposition. A formula for the nutrient budget may be written as follows:

$$P + R + O + G + A - Q - S - B = 0$$

where:

P = point source contributions
R = surface runoff contributions
O = organic sediment contributions
G = groundwater contributions
A = atmospheric contributions
Q = losses to streamflow
S = losses to groundwater
B = losses to organisms and sediment burial

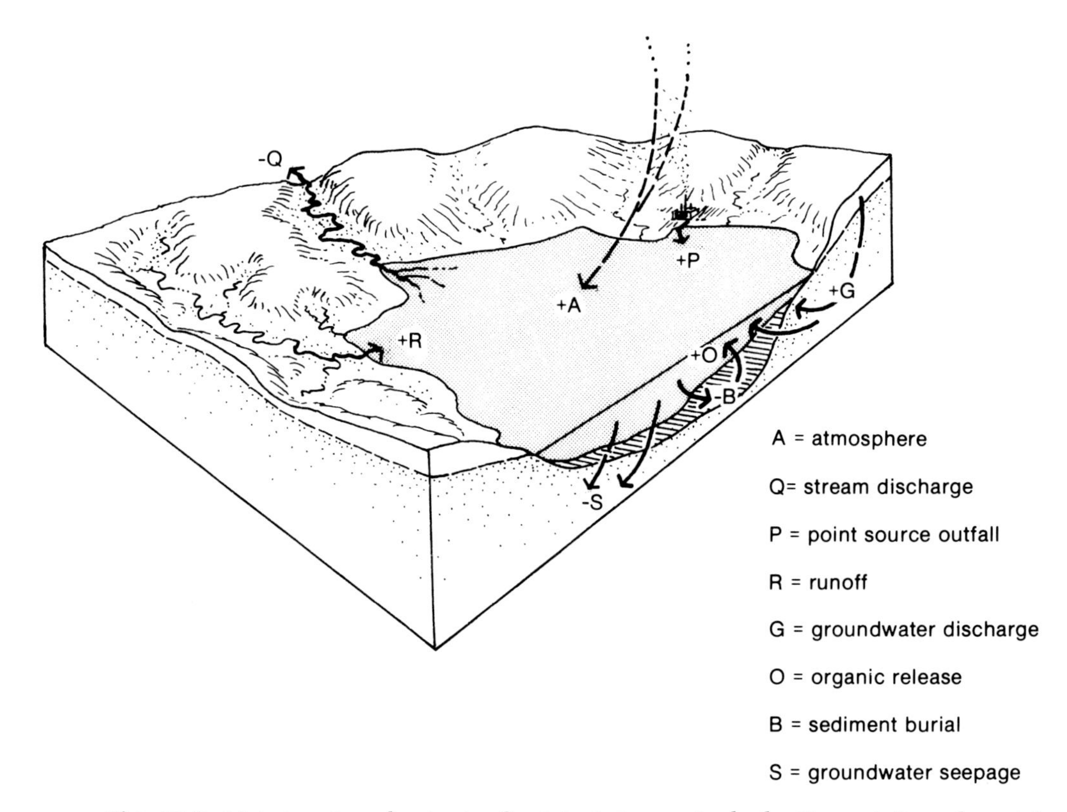

Fig. 10.3 Main inputs and outputs of nutrients to a water body. Computation of a nutrient budget over some time period tells us the trend of the nutrient concentrations in the water.

While the nutrient budget is easy to describe, it has proven very difficult to compute accurately for most water bodies. The primary reasons for this are: (1) some of the pertinent data, such as nutrient losses to groundwater seepage, are often difficult to generate; and (2) exchanges of nutrients among water, organisms, and organic sediments are difficult to gauge. As a result, most nutrient budgets are based on a limited set of data, usually those representing streamflow, septic drainfield seepage, point sources, and atmospheric fallout. Recently, however, data have been produced relating land use and surface cover to the nutrient content of runoff.

LAND USE AND NUTRIENT LOADING OF RUNOFF

In a study of the nitrogen and phosphorus contents of United States streams, the United States Environmental Protection Agency and various state water quality programs made several interesting findings. First, the export of these nutrients from the land by streams tends to vary widely for different runoff events and for different watersheds with similar landuses. Second, although regional variations in nutrient export do appear in the United States for small drainage areas, they do not correlate very well with rock and soil type. Instead, nutrient export by small streams tends to correlate best with land use and cover, in particular to the proportion of agricultural and urban land in a watershed. Nitrogen and phosphorus concentrations in streams draining agricultural land, for example, are typically five to ten times higher than those draining forested land.

Based on these findings, it is possible to estimate the nitrogen and phosphorus loading of streams and lakes in most areas. The loading values are applicable to surface runoff, principally channel flow. The values in Table 10.3 are given in kilograms per square kilometer for six basic land use/cover types. (The definition of each land use/cover type is given in the next section.) Figure 10.4 gives the loading values in milligrams per liter of

Table 10.3 Nutrient Loading Rates for Six Land Cover/Use Types

Cover/Use	*Nitrogen (kg/km²/yr.)*	*Phosphorus (kg/km²/yr.)*
Forest	440	8.5
Mostly forest	450	17.5
Mostly urban	788	30.0
Mostly agriculture	631	28.0
Agriculture	982	31.0
Mixed	552	18.5

From J. Omernik, *The Influence of Land Use on Stream Nutrient Levels,* U.S. EPA, 1977.

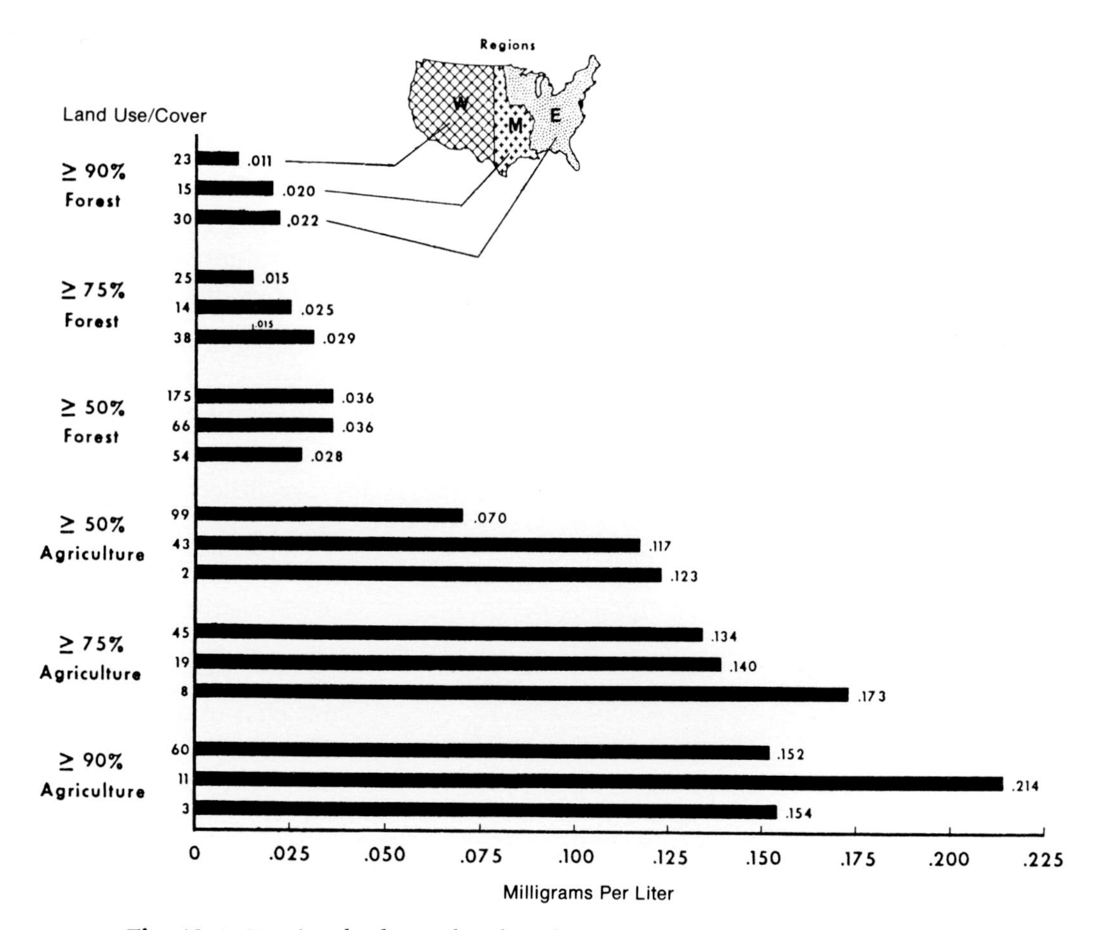

Fig. 10.4 Nutrient loading values based on agricultural and forest land use/cover for the coterminous United States. (From J. M. Omernik, *Non-point Source—Stream Nutrient Level Relationships: A Nationwide Study,* U.S. EPA, 1977.)

water (runoff) for agriculture and forest use/cover types for the eastern, middle, and western regions of the coterminous United States. For comparative purposes, examine these values with those representative of the water systems listed in Table 10.4.

Table 10.4 Representative Levels of Phosphorus and Nitrogen in Various Waters

Water	*Total P, mg/l*	*Total N, mg/l*
• Rainfall	0.01–0.03	0.1–2.0
• Lakes without algal problems	less than 0.025	less than 0.35
• Lakes with serious algal problems	more than 0.10	more than 0.80
• Urban stormwater	1.0–2.0	2.0–10
• Agricultural runoff	0.05–1.1	5.0–70
• Sewage plant effluent (secondary treatment)	5–10	more than 20

Sources include John W. Clark et al., *Water Supply and Pollution Control*, 3rd ed., New York: IEP/Dun-Donnelley, 1977); and American Water Works Association, "Sources of Nitrogen and Phosphorus in Water Supplies," *Journal of the American Water Works Association*, vol. 59, 1967, pp. 344–366.

ESTIMATING NUTRIENT LOADING

To determine the nutrient contributions to a water body, one must first delineate the drainage system and define its drainage areas. This may be a difficult task in developed areas because of the complexity of the land uses and the man-made drainage patterns that are superimposed on the natural drainage system. As a first step, it is instructive to identify the various land uses and major cover types in the area, noting their relationship to the drainage system and water features, what kinds of pollutant they are apt to contribute (both nutrients and other types), and the locations of critical entry points. The remaining steps are as follows:

- Determine the percentage of each drainage area occupied by forest, agriculture, and urban development.
- Classify each area according to the relative percentages of forest, agriculture, and urban land uses based on the following percentages:

Forest:	>75% forested
Mostly Forest:	50–75% forested
Agriculture:	>75% active farmland
Mostly Agriculture:	50–75% active farmland
Mostly Urban:	>40% urban development (residential, commercial, industrial, institutional)
Mixed:	Does not fall into one of the above classes; for example, 25% urban, 30% agriculture, and 45% forest.

- Using the nutrient-loading values given in Table 10.3, multiply the appropriate value for each area by its total area in square kilometers.
- To calculate the loading potential from septic drain field seepage, count the number of homes within one hundred yards of the shore or

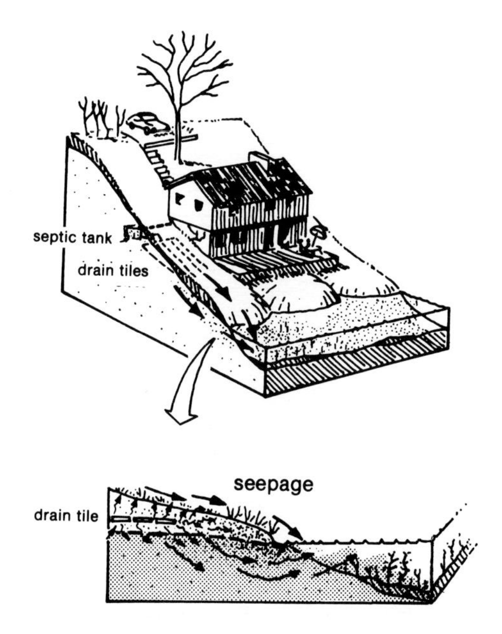

streambank for each drainage area and multiply this number by the following nutrient loading rates:[1]

Phosphorus/Year	*Nitrogen/Year*
0.28 kg/home	10.66 kg/home

- Combine the two totals for each drainage area for the total input from the watershed. If the *grand* total input from all sources is called for, then atmospheric contributions must also be considered. Atmospheric input need only be considered for water surfaces, since it is already included in the values given for land surfaces. Given a fallout rate for the area in question, this value (in mg/m^2) should be multiplied by the area of the water body; this quantity would be added to those from the watershed to obtain the grand total input (less groundwater contributions, if any) to the water body.

CASE STUDY

Wetlands and Community Wastewater Treatment

Donald Tilton

The recognition of the decline in local water resources and the high cost of conventional wastewater treatment techniques in the United States were the impetus behind several efforts to develop alternative wastewater treatment techniques. One of these techniques was the utilization of freshwater wetlands (marshes, swamps, bogs, etc.) for the treatment of municipal wastewater. Research projects on this technique were conducted in California, New York, Wisconsin, Minnesota, Michigan, New Jersey, Florida, and Louisiana, and there continues to be a need for additional research.

[1]From *National Eutrophication Survey Methods for Lakes Sampled in 1972: Working Paper No. 1*, Washington, D.C.: U.S. EPA, 1973.

CASE STUDY (cont.)

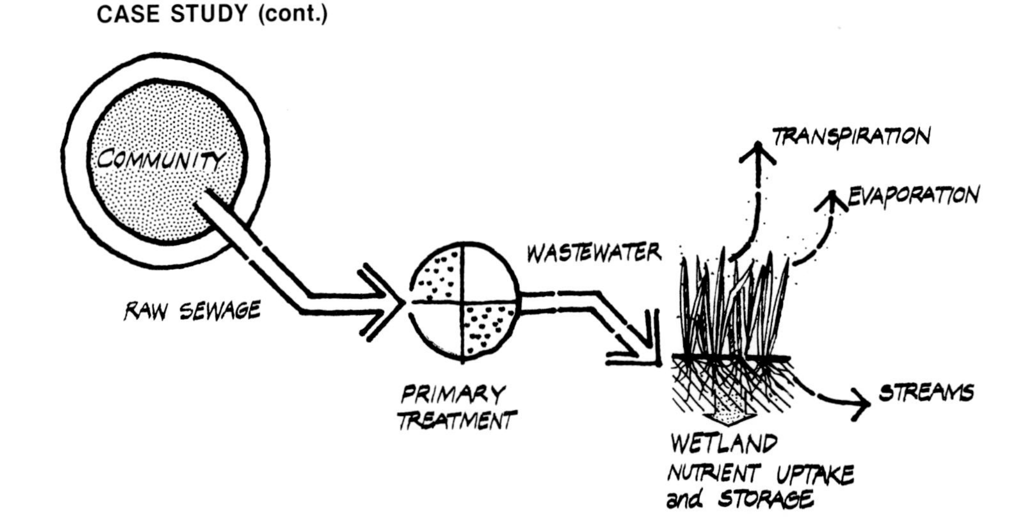

The treatment of municipal wastewater by irrigating wetlands relies on the ecological, biological, and physical processes that occur naturally in a wetland. The first stage of this wastewater treatment method is to process the untreated wastewater by conventional engineering technology. The product of this treatment process is a wastewater contaminated with bacteria and dissolved nutrients, which, if discharged to a river or lake, could cause serious pollution and environmental degradation. When this same wastewater is discharged to certain types of wetlands, however, the nutrients are absorbed by the plants, converted by microbial processes into gaseous forms, or trapped in the soil with a variable degree of environmental degradation.

Unfortunately, it is not clearly understood what environmental harm can result from the discharge of wastewater to different types of wetlands. At certain project sites there has been a significant deterioration of the habitat after only a few years of wastewater discharge, while at other sites, particularly California, the wastewater was managed in such a way so as to create valuable wildlife habitat. There is a third class of project where the wastewater has had some impact on the original communities of plants and animals, but the impacts were not severe enough to reduce the use of the site by wildlife.

Besides variation in the degree of environmental impact at various sites, there is also variation in the degree of wastewater treatment that can be achieved in different wetlands. The project sites studied to date have recorded nutrient removals of 40 to 90 percent, but there is no formula for predicting what level of nutrient removal efficiency will be achieved at a particular site. The issue of treatment efficiency is further complicated by the fact that some wetland ecosystems temporarily store nutrients during the growing season with a nutrient discharge in the fall and winter.

In three small Michigan communities near Houghton Lake, for example, community wastewater is collected from residents, given primary treatment, and discharged to a large wetland of approximately 1700 acres. Several years of research by environmental scientists and engineers preceded a full-scale discharge of secondary wastewater in 1979. Studies performed in conjunction with that discharge showed that significant levels of nutrients, particularly nitrate-nitrogen and phosphorus, were removed from the wastewater pumped onto the wetland surface. This particular wetland treatment system has demonstrated satisfactory nutrient removals, but some new plant species have invaded the plant communities closest to the site of wastewater discharge. Only long-term studies can document the final impact of the wastewater on the wetland environment, but early indications suggest that under proper conditions this wetland will provide a natural means of treating community wastewater with acceptable environmental impact.

The use of wetland ecosystems for the treatment of municipal wastewater may provide communities with an efficient alternative to conventional wastewater treatment. The challenge to community planners, environmental scientists, and engineers, however, is to insure that during the treatment of municipal wastewater the wetland environment is not irreparably degraded. This means that not only must the capacity of the biological systems be maintained, but that the quality of the wetland as an environment be maintained at some level. In approaching the issue of wetland quality, it is necessary to recognize the reality of environmental tradeoffs in master planning. Simply stated, trade-offs are based on the understanding that not all types of environments in a community can be maintained at the highest quality level, and that a choice must often be made between those to be protected and those to be altered and degraded by land use activities. In the case of wastewater disposal, it often comes down to a choice between a stream or lake on one hand and a wetland on the other. The choice ultimately rests largely on the social and cultural values of the community.

Donald Tilton is a senior environmental analyst with Smith, Hinchman, Grylls Associates, an architecture, engineering, and planning firm.

PLANNING FOR WATER QUALITY MANAGEMENT IN SMALL WATERSHEDS

Planning for residential and related land uses near water features presents a fundamental dilemma: People are attracted to water, yet the closer to it they build and live, the greater the impact they are apt to have on it. As impacts in the form of water pollution and scenic blight rise, the value of the environment declines, which in turn usually lowers land values (Fig. 10.5). Moreover, the greater the proximity of development to water features, generally the greater the threat to property from floods, erosion, and storms. In spite of these well-known problems, the pressure to develop near water features has not waned in recent years; in fact, it seems to have increased, and in areas where there are few natural water features, developers are often inclined to build artificial ones to attract buyers.

In planning and management programs aimed at water quality, basically two approaches may be employed: preventive and corrective. The corrective approach is used to address an unsatisfactory condition that has already developed; for example, cleaning beaches and surface water after an oil spill. On small bodies of water with weed growth problems, corrective measures may involve treatment with chemicals that inhibit aquatic plants, or basin dredging to remove organic sediments. Such measures are generally used as a last resort for water bodies with serious nutrient problems and/or advanced states of eutrophication.

A preventive approach is generally preferred for most bodies of water, though it may actually be more difficult to employ successfully.

Fig. 10.5 Environmental damage along a water body leading to deterioration of land values.

This approach involves limiting or reducing the contributions of nutrients and other pollutants from the watershed by controlling on-site sources of pollution, limiting the transport of pollutants from the watershed to the lake, or both. Measures to control nutrient sources include improving the performance of septic drainfields for sewage disposal, replacing septic drainfields with community sewage treatment systems, reducing fertilizer applications to cropland and lawns, controlling soil erosion, and eliminating the burning of leaves and garbage. Measures to limit nutrient transport to the lake include on-site retention of stormwater, abolition of stormsewers in site drainage, diversion of low-quality water to sump basins, and the maintenance of wetlands, floodplains, and natural stream channels.

The formulation of a water quality management program for a water body usually begins with an estimate of the nutrient budget. Nutrient contributions are placed into two categories: those that can be managed using available technology and funding, and those that cannot. The latter includes groundwater and atmospheric contribution, whereas the former includes primarily surface and near surface runoff, that is, stormwater, septic drainfield seepage, and the like.

The second step entails defining management goals, such as "to slow the rate of eutrophication," or "to improve on the visual character" of a certain part of the water body. This is an important process because the goals must be realistic and attainable. Generally speaking, the more ambitious the goals (for example, "to reverse the trend in eutrophication"), the more difficult and expensive they will be to achieve. Once goals are established, a plan (or set of plans) is formulated that identifies the actions and measures that need to be implemented. Some actions, such as a prohibition on the use of phosphorus-rich fertilizers, may apply to the entire watershed, while others may apply only to a specific subarea that the nutrient budget data show to be a large contributor. In addition, measures are proposed to limit nutrient production and transport related to future development by defining appropriate development zones, and enacting guidelines on sediment control, stormwater drainage, and sewage disposal.

Finally, after the plan is implemented, the water body must be monitored and the results weighed against the original goals, financial costs, and public support. This is a difficult task because it often requires comparing the results with forecasts about what conditions would have been without the plan. Uncertainty often arises relating to the validity of the original forecasts and what the effects of natural perturbations in climatic, hydrologic, or biotic systems may have had in masking or enhancing the efforts of the plan.

THE SPATIAL FRAMEWORK FOR PLANNING

We should not end this chapter without underscoring the need in water quality planning to understand the physical system we are dealing with. As we noted in Chapter 7, every water body is supported by a watershed, which is comprised of many water systems including stream networks, groundwater, and precipitation and evaporation. The runoff system, particularly that on the surface, provides the spatial framework within which land use planning related to water quality management takes place.

The watershed in Fig. 10.6 is typical of that around most impoundments. The bulk of the area is taken up by *subbasins* that drain all but the narrow belt of land along the shore, called *shoreland*. Subbasins are usually emptied via streams, whereas shorelands, being too narrow to develop streams, are drained directly to the water body by overland flow and interflow. Land uses in the watersheds of impoundments tend to correlate with these two types of drainage areas. Water-oriented development (recreational activities and residence uses) are concentrated in the shoreland, whereas nonwater-oriented development (agriculture, suburban residential, commercial, and so on) is located in the subbasins.

Management aimed at the shoreland zone usually involves dealing with individual property owners because each site drains directly into the water body. In the subbasins, on the other hand, many land uses are integrated by a runoff system that connects to the water body at a single point. Management of a subbasin, therefore, is in one respect a more difficult task because many players are often involved and the system is larger and more complex. Subbasins do, however, provide management opportunities that shorelands do not; for example, runoff can be collected at central locations and processed by natural or artificial means to reduce nutrients, sediment, and other impurities.

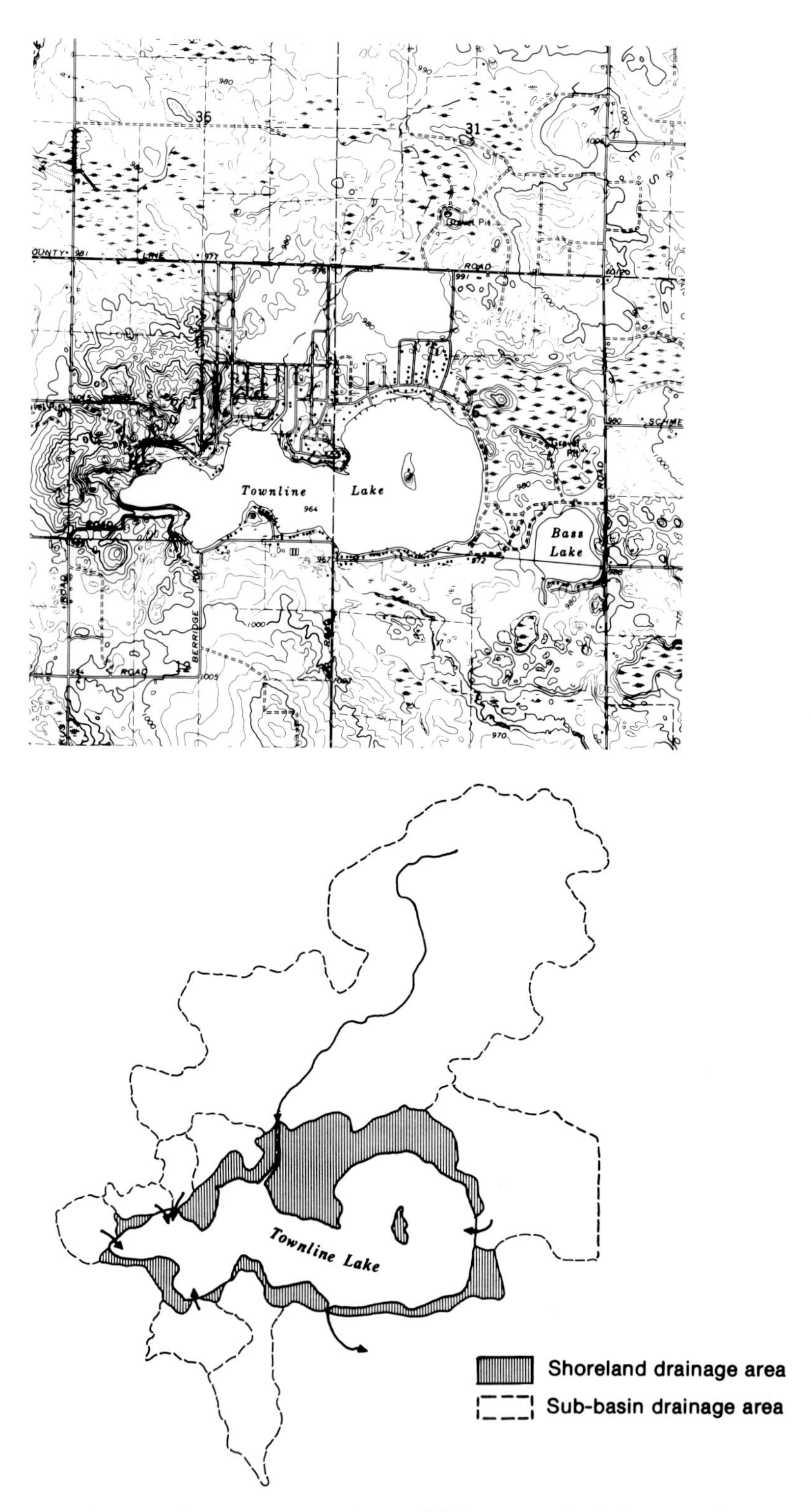

Fig. 10.6 The spatial organization of a small lake watershed showing the two types of drainage subareas: shoreland and subbasin.

PROBLEM SET

I. The following map and aerial photograph show a segment of a river and a local area that drains into it.

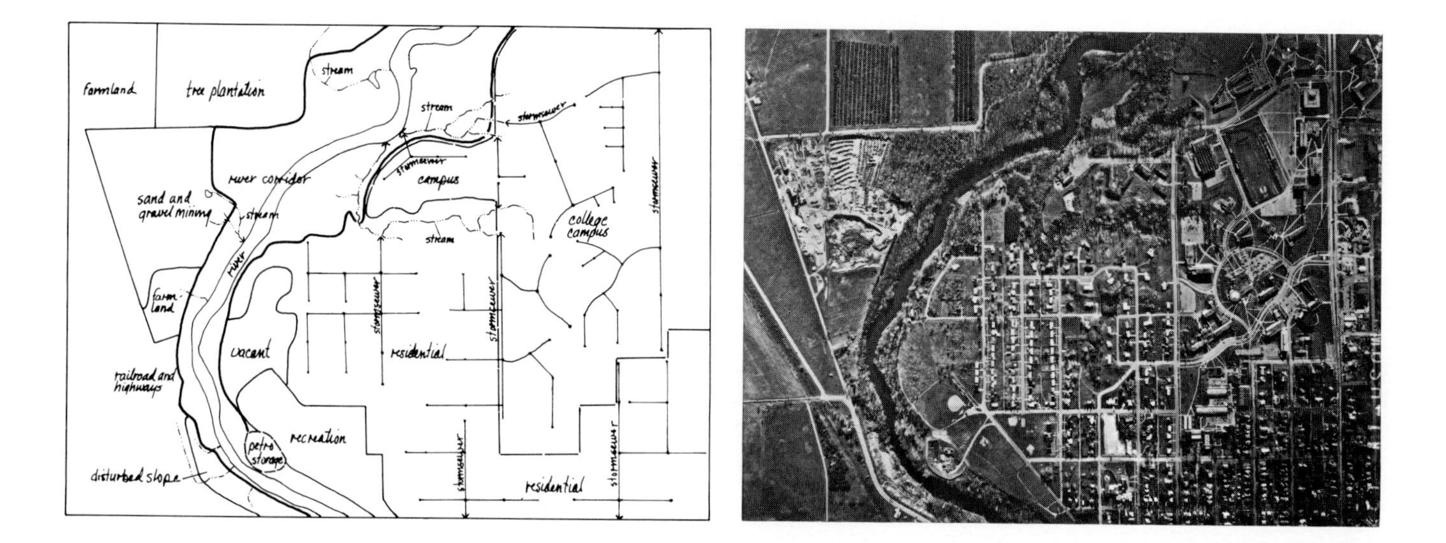

1. Locate the apparent entry points for stormwater, and, based on the land uses in the areas that produce these discharges, identify which of the following pollutants might be significant in each: sediment, phosphorus, nitrogen, petroleum residues, insecticides, disease-bearing agents such as harmful bacteria.

2. Identify some locations where those "difficult to detect" nonpoint pollutants *could* be entering the river.

II. The map on page 198 shows a lake watershed and its drainage subareas. The land use/cover type and area for each subarea are as follows:

Subarea	*Use/Cover*	*Area, km²*
1	agriculture	.92
2	mostly agriculture	1.81
3	forest	1.63
4	mostly urban	.77
5	mixed	1.22
6	mostly urban	.28
7	mostly urban	.50
8	mostly forest	.17
9	mixed	.15
10	mixed	.15

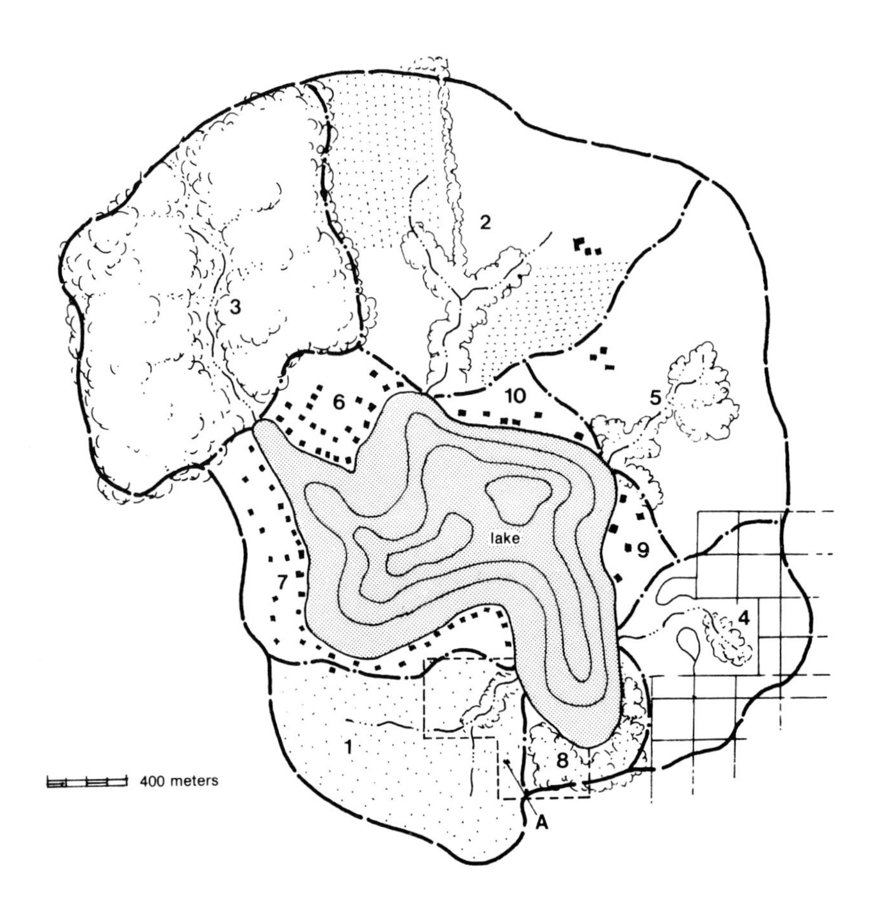

1. Determine the contributions of phosphorus and nitrogen to the lake from these subareas.
2. Assuming the residences shown rely on septic drainfields for sewage disposal, determine the additional contributions of phosphorus and nitrogen from those residences within one hundred yards of the lake shore. Considering the two types of drainage subareas (shorelands and subbasins), which present the most serious source of nutrient input to the lake?
3. Name several strategies that could be employed in each drainage subarea to reduce phosphorus contributions to the lake. Try to relate these to the land use/cover and the drainage system in each.
4. List a set of guidelines for the development of a new subdivision (at site A) that would help prevent additional nutrient contributions to the lake. Suggest some locations for future development that would be more appropriate than site A from a water quality standpoint.

SELECTED REFERENCES FOR FURTHER READING

Clark, John W.; Viessman, Warren, Jr.; and Hammer, Mark J. *Water Supply and Pollution Control.* 3rd ed. New York: IEP/Dun-Donnelley, 1977, 857 pp.

Dillon, P. J., and Vollenweider, R. A. *The Application of the Phosphorus Loading Concept to Eutrophication Research.* Burlington, Ontario: Environment Canada; Center for Inland Waters, 1974, 42 pp.

Marsh, William M., and Borton, T. E. *Inland Lake Watershed Analysis: A Planning and Management Approach.* Lansing, Mich.: Michigan Department of Natural Resources, 1976, 88 pp.

Millar, John B. *Wetland Classification in Western Canada.* Ottawa: Environment Canada; Wildlife Service, 1976, 38 pp.

National Academy of Sciences. *Eutrophication: Causes, Consequences, and Correctives.* Washington, D.C.: National Academy of Sciences and the National Research Council, 1969, 463 pp.

Omernik, James M. *Nonpoint Source—Stream Nutrient Level Relationships: A Nationwide Study.* Corvallis, Ore.: U.S. Environmental Protection Agency, 1977, 150 pp.

Soil Conservation Service. *Ponds For Water Supply and Recreation.* Washington, D.C.: U.S. Dept. of Agriculture, 1971, 55 pp.

Tilton, Donald L., and Kadlec, R. H. "The Utilization of a Fresh-Water Wetland for Nutrient Removal from Secondarily Treated Waste Water Effluent." *Journal of Environmental Quality,* vol. 8, no. 3, 1979, pp. 328–334.

Vallentyne, John R. *The Algal Bowl: Lakes and Man.* Ottawa: Environment Canada; Fisheries and Marine Service, 1974, 186 pp.

11

TOPOGRAPHY, SLOPES, AND LAND USE PLANNING

INTRODUCTION

In searching for a building site, an architect or landscape architect will often be guided by topographic considerations. Depending on the sort of building he or she is designing and the activities associated with it, a level site, sloping site, or hilly site may be desired. Level or gently sloping sites are usually preferred for industrial and commercial buildings, whereas hilly sites are preferred for fashionable suburban residences.

The influence of slope and topography can be extended to other land uses as well. Cropland today is generally limited to slopes of less than 10 degrees (18 percent), because of the performance and safety restrictions posed by the operation of tractors and field machinery. In the era of horse and oxen power, slopes of 15 degrees (26 percent) or steeper could be used for cultivated crops. The influence of slopes and topography on the alignments of modern roads depends on the class of the road; the higher the class, the lower the maximum grades allowable. Interstate class expressways (divided, limited access, four or six lanes) are designed for high speed, uninterrupted movement and limited to grades of 4 percent; that is, four feet of rise per one hundred feet of distance. On city streets, where speed limits are 20 to 30 mph, grades may be as steep as 10 percent, whereas driveways may be as steep as 15 percent.

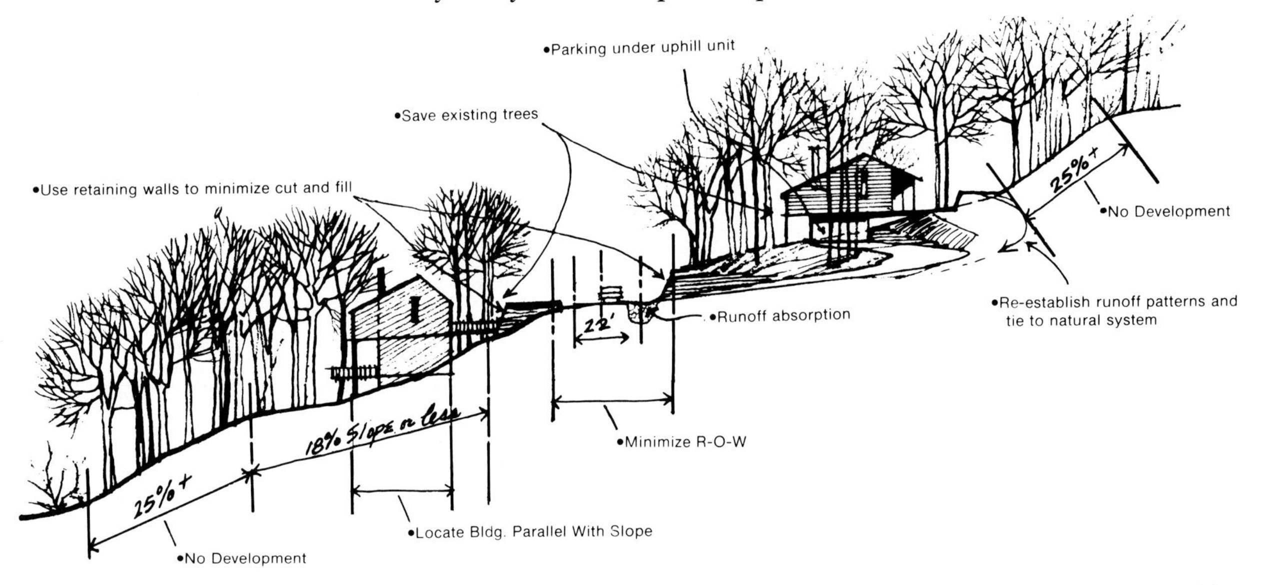

A design concept for residential development on a hillside. (By Carl D. Johnson. Used by permission.)

SLOPE PROBLEMS

The need to consider topography in planning is an outgrowth of widespread realization not only that land uses have slope limitations but that slopes have been misused in modern land development. The misuse arises from two types of practices: (1) the placement of structures and facilities on slopes that are already unstable or potentially unstable; and (2) the disturbance of stable slopes resulting in failure, accelerated erosion, and/or ecological deterioration of the slope environment.

The first type can result from inadequate survey and analysis of slopes in terrain that has a history of slope instability; more infrequently, however, it probably results from inadequate planning controls (for example, zoning and environmental ordinances) on development. In some instances, admittedly, surveys reveal no evidence of instability and failure of a slope catches inhabitants completely unawares.

A small example of improper siting of facilities on a slope. (Photograph courtesy of the U.S. Soil Conservation Service.)

Disturbance of slope environments is unquestionably the most common source of slope problems in North America. Three types of disturbances stand out:

- *Mechanical cut and fill* in which slopes are reshaped by heavy equipment. This often involves steepening and straightening, resulting in a loss of the equilibrium associated with natural conditions; in Canada and the United States this is best exhibited in mining areas and along major highways.
- *Deforestation* in a hilly terrain by lumbering operations, agriculture, and urbanization. This not only results in a weakened slope because of the reduced stabilizing effect of vegetation, but also increases stress from runoff and groundwater.
- *Improper siting and construction* of buildings and related facilities, leading to an upset in the slope equilibrium because of the alteration of vegetation, slope materials, and drainage (Fig. 11.1).

MEASUREMENT OF SLOPES AND TOPOGRAPHY

Years ago the configuration of the terrain could be measured only by field surveying. In its simplest form, this involved projection of a level line into the terrain from a point of known elevation and then measuring the distances above and below the line to various points in the terrain. Once elevation points were known, a contour map could be constructed.

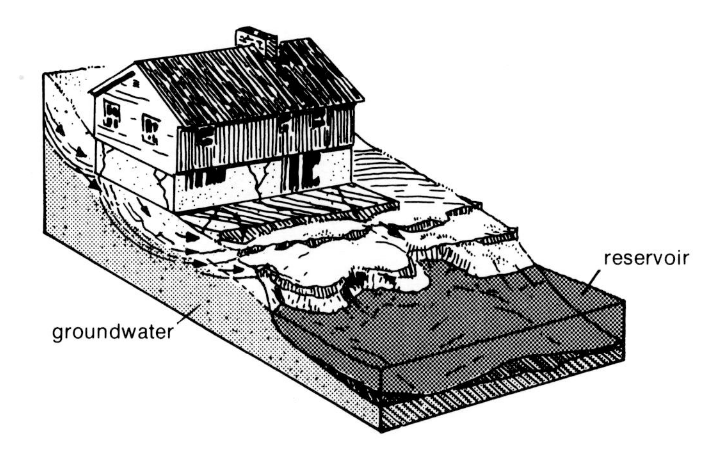

Fig. 11.1 Slope failure resulting from alteration of drainage, vegetation, and soil in residential development.

Contour maps are comprised of lines, called *contours,* connecting points of equal elevation. In modern mapping programs, such as the one practiced by the U.S. Geological Survey, the contours are drawn from specially prepared sets of aerial photographs. These photographs and the optical apparatus used to view them enable the mapper to see an enlarged, three-dimensional image of the terrain, and based on this image, the mapper is able to trace a line, the contour, onto the terrain at a prescribed elevation. The contour elevation is calibrated on the basis of survey markers, called *bench marks,* placed on the land by field survey crews (see Chapter 16, *Maps and Map Reading*).

To determine the inclination of a slope from a topographic contour map, one must know the scale of the map and the contour interval. With these, the change in elevation over distance can be measured, and in turn a percentage can be calculated:

$$\text{percent slope} = \frac{\text{change in elevation}}{\text{distance}} \times 100$$

This is one of the two conventional expressions for slope, the other being degrees. Conversions can be made for degrees or percentages with the aid of the diagram in Fig. 11.2.

SLOPE MAPPING FOR LAND USE PLANNING

To avoid costly damage to the environment or to structures and utilities, it is necessary to make the proper match between land uses and slopes. In most instances this is simply a matter of assigning to the terrain uses that would: (1) not require modification of slopes to achieve satisfactory performance, and (2) not themselves be endangered by the slope environment and its processes. Generally speaking, topographic contour maps alone do not provide information in a form suitable for most planning problems. The contour map must instead be translated into a map made up of slope classes tailored to planning problems. The utility of such slope maps is a function of: (1) the criteria that are used to establish the slope classes, and (2) the scale at which the mapping is undertaken.

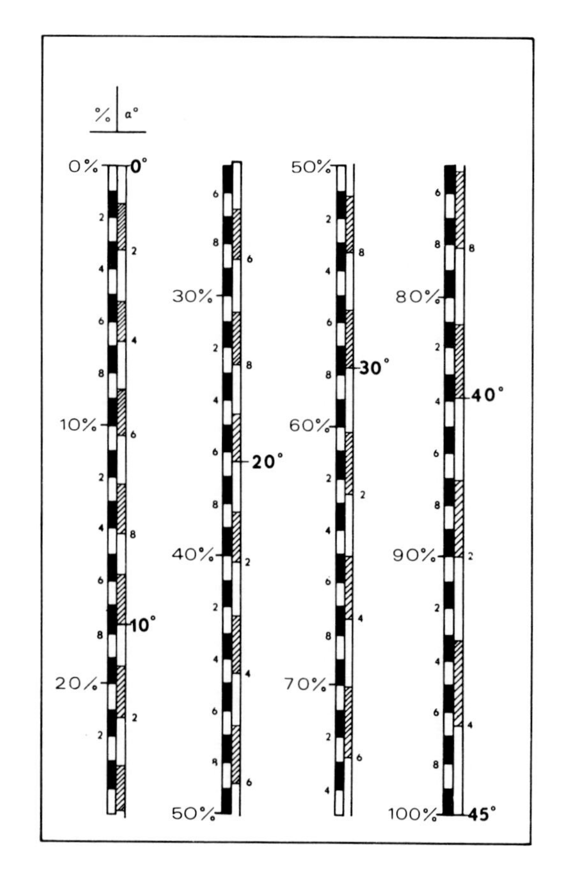

Fig. 11.2 Degree equivalence of percent slope up to 100 percent.

The scale of mapping and the level of detail that are obtainable are strictly limited by the scale and contour interval of the base map. In areas where maps of two or three different scales are available, the scale chosen should be the one that best suits the scale of the problem for which it is intended, such as site plan review, master planning, or highway planning.

The criteria used to set the slope classes are dependent foremost on the problems and questions for which the map will be employed (Fig. 11.3). For areas under the pressure of suburban development, the maximum and minimum slope limits of the various community activities would be one set of criteria. Another would be the natural limitations and conditions of the slopes themselves, which are taken up in the next section. With respect to land use activities, one would want to know the optimum slopes for parking lots, house sites, residential streets, playgrounds and lawns, and so on (Table 11.1).

In addition to the selection of slope classes and the appropriate base map, the preparation of a slope map also involves:

1. *Definition of the minimum size mapping unit.* This is the smallest area of land that will be mapped, and it is usually fixed according to the base map scale, the contour interval, and the scale of the land uses involved. For 7.5-minute U.S. Geological Survey quadrangles (1:24,000), units should not be set much smaller than 10 acres, or 660 feet square.
2. *Construction of a graduated scale* on the edge of a sheet of paper, representing the spacing of the contours for each slope class. For exam-

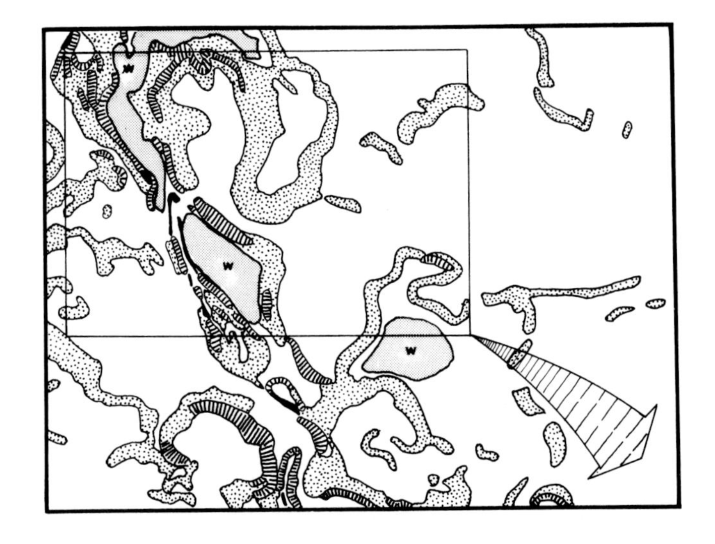

Very Steep (greater than 25°)
Steep (15-25°)
Moderate (5-15°)
Gentle (less than 5°)
Soil: loam

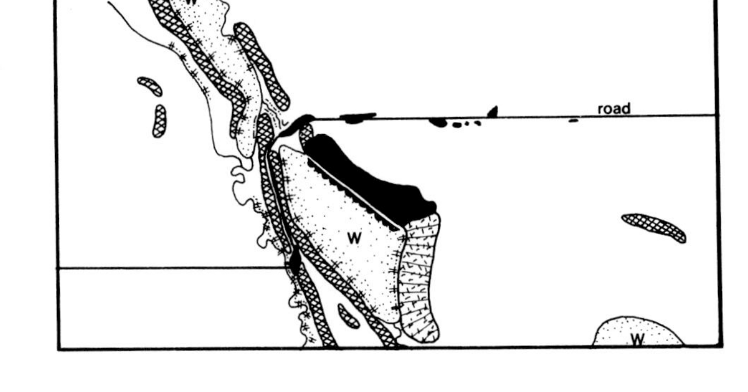

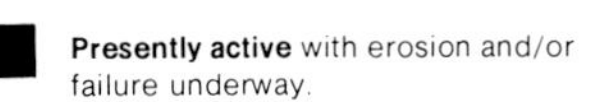

Presently active with erosion and/or failure underway.

Highly susceptible to failure and erosion should forest be removed.

Erosion imminent under present use.

Least susceptible to failure and erosion under agricultural, residential and related uses.

Natural waters presently influenced by sedimentation from eroded slopes.

Natural waters highly susceptible to sedimentation should nearby slopes be activated.

Fig. 11.3 Slope classification for the purpose of identifying slopes prone to erosion and failure. (From *Environmental Analysis for Land Use and Site Planning*, by W. M. Marsh. Copyright © 1978, McGraw-Hill, New York. Used with the permission of McGraw-Hill Book Company.)

ple, on the 7.5-minute quadrangle, where 1 inch represents 2,000 feet, a 10-percent slope would be marked by contour every 1/20 inch.

3. *Next, the scale should be placed on the map* in a position perpendicular to the contours to delineate the areas in the various slope classes (Fig. 11.4).
4. *Finally*, each of the areas delineated should be coded or symbolized according to some cartographic scheme.

INTERPRETING STEEPNESS AND FORM

In addition to land use requirements, one must also know what sort of material comprises the slope to interpret accurately the meaning of different inclinations. For any earth material, there is maximum angle, called the

Table 11.1 Slope Requirements for Various Land Uses

Land Use	*Maximum*	*Minimum*	*Optimum*
House Sites	20–25%	0%	2%
Playgrounds	2–3%	.05%	1%
Public Stairs	50%	—	25%
Lawns (mowed)	25%	—	2–3%
Septic Drainfields	15%*	0%	.05%
Paved Surfaces			
Parking lots	3%	.05%	1%
Sidewalks	10%	0%	1%
Streets and Roads	15–17%	—	1%
20 mph	12%		
30	10%		
40	8%		
50	7%		
60	5%		
70	4%		
Industrial Sites			
Factory Sites	3–4%	0%	2%
Lay Down Storage	3%	.05%	1%
Parking	3%	.05%	1%

*Special drainfield designs are required at slopes above 10 to 12 percent.

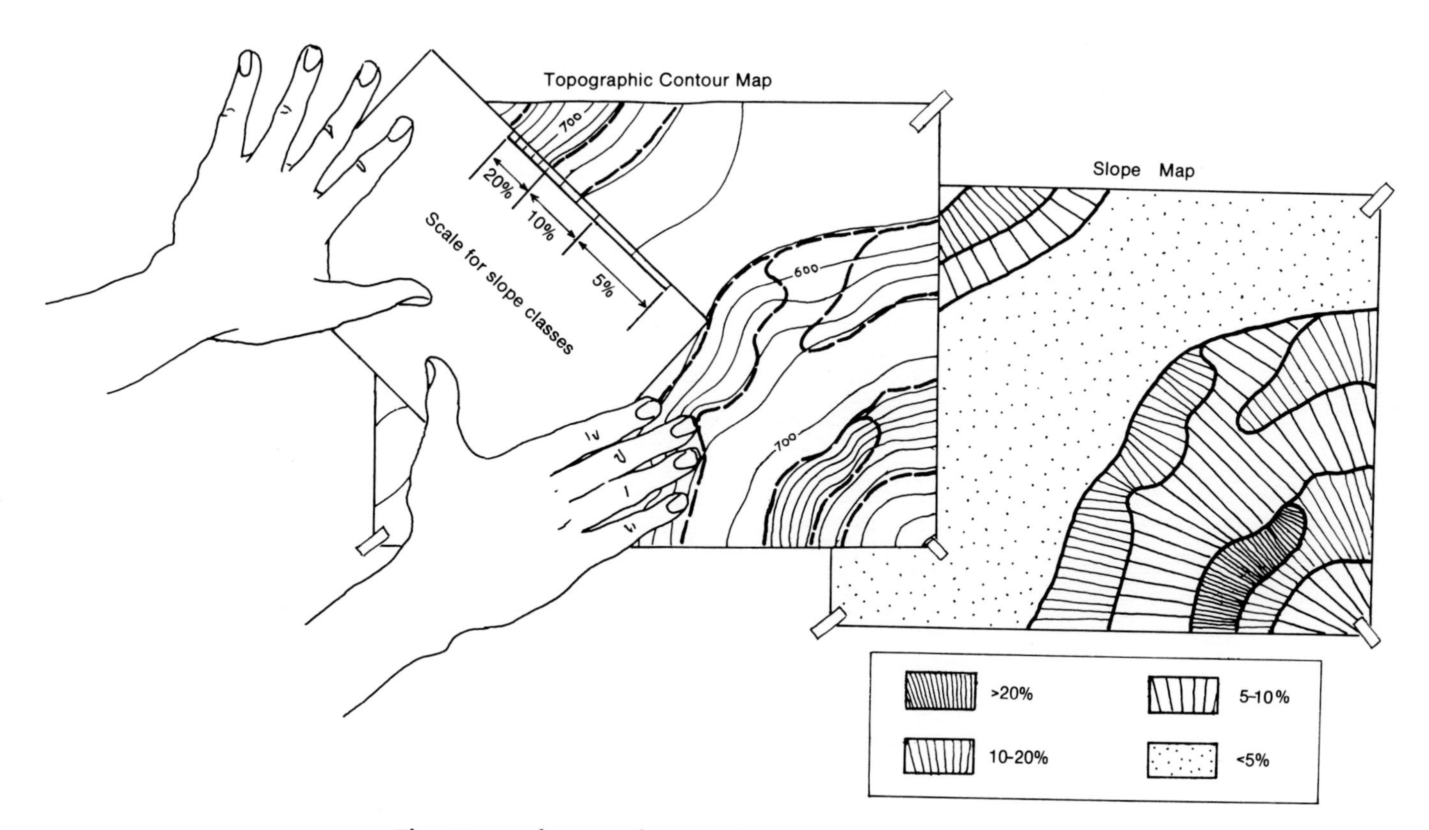

Fig. 11.4 Schematic diagram showing the use of a scale for mapping slope. The lower map shows the results.

angle of repose, at which it can be safely inclined and beyond which it will fail. The angle of repose varies widely for different materials, from 90 degrees in strong bedrock to less than 10 degrees in some unconsolidated materials. Moreover, in unconsolidated material it may vary substantially with changes in water content, vegetative cover, and the internal structure of the particle mass. This is especially so with clayey material: A poorly compacted mass of saturated clay may give way at angles as low as 5 percent, whereas the same mass of clay with high compaction and lower water content may be able to sustain angles greater than 100 percent. Coarse materials, such as sand, pebbles, cobbles, boulders, and bedrock itself are less apt to vary with changes in compaction and water content, and therefore it is possible to define some representative angles of repose for them (Fig. 11.5). Beyond these angles, these materials are susceptible to failure in which the ground ruptures and slides, slumps, or falls.

The influence of vegetation on slope is highly variable depending on the type of vegetation, the cover density, and the type of soil. Vegetation with extensive root systems undoubtedly imparts added stability to slopes comprised of soils of clay, silt, sand, and gravel, but for very coarse materials such as cobbles, boulders, and bedrock, the influence of vegetation is probably insignificant. On sandy slopes the presence of woody vegetation can increase the angle of repose by 10 to 15 degrees; accordingly, loss of the cover on such slopes is almost certain to trigger failure.

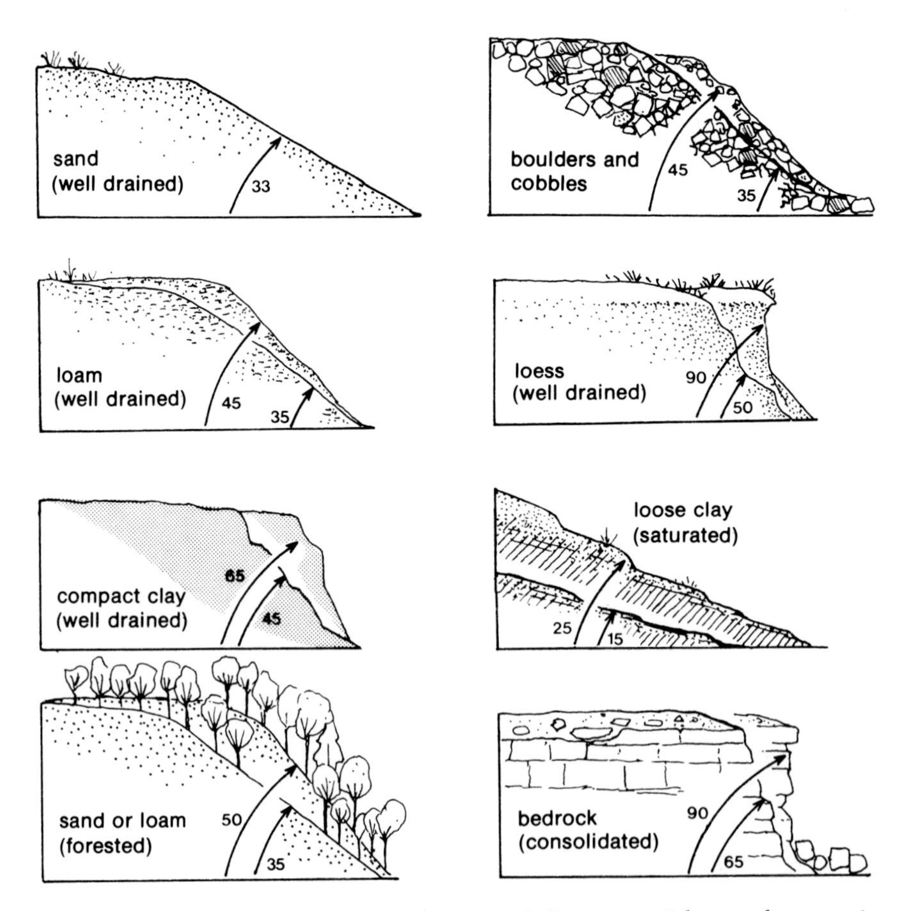

Fig. 11.5 Angles of repose for various types of slope materials. Angles are given in degrees.

In addition to the overall angle, shape can also be an important factor in slope analysis. Shape is expressed graphically in terms of a slope profile, which is basically a silhouette of a slope drawn to known proportions with distance on the horizontal axis and elevation on the vertical axis. The vertical axis is often exaggerated to ease construction and accentuate topographic details (Fig. 11.6).

Five basic slope forms are detectable on contour maps: straight, *S*-shape, concave, convex, and complex (Fig. 11.7). Understanding these forms for problems of land use planning and landscape management usually requires understanding of local geologic, soil, hydrologic, and vegetative conditions. In areas where slopes are comprised of unconsolidated materials (soils and various types of loose deposits), and bedrock is found at great depths, slope form often varies with the vegetative cover, slope (soil) composition, and the recent and past occurrence of events such as undercutting by rivers, excavations by humans, landslides induced by earthquakes, and erosion associated with deforestation.

Smooth *S*-shapes usually indicate long-term slope stability. Such slopes rarely exceed 45 degrees inclination and are usually secured against heavy erosion by a substantial plant cover. Concavities in otherwise straight or *S*-shaped slopes are often signs of failures, such as slides or slumps, and may

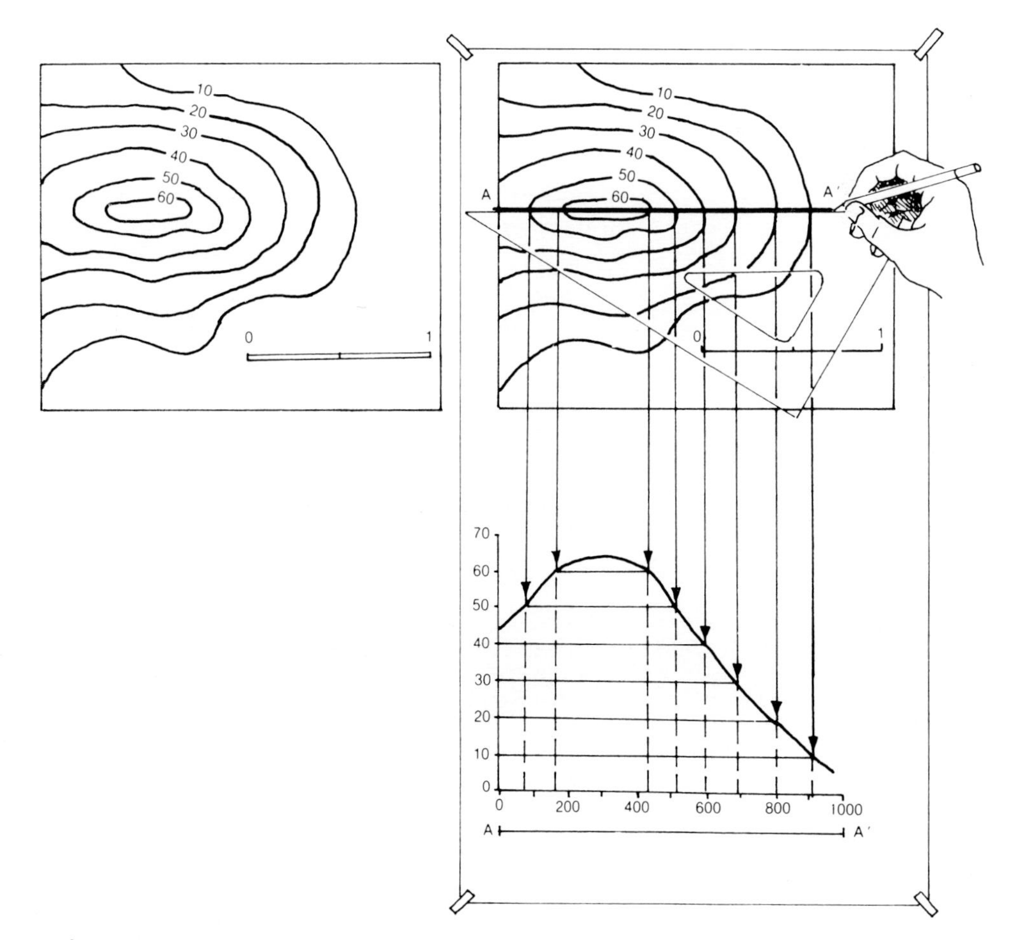

Fig. 11.6 Construction of a slope profile from a topographic contour map.

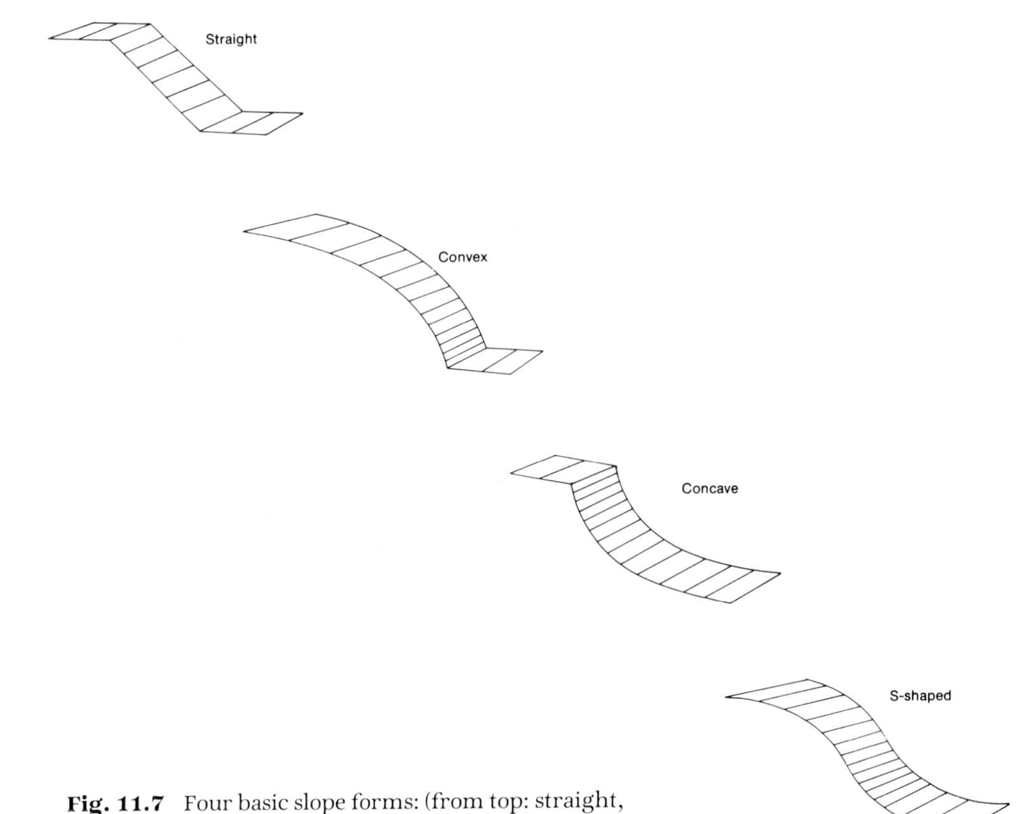

Fig. 11.7 Four basic slope forms: (from top: straight, convex, concave, *S*-shaped).

suggest a general susceptibility to failure for slopes of this class and situation. From the air, scars caused by slope failure or severe erosion can often be identified on the basis of breaks in the plant cover. Fresh scars are marked by light tones on aerial photographs; older scars by less mature vegetation or different plant species (Fig. 11.8).

ASSESSING SLOPES FOR STABILITY

Many criteria should be taken into account in assessing the susceptibility of slopes to failure. At the top of the list are the angle and composition of the slope (both soil and bedrock) and the history of slope activity. Steep slopes with a record of instability stand a greater chance of failure when subjected to development because construction activity, loss of vegetation, and changes in drainage typically lower the stability threshold. If a rock formation or sediment layer of known instability (or high erodibility) is situated within the slope, the potential for failure is even greater.

Plant cover is another important criterion inasmuch as devegetated slopes show a much greater tendency to fail than fully vegetated ones. Studies of slopes in western United States show that clearcut slopes fail more frequently under the stress of heavy precipitation than do fully forested ones.

Fig. 11.8 Slope failure scars marked by breaks in the vegetative cover in the walls of a large gully. (Photograph by U.S. Soil Conservation Service.)

Slope undercutting and earthquake activity are also significant. Active erosion at the foot of a slope by waves, rivers, or human excavation produces steeper inclinations and less confining pressure on the lower slope, thereby increasing the failure potential. When earthquakes jar rock and soil material, interparticle bonds may be weakened and the material's resistance to failure reduced. Some of the worst disasters from slope failure have been triggered by earthquakes.

Finally, drainage must be considered. Though often difficult to evaluate, soil water and groundwater can have a pronounced influence on slope stability: (1) the addition of water to clayey soils can transform them from solid to plastic and liquid states, thereby reducing their resistance to displacement; (2) groundwater seepage can produce undermining of slopes by sapping and piping; and (3) pore-water pressure near seepage zones may weaken the skeletal strength of soil materials within a slope (Fig. 11.9).

Unfortunately, the means to integrate analytically all these variables have not been developed, and for large areas, field and laboratory testing are not economically feasible. Therefore, evaluations of slope stability for purposes of land use planning and environmental management must be based on some sort of systematic review of the previous criteria. Whether all these criteria can be used depends on the availability and reliability of

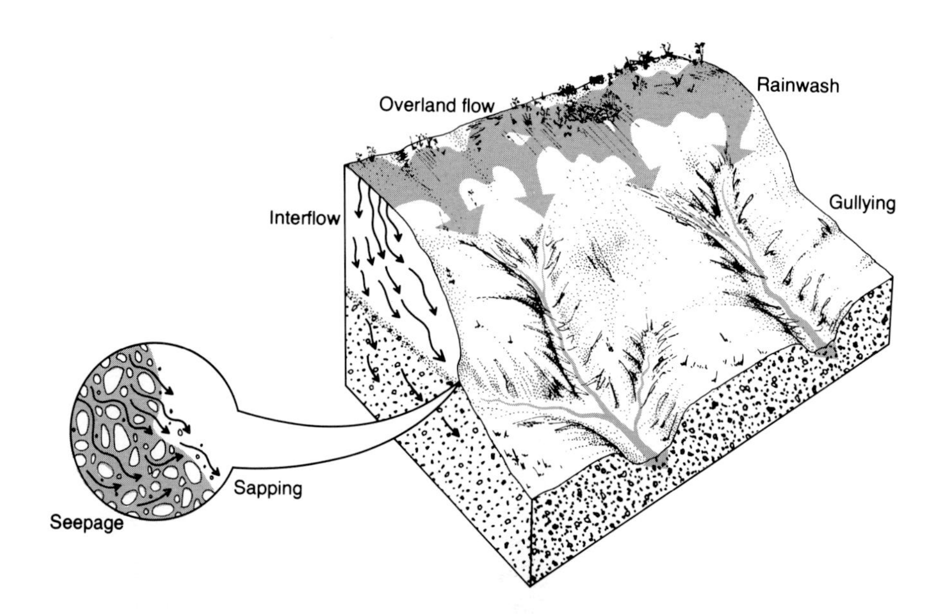

Fig. 11.9 Schematic diagram showing the influence of drainage on slope stability. (From W. M. Marsh and J. Dozier, *Landscape: An Introduction to Physical Geography*, © 1981, Addison-Wesley, Reading, Massachusetts. Fig. 26.4. Reprinted with permission.)

data sources. Generally, topographic contour maps, aerial photographs, geologic maps, siesmic maps, and soil maps are used as data sources, but they require interpretation and adaptation because of differences in scale, resolution, and units of measurement. When the desired criteria can be employed, the map overlay technique has been used to delineate areas of different levels of slope stability. The result is usually a map such as the one shown in Fig. 11.10.

APPLICATIONS TO COMMUNITY PLANNING

In the absence of effective planning controls on development and the base information necessary to help guide decision making, many American and Canadian communities have in the past several decades seen development creep over the terrain, enveloping sloping and flat ground alike. Unlike development of an earlier era when the decision-maker, developer, and resident were one and the same, in modern development these roles are often played by different and separate parties. As a result, in the modern, large-scale development project poor decisions on siting by the developer are upon sale of the property inherited by the buyer, and when a problem eventually arises, it becomes the responsibility of the owner resident. Complicating matters further, the property may be in the second or third owner by the time the problem actually surfaces. The social and financial costs for remedial action are typically high and must be borne by the current property owner, by the community, by some higher level of government, or by all three parties. This harsh lesson has taught communities that the tendency to push development beyond the terrain's capacity must be curbed by land use regulations.

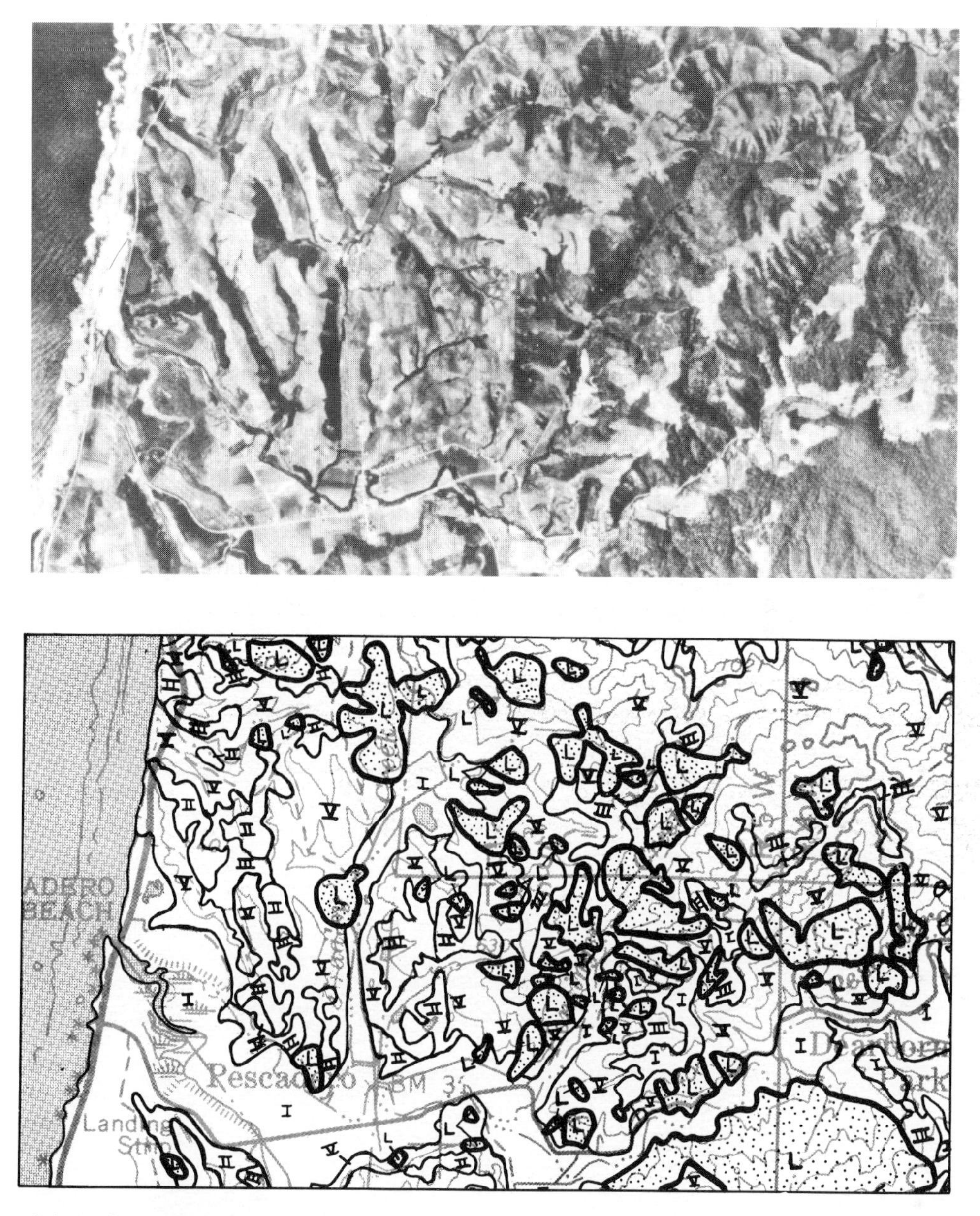

I Least Susceptibility to Landsliding
II Low Susceptibility
III Moderate Susceptibility
IV Moderately High Susceptibility
V High Susceptibility
VI Very High Susceptibility
L Highest Susceptibility

Fig. 11.10 An example of a slope stability map (A) built from multiple sources of mapped data for the area shown in B. (From E. E. Brabb, and E. H. Pampeyan, "Preliminary Geologic Map of San Mateo County, California," *U.S. Geological Survey Miscellaneous Field Studies Map MF-328*, 1972.)

The formulation of policies to guide local development has been aimed at avoiding terrain that: (1) poses a threat of some sort to development, or (2) is valued by the community for its ecological and/or aesthetic character. Implementation of such policies necessitates building a map of suitable and unsuitable terrain and devising a procedure to review and evaluate development proposals and site plans. In slope mapping, two strategies can be used: one is based on development density and average slope inclination within specified land areas or units. Terrain units (such as valley floors or foothills) comprising many individual but similar slopes are assigned a maximum allowable density of development according to the average slope inclination (Fig. 11.11). The percentage of ground to be left undisturbed as required by three California communities is shown in Table 11.2. In Chula Vista, for example, terrain with slopes averaging above 30 percent requires that for each acre of development, nine acres be left in an undisturbed state.

Fig. 11.11 An example of terrain units in an area of mountainous topography. Highest development densities would be allowed in class I where slopes average 10 percent or less. (Photograph courtesy of Pima County Planning and Zoning Dept., Tucson, Arizona.)

Table 11.2 Undisturbed Area Requirements for Sloping Ground in Three California Communities

Percent Slope (avg.)	*Chula Vista*	*Pacifica*	*Thousand Oaks*
10	14%	32%	32.5%
15	31%	36%	40%
20	44%	45%	55%
25	62.5%	57%	70%
30	90%	72%	85%
35	90%	90%	100%
40	90%	100%	100%

The second mapping strategy is slope or site specific and follows the procedure discussed earlier in the chapter. This entails mapping individual slopes and related features that are critical to development; for example, slopes above 25 percent, failure-prone slopes, heavily forested slopes, and unstable soils. Development plans are compared directly to the specific distributions of each factor. This strategy is commonly employed by communities in nonmountainous regions, whereas the strategy based on average slope and terrain units is more common in mountainous regions.

PROBLEM SET

Years ago, when rural communities in the United States and Canada were setting up their systems of government, large areas of undeveloped land were often declared part of the community by virtue of the exaggerated alignment chosen for the town or village limits. These areas were often drawn into maps of the community with little or no attention to the nature of the terrain, and, in most cases, no realistic expectation on the part of the town fathers that the community itself would ever expand into them. Some communities even went so far as to plat these areas; that is, subdivide them on paper into residential lots, city blocks, streets, and neighborhoods. In recent decades, many of these communities have expanded at unprecedented rates, rapidly taking up outlying areas. Where these areas are already platted, the communities often have little control on where development takes place. They must formulate planning ordinances to gain the necessary control. In areas of hilly terrain, an ordinance based on slope inclination can be used to control development, but such an ordinance requires documentation in the form of a slope map. In this case, the area under study is platted for residential development. Neither streets nor utilities currently serve the area. In addition, it is not served by a sanitary sewer system; therefore, each house would have its own septic tank and drainfield.

I. The following map shows the layout of a coastal community as it appeared in a 1943 city map. The hills are forested and sandy with springs seeping from the base of the slopes at selected points.

1. Based on the slopes alone, designate the nonbuildable zones.

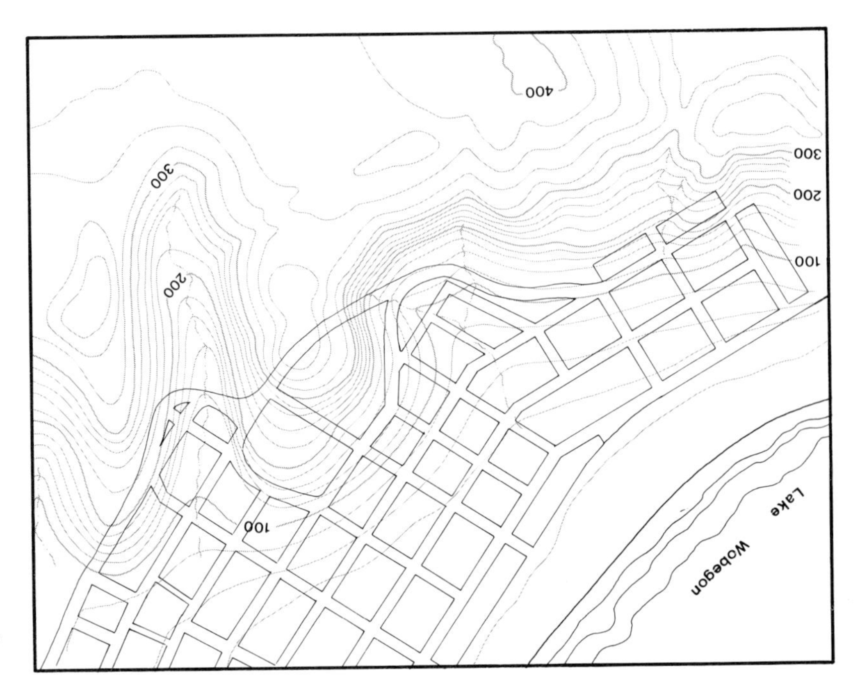

Scale: 1 inch = 2150 ft
(1:26,000)

2. Draw an alternative layout pattern for streets.

3. Based on shape and steepness, identify those slopes that would be most prone to failure if deforested. Construct a representative profile of such a slope.

II. Examine the platted area in the map in problem I of Chapter 5 (page 108) and determine the range (extremes) of slope inclination among the sites (lots) shown.

1. Set up three slope classes: (a) suitable for residential development; (b) moderately suitable for residential development (this class, for example, would include ground where special construction techniques, architecture, and wastewater disposal would be called for); and (c) unsuitable for residential development. Assume that the development would be of a conventional residential variety with single-family detached homes served by on-site sewage disposal (drainfield).

2. Using the procedure outlined on pages 205–207, construct a slope map for the area. The lots measure 100 ft. by 200 ft. (30 m × 60 m); use an area of one-half lot (100 ft. × 100 ft.) as the minimum size mapping unit. Color or hachure the areas in each class.

3. Provide a legend for the map and an explanation of what each slope class represents. Give the slope inclinations in both percentages and degrees.

4. Count the number of suitable lots in this area. Then, using a piece of tracing paper, try to design a pattern of streets and lots that would accommodate a greater number of suitable lots than the current pattern does. In addition to the number of developable lots, can you identify other disadvantages to the rectilinear design scheme?

SELECTED REFERENCES FOR FURTHER READING

Bailey, Robert G. *Land Capability Classification of the Lake Tahoe Basin, California-Nevada*, Washington, D.C.: Forest Service, U.S. Department of Agriculture, 1974, 32 pp.

Briggs, Reginald P., et al. "Landsliding in Allegheny County, Pennsylvania." *Geological Survey Circular 728*, 1975, 18 pp.

Carson, M. A., and Kirkby, M. J. *Hillslope Form and Process.* Cambridge, United Kingdom: Cambridge University Press, 1972, 475 pp.

Schuster, Robert L., and Krizek, Raymond J. *Landslides: Analysis and Control.* Washington, D.C.: National Academy of Sciences, 1978, 234 pp.

Utgard, R. O., et al. *Geology in the Urban Environment.* Minneapolis: Burgess, 1978, 355 pp.

Vitek, John D., and Marsh, William M. "Landslide Hazard Mapping for Local Land Use Planning." In *Environmental Analysis for Land Use and Site Planning.* New York: McGraw-Hill, 1978, 292 pp.

Way, Douglas S. *Terrain Analysis: A Guide to Site Selection Using Aerial Photographic Interpretation.* Stroudsburg, Penn.: Dowden, Hutchinson and Ross, 1973, 392 pp.

12

SOIL EROSION, LAND USE, AND STREAM SEDIMENTATION

INTRODUCTION

Soil erosion represents one of the most serious depletions of natural resources in the world today. Loss of top soil to runoff each year substantially reduces the food production capability of virtually every agricultural nation (Fig. 12.1). To offset this loss, more land must be opened for farming, and fertilizer application must be increased, both of which add to the overall cost of food production in both the short and long run.

This is not the only issue associated with soil erosion, however, for the soil material that enters the water systems also causes environmental degradation. In particular, the water quality of the receiving streams and lakes is reduced because of added turbidity (muddiness) and chemicals such as nitrogen and phosphorus, which are washed in with soil particles. Moreover, the heavier sediments glut stream channels and reservoirs, decreasing their capacities to contain large flows.

SOIL EROSION AND LAND USE

Over the past several thousand years, deforestation, crop agriculture, and grazing have promoted the greatest soil erosion, and these activities continue to hold that dubious distinction on a worldwide basis today. In the past century or so, however, urbanization has become important in the soil erosion issue, especially as it is practiced in North America. The general sequence of land use change that ends in urban development and the corresponding rates of soil erosion are given in Fig. 12.2. A description of the process begins with forest clearing and agricultural development sometime in the 1800s for most areas. Erosion rose sharply with the destruction of natural vegetation and the establishment of cropland and pasture. This trend continued until the first half of the twentieth century when agriculture declined, and as abandoned farmland was taken over by weedy vegetation, soil erosion probably declined somewhat as well.

Erosion increased dramatically with urban sprawl in the second half of the twentieth century when farmland, abandoned farmland, and forest were cleared for development. Wholesale exposure of soil during the construction phase of development gave rise to erosion rates as high as 200 tons per acre per year. But this trend is shortlived because the soil is quickly secured under buildings, roads, and landscaped surfaces, as development is completed. Under full urbanization, erosion rates appear to drop to levels less than those associated with twentieth century agriculture (Fig. 12.3).

FACTORS INFLUENCING SOIL EROSION

At least four factors must be taken into account in any attempt to forecast soil erosion rates: vegetation, soil type, slope size and inclination, and the frequency and intensity of rainfall. Tests show that intensive rainfalls such as those produced by thunderstorms promote the highest rates of erosion.

Pottawattamie County, Iowa, 1956

Ottawa County, Michigan, 1949

Whitman County, Washington, 1956

Hernando County, Florida, 1960

Baca County, Colorado, 1935

Fig. 12.1 Examples of soil erosion from various parts of North America in this century. (Photographs courtesy of the U.S. Soil Conservation Service.)

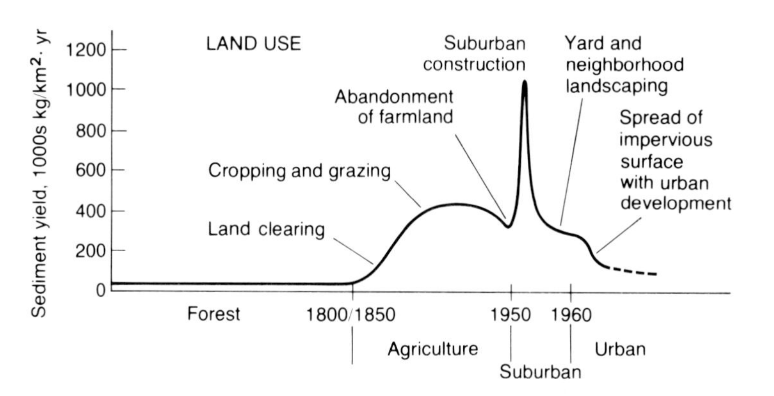

Fig. 12.2 The trend of soil erosion related to land use based on the North American experience. (Based on M. G. Wolman, "A Cycle of Sedimentation and Erosion in Urban River Channels," *Geografiska Annaler* 49A, 1967.)

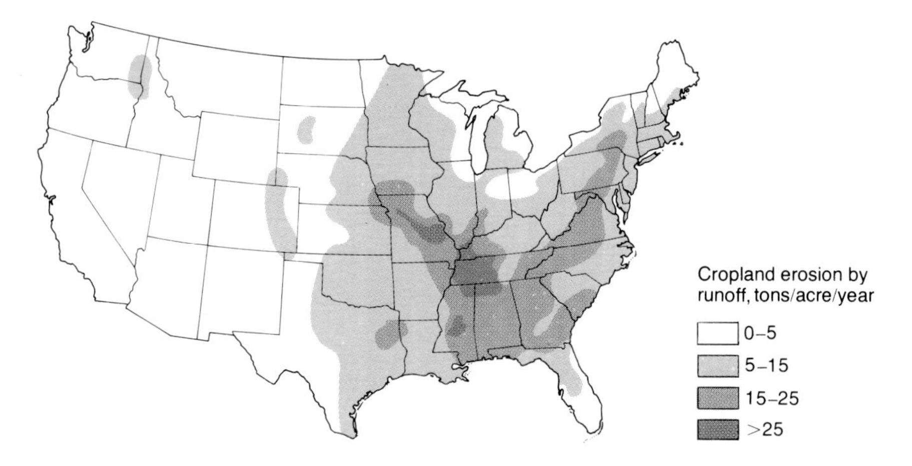

Fig. 12.3 Erosion of cropland by runoff in the United States, 1975. (From "Cropland Erosion," Soil Conservation Service, U.S. Department of Agriculture, 1977.)

Accordingly, the incidence of such storms together with the total annual rainfall can be taken as a reliable measure of the effectiveness of rainfall in promoting soil erosion. The U.S. Soil Conservation Service has translated this into a *rainfall erosion index* that represents the erosive energy delivered to the soil surface annually by rainfall (Fig. 12.4). The index values vary appreciably over the United States, and in some regions, from one side of a state to another. For instance, values decline from 250 to less than 100 from the southeastern to the northwestern corner of Kansas, meaning that on the average erosion on comparable sites should be more than 2.5 times greater in the southeast.

On most surfaces, vegetation appears to be the most important single control on soil erosion. Foliage intercepts raindrops, reducing the force at which they strike the soil surface. Organic litter on the ground further reduces the impact of raindrops, and plant roots bind together aggregates of soil particles, increasing the soil's resistance to the force of running water. The one feature of vegetation that appears to have the greatest influence on erosion is cover density; the heavier the cover, either in the form of ground cover or tree canopy, the lower the soil loss to runoff.

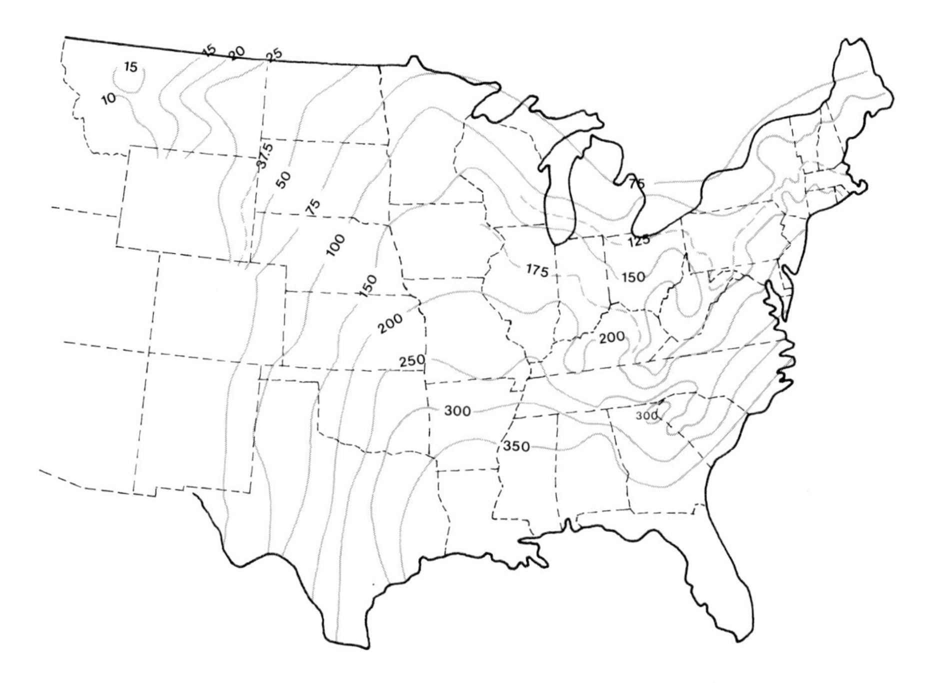

Fig. 12.4 Rainfall erosion index for the United States and southeastern Canada. The values are based on the erosive energy of total annual rainfall times the average annual maximum intensity of the thirty-minute rainfall. (From U.S. Soil Conservation Service, 1975.)

If running water is applied to soils of different textures, sand will usually yield (erode) first. In order to erode clay, the velocity of the runoff would have to be increased to create sufficient stress to overcome cohesive forces that bind the particles together. Similarly high velocities would also be needed to move pebbles and larger particles because their masses are so much greater than those of sand particles. Thus, in considering the role of soil type in erosion problems, it appears that intermediate textures tend to be most erodible, whereas clay and particles coarser than sand are measurably more resistant (Fig. 12.5). Other soil characteristics, such as compactness and structure, also influence erodibility, but in general texture can be taken as the leading soil parameter in assessing the potential for soil erosion.

The velocity that runoff is able to attain is closely related to the slope of the ground over which it flows. In addition, slope also influences the quantity of runoff inasmuch as long slopes collect more rainfall and thus generate a larger volume of runoff, other things being equal. In general, then, slopes that are both steep and long tend to produce the greatest erosion because they generate runoff that is high in both velocity and mass. But this is true only for slopes up to about 50 degrees, because at steeper angles, the exposure of the slope face to rainfall grows rapidly smaller, vanishing altogether for vertical cliffs. In land use problems, however, greatest consideration is usually given to slopes less than 50 degrees, particularly in so far as urban development, residential development, and agriculture are concerned. Table 12.1 gives relative values for soil erosion, or the potential for it, based on slope steepness and length for slopes up to 50 percent inclina-

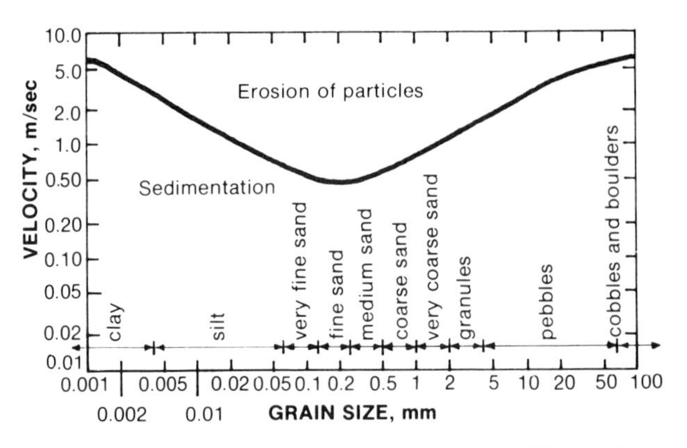

Fig. 12.5 Erosion thresholds of clay-, silt-, sand-, and pebble-sized particles under the force of running water. Sand gives way at the lowest velocities, whereas much higher velocities are required to dislodge clay and very coarse particles. (After F. Hjulström, "Transport of Detritus by Moving Water," in *Recent Marine Sediments: A Symposium* (P. Trask, ed.), Tulsa, Okla.: Amer. Assoc. Petroleum Geologists, 1939).

Table 12.1 Slope Geometry Factor Based on Steepness and Length

Slope Length in Feet	*Slope Steepness in Percent*														
	4	*6*	*8*	*10*	*12*	*14*	*16*	*18*	*20*	*25*	*30*	*35*	*40*	*45*	*50*
50	.3	.5	.7	1.0	1.3	1.6	2.0	2.4	3.0	4.3	6.0	7.9	10.1	12.6	15.4
100	.4	.7	1.0	1.4	1.8	2.3	2.8	3.4	4.2	6.1	8.5	11.2	14.4	17.9	21.7
150	.5	.8	1.2	1.6	2.2	2.8	3.5	4.2	5.1	7.5	10.4	13.8	17.6	21.9	26.6
200	.6	.9	1.4	1.9	2.6	3.3	4.1	4.8	5.9	8.7	12.0	15.9	20.3	25.2	30.7
250	.7	1.0	1.6	2.2	2.9	3.7	4.5	5.4	6.6	9.7	13.4	17.8	22.7	28.2	34.4
300	.7	1.2	1.7	2.4	3.1	4.0	5.0	5.9	7.2	10.7	14.7	19.5	24.9	30.9	37.6
350	.8	1.2	1.8	2.6	3.4	4.3	5.4	6.4	7.8	11.5	15.9	21.0	26.9	33.4	40.6
400	.8	1.3	2.0	2.7	3.6	4.6	5.7	6.8	8.3	12.3	17.0	22.5	28.7	35.7	43.5
450	.9	1.4	2.1	2.9	3.8	4.9	6.1	7.2	8.9	13.1	18.0	23.8	30.5	37.9	46.1
500	.9	1.5	2.2	3.1	4.0	5.2	6.4	7.6	9.3	13.7	19.0	25.1	32.1	39.9	48.6
550	1.0	1.6	2.3	3.2	4.2	5.4	6.7	8.0	9.8	14.4	19.9	26.4	33.7	41.9	50.9
600	1.0	1.6	2.4	3.3	4.4	5.7	7.0	8.3	10.2	15.1	20.8	27.5	35.2	43.7	53.2
650	1.1	1.7	2.5	3.5	4.6	5.9	7.3	8.7	10.6	15.7	21.7	28.7	36.6	45.5	55.4
700	1.1	1.8	2.6	3.6	4.8	6.1	7.6	9.0	11.1	16.3	22.5	29.7	38.0	47.2	57.5
750	1.1	1.8	2.7	3.7	4.9	6.3	7.9	9.3	11.4	16.8	23.3	30.8	39.3	48.9	59.5
800	1.2	1.9	2.8	3.8	5.1	6.5	8.1	9.6	11.8	17.4	24.1	31.8	40.6	50.5	61.4
900	1.2	2.0	3.0	4.1	5.4	6.9	8.6	10.2	12.5	18.5	25.5	33.7	43.1	53.5	65.2
1000	1.3	2.1	3.1	4.3	5.7	7.3	9.1	10.8	13.2	19.5	26.9	35.5	45.4	56.4	68.7

tion and 1000 feet length. In using this table, note that the rate of change with slope steepness is not linear, whereas it is with slope length. For example, for a 500-foot slope of 10 percent the value is 3.1, whereas for the same length at 20 percent, the value is 9.3.

COMPUTING SOIL LOSS FROM RUNOFF

An estimate of soil loss to runoff can be computed by combining all four of the major factors influencing soil erosion. For this we use a simple formula called the *universal soil loss equation,* which gives us soil erosion in tons per acre per year:

$$A = R \cdot K \cdot S \cdot C$$

where:

A = soil loss, tons/acre/year
R = rainfall erosion index
K = soil erodibility factor
S = slope factor, steepness and length
C = plant cover factor

In problems involving agricultural land, a fifth factor, cropping management, is also included, but for problems involving nonagricultural land, abandoned farmland, and urban land types, it is not applicable.

In order to interpret the universal soil loss equation reliably, it is necessary to understand that the computed quantity of soil erosion represents only the displacement of particles from their original positions. This may or may not be equated to loss from a study site, depending on the size and configuration of the site. For a site such as the one shown in Fig. 12.6, for example, much of the soil lost from the upper slope will be deposited near the foot of the slope. Therefore, in computing the soil loss for the lower surface, one should take into account sediment added from deposition and solve for the net change in the soil mass.

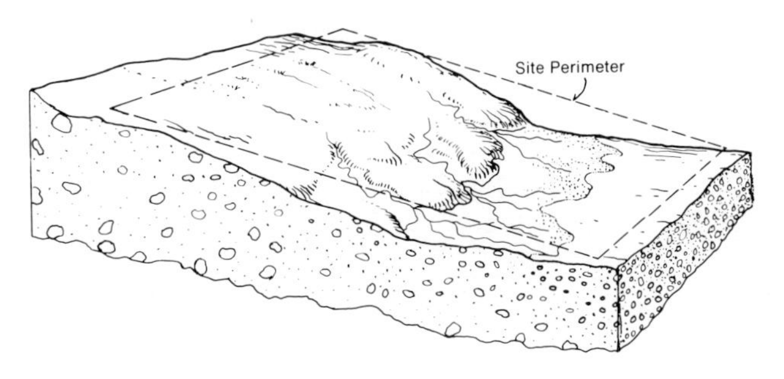

Fig. 12.6 An example of a site comprised of two slope classes, one that yields sediment and the other that accumulates it. Although there may be no net loss of sediment from the site, appreciable erosion occurs on the upper slope. A reliable assessment of erosion on this site must recognize this arrangement.

The data needed to make a soil loss computation can usually be obtained from topographic sheets, county soil reports, and a few additional maps and tables. Field inspection of the site is recommended to determine plant cover and erosion and deposition patterns. If that is not possible, then ground and aerial photographs may be used instead.

Soil type is expressed in terms of an erodibility factor, or *K factor*, which is a measure of a soil's susceptibility to erosion by runoff. This factor is derived from tests conducted on field plots for each soil series in a state and is given as a dimensionless number between 0 and 1.0. The higher the number, the greater the susceptibility to erosion. In some cases, two numbers are given for a soil, one for disturbed and one for undisturbed ground. Disturbed includes filled and rough graded ground. *K* factors are usually available from county or state offices of the U.S. Soil Conservation Service.

The rainfall erosion values for most states may be read from the map in Fig. 12.4. Those for the western states have not been generated as yet. The slope factor can be read from Table 12.1, and the plant cover factor can be approximated from Table 12.2 based on ground cover and canopy density.

Table 12.2 Plant Cover Factors

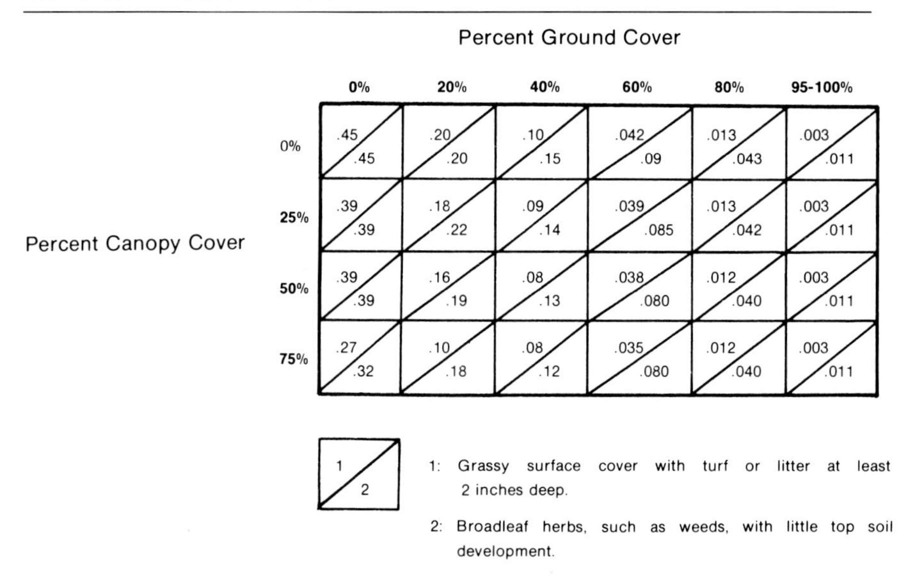

Percent Canopy Cover	Percent Ground Cover 0%	20%	40%	60%	80%	95-100%
0%	.45 / .45	.20 / .20	.10 / .15	.042 / .09	.013 / .043	.003 / .011
25%	.39 / .39	.18 / .22	.09 / .14	.039 / .085	.013 / .042	.003 / .011
50%	.39 / .39	.16 / .19	.08 / .13	.038 / .080	.012 / .040	.003 / .011
75%	.27 / .32	.10 / .18	.08 / .12	.035 / .080	.012 / .040	.003 / .011

1 / 2

1: Grassy surface cover with turf or litter at least 2 inches deep.

2: Broadleaf herbs, such as weeds, with little top soil development.

The reliability of soil erosion computations based on the universal soil loss equation depends not only on the accuracy of the data used, but on the way in which the problem is set up. For sites in areas of diverse terrain, the site should be divided into subareas within which soil, slope, and vegetation are reasonably uniform. Soil loss should be computed for each subarea and then adjusted for any deposition that may be received from adjacent subareas. In making the adjustment for deposition, it is important to consider not only the potential contribution from upslope areas, but the patterns of gullies, streams, and swales through which runoff and sediment can be funnelled across flat areas and discharged onto low ground or into streams, lakes, and wetlands.

To summarize, the procedure for computing soil loss from a site may be described in six steps:

1. Define the site (problem area) boundaries on a large-scale base map such as a topographic sheet or soil map.
2. Using aerial photographs, soil maps, topographic maps, and whatever other sources of data are available, divide the site into subareas. If the site is essentially uniform throughout, this is not necessary.
3. For each subarea, assign a *soil erodibility factor* (based on SCS soil series K-factor designation), a *slope factor* (read from Table 12.1 based on steepness and length data taken from topographic map), and a *plant cover factor* (read from Table 12.2 based on aerial photographs or field observation).
4. Multiply the appropriate value from the *rainfall erosion* map in Fig. 12.4 times the three factors assigned to the subarea in step 3. The answer is in tons per acre per year.
5. Examine the relations between slopes and drainage patterns in each subarea, identify where sediment would be expected to accumulate, and, if possible, adjust the quantity of soil loss computed in step 4.
6. Determine the total soil loss for the entire site by summing the net amounts of soil loss for each subarea. If this is not possible because of uncertainty over the relationship among various subareas, then the gross soil loss from the subareas should be summed for a gross site total.

APPLICATIONS TO LAND PLANNING AND ENVIRONMENTAL MANAGEMENT

An important task of most local planning agencies is the review and evaluation of land development proposals for housing, industrial, and commercial projects. Proposals are evaluated according to a host of criteria including soil erosion, or the potential for it. Consideration of soil erosion stems not only from a concern over the loss of top soil and depletion of the soil resource in general but also from the impact of sedimentation on terrestrial vegetation, wetlands, river channels, and drainage facilities such as stormsewers. To gain the necessary information, developers are often asked to prepare a site plan and respond to questions such as these:

- What percentage of site lies in slopes of 15 percent or greater, and of this area how much (a) is proposed for development, and (b) if developed, will be affected by construction?
- What percentage of this site is forested, grass covered, and shrub covered? How much of each of these covers will be destroyed as a result of development?
- What are the minimum distances between the proposed development zone and (a) water features (streams, ponds, reservoirs, and wetlands), and (b) existing drainage facilities (stormsewers, stormwater retention ponds, and ditches)?

- What erosion and sedimentation control measures are proposed during (a) the construction phase, and (b) the operational phase of the proposed project?
- What is the anticipated length of the construction period, and which months in the year are proposed for (a) land clearing, (b) excavation and grading, (c) construction of building and facilities, and (d) regrading and landscaping (Fig. 12.7)?

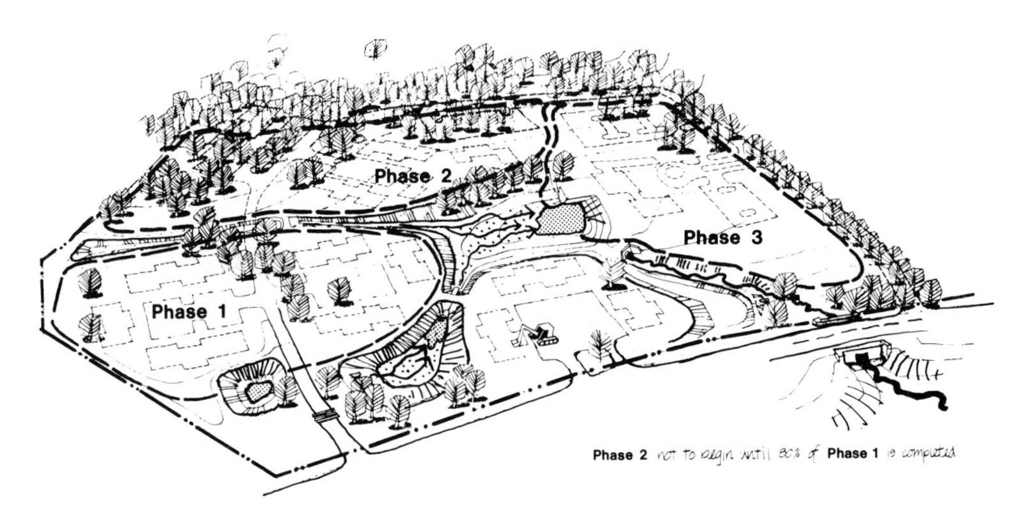

Fig. 12.7 Phasing of development is necessary in large projects in order to minimize soil erosion and sedimentation of local water features. In this project, 25 to 30 percent of the site would be opened at any time. (Original illustration by Howard Deardorff. Used by permission.)

Control of erosion and sedimentation during the construction phase of development is viewed very seriously in many communities—so much so, in fact, that many have legislated erosion control ordinances. At the heart of such ordinances are the control measures, or *mitigation measures*, as they are often called. These include the placement of berms around construction zones, the use of fiber nets on slopes, and the use of sedimentation basins to protect drainage features (Fig. 12.8).

CONSIDERATIONS IN WATERSHED MANAGEMENT

The geographer's and planner's perspectives on soil erosion and stream sedimentation must extend beyond the site scale of observation. Every site is part of a larger drainage area linked together by a network of drainage channels. These channels are collectors for runoff and sediment. In most natural drainage systems, the size of the channel is adjusted to the size of flows it carries, in particular to certain flows of larger than average magnitudes. We reason, therefore, that when two streams join in the network, the size of resultant channel should approximate the sum of the two tributary channels. (Actually it is the *capacity* of the channel, defined by the magnitude of discharge that it can accommodate, that increases with the merger of tributaries.)

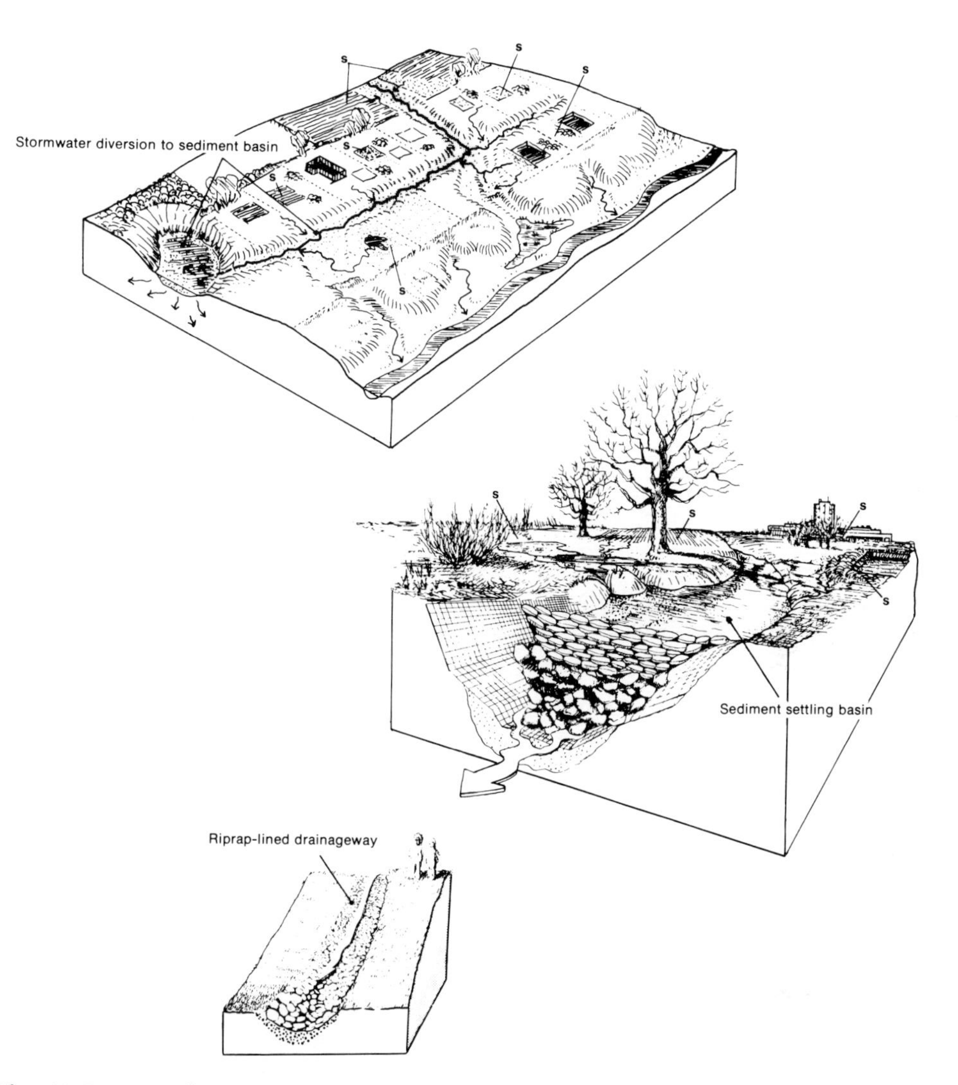

Fig. 12.8 Examples of measures used in construction site management to control erosion and avert sedimentation of water features. Sediment sources are denoted by "s."

CASE STUDY

Erosion and Sediment Control Planning for a Strip Mining Operation

Phil Shumaker

This project calls for the strip mining of two shallow coal seams. Operational plans call for disturbance of approximately sixty acres of land in the Pleasant Valley Run Watershed, situated in Garrett County, Maryland. The area lies within the Allegheny Plateau Physiographic Section of the Appalachian Mountains.

Both coal seams outcrop along the northern side of the Pleasant Run Valley and dip northwestward at angles of 4 to 5°. Core samples indicate that the overburden resting on the lower seam is primarily composed of a soft yellow clay with a thin bed of dark-grey shale between the clay and the coal seam. The strata overlying the upper seam consist of thinly bedded

(cont.)

CASE STUDY (cont.)

shale and sandstone. The analysis of overburden material indicates that the shale near the coal seam has a high pyrite content and, therefore, a high acid forming potential. This requires treatment of all water that is pumped from the pit and the burial of all acid shale away from the surface and the pit floor.

Preliminary soil analysis indicates that topsoil material (defined in this case as that portion of the soil that is capable of supporting vegetation with little or no special preparation) is extremely thin in the mining area. On the whole, the soils, even before mining, have a high runoff potential. Therefore, they will be graded in such a way as to insure maximum infiltration. Topsoil material will be placed on the surface whenever possible, and further supplemented by the high alkaline, potassium, and phosphorus-rich yellow clay.

In designing erosion and sediment control structures, and in scheduling the planting program, climatic data on rainfall intensity and freezing must be carefully analyzed. The data indicate that the intensity of a one-hour rainfall expected once in ten years is 2.25 inches. Accordingly, each structure is designed to handle runoff from such a storm with provisions for an emergency spillway to handle a twenty-five-year storm. Freeze data define the growing season as averaging only 122 days between late spring and early fall. Planting of all vegetative species will be conducted during this period to insure maximum survival and early growth.

The drainage area of Pleasant Valley Run is 2.12 square miles, and of this area, 63.6 acres (7.1 percent) will be disturbed by mining and associated operations. The desire to protect Cunningham Lake and reduce the amount of runoff and sediment pollution directly entering Pleasant Valley Run will require directing runoff away from the valley and lake. Owing to the dip of the underlying strata, it will be feasible to direct the runoff from the mined area into the pit and then pump it into a sediment pond, where it will be discharged into Cunningham Swamp. Cummingham Swamp is intended to act as a natural filter area for the runoff from the mined area and the haul-

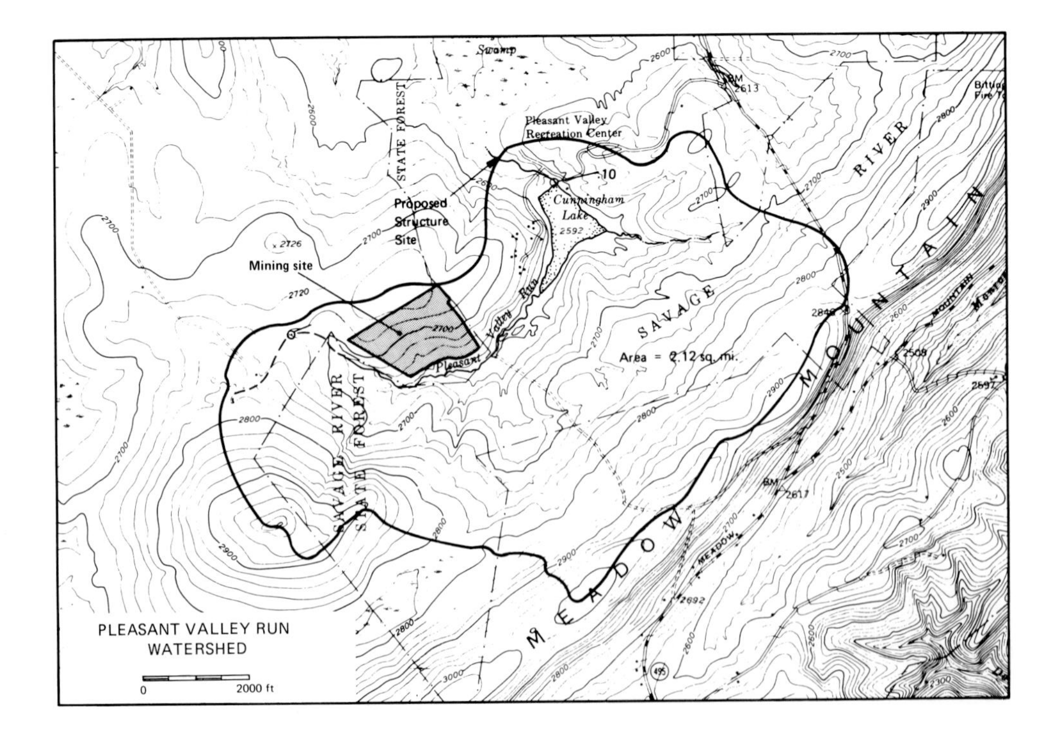

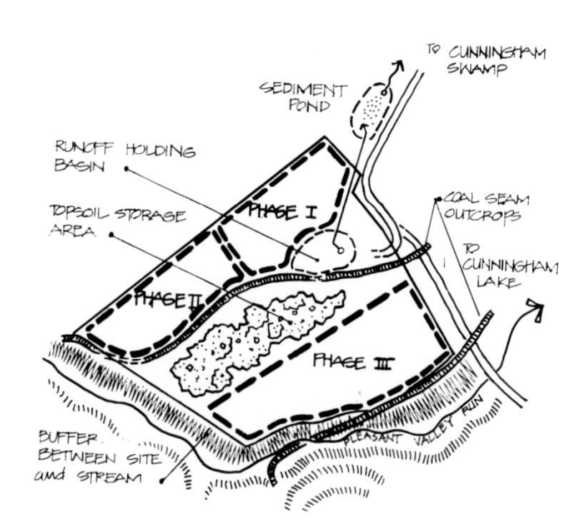

age road. Observation indicates that vegetated strips between the mined area and adjacent streams as well as the area along the haulage road and below the outlet of the sediment pond possess good stands of erosion-resisting plant species, which can act as effective filters for sediment.

The mining operation will consist of three separate phases. All three phases will use the same haulage road and sediment pond. In all three cases, backfilling will be to the original contour and done in such a manner as to encourage maximum absorption of precipitation, thereby preventing accumulations of water on backfilled areas. The backfill shall also be graded in such a manner as to permit runoff water to flow into the unaffected areas gradually and in small quantities, thereby not carrying soils and sediment deposits into the surrounding areas.

The planting program will begin with a temporary cover of quick-growing grasses (oats or rye) to guard against erosion in the short term. This will be followed by seeding of the entire site in permanent grasses. After one year, selected tree species will be introduced with the goal of eventually reforesting the site in manner compatible with the tree cover of nearby Savage State Forest.

Phil Shumaker is an engineer with Delta Mining, Inc., Grantsville, Maryland.

Sediment transport capacity also increases with the magnitude of flow. Therefore, in assessing what could happen when sediments are released to streams from an erosion site, it is important to consider the size of the receiving channel and the relative position of the channel in the drainage network. Massive loadings in small streams located in the upper parts of a watershed imply that: (1) channels may become glutted, thereby reducing their capacities to carry large flows; and (2) sediment in the streambed will remain there for a long time, owing to the limited capacity of the stream to carry it away.

In a watershed where several sites are releasing sediment to a stream network at the same time, the sediment bodies will move through the channel network at uneven rates, making it difficult to assess the impact on the system as a whole. Clays will move through the entire system rapidly, whereas sand and silts will not only move more slowly but build up at selected points (Fig. 12.9). Where in the drainage system sediment is likely to accumulate can often be estimated based on sediment size and channel

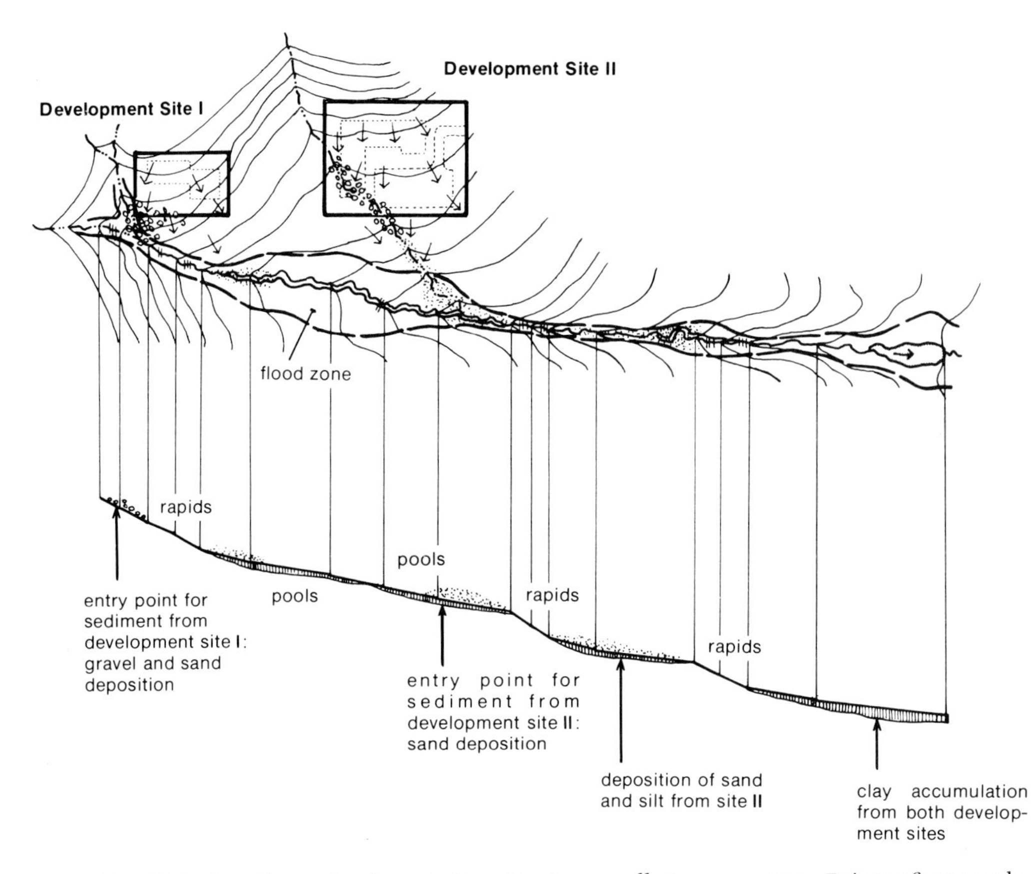

Fig. 12.9 Locations of sedimentation sites in a small stream system: Points of accumulation vary with sediment size, channel shape and gradient, and proximity to sediment source areas.

shape and gradient. Broad reaches of slow-moving water are the favored deposition sites for sands and silts. Among these, wetlands and impoundments such as reservoirs and lakes are generally considered the most critical from a watershed management standpoint, and it is often necessary to estimate the amount of sediment that would be added to them based on land use practices in the drainage basin.

Sediment loading of a drainage basin requires a yearly calculation of sediment production on a site-by-site basis for the entire watershed. The locations represented by each of these quantities must be pinpointed in the watershed and the drainage network, taking special care to determine where sediment would actually enter the network. In most cases where a site falls on a drainage divide within the watershed, the site may contribute sediment to two or more channels, as is illustrated by sites 1 and 4 in Fig. 12.10. Once these relationships are known, the receiving water bodies downstream can be identified and a loading rate can be calculated.

Finally, the sediment loads must be converted to volumetric units to determine how much space in the channel or impoundment will actually be taken up by the sediments. The following sediment densities are recommended:

clay	= 60–80 lbs/ft^3
clay/silt/sand mixture	= 80–100 lbs/ft^3
sand and gravel	= 95–130 lbs/ft^3

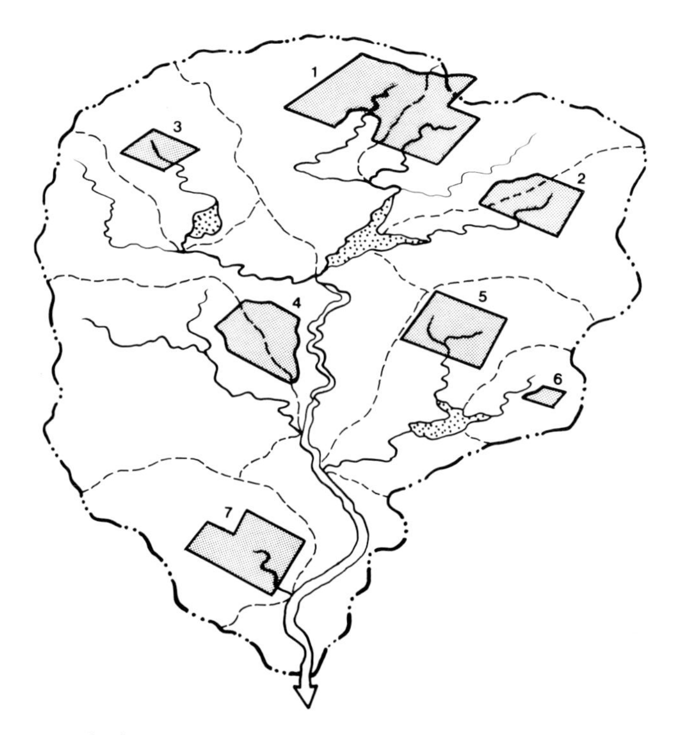

Fig. 12.10 A watershed in which forest has been cleared at seven sites, each of which is related to a different part of the drainage network.

Using an intermediate value of 90 pounds per cubic foot, a reservoir receiving 10,000 tons of sediment per year would lose 222,000 ft^3 (8200 yds^3) of volume each year. Given a reservoir capacity of 2,000,000 ft^3 (one, for example, with dimensions of 200 feet wide, 1000 feet long, and 10 feet deep) the life of the reservoir would be only nine years.

For impoundments that have low residence times, that is, those that exchange water in only a matter of hours or several days rather than months or years, most of the clays will not settle out. Owing to their small sizes, clay particles settle exceedingly slowly, allowing most to pass through small impoundments. Impoundments with long residence times, which may be as great as ten to twelve years in some lakes, retain fine sediments and deposit them over the lake floor, burying or coating bottom organisms.

PROBLEM SET

I. A site of eighteen acres has been cleared for construction of a shopping plaza, but because of financing difficulties construction is held up for a year. The site is located in central South Carolina and has these characteristics:

- All ground slopes in one direction with drainage routes focusing on one corner of the site.
- Three subareas are definable:

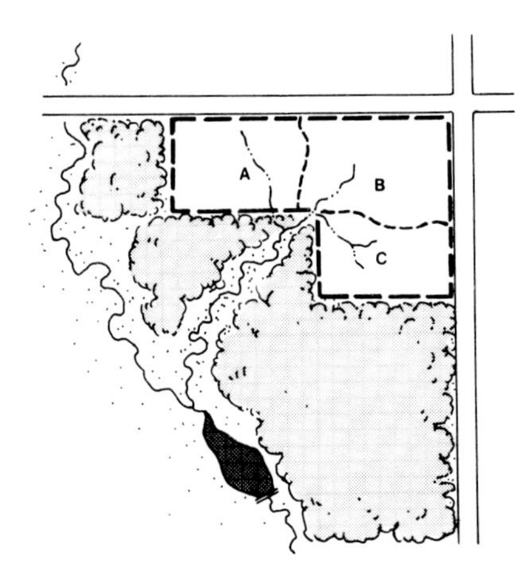

	Subarea A	*Subarea B*	*Subarea C*
• area	5.4 ac	7.4 ac	5.2 ac
• slope	8.0%	12.5%	9%
• average slope length	300 ft	410 ft	220 ft
• *K* factor	0.32	0.24	0.28
• plant cover factor			
• rainfall erosion index			
• total sediment yield			

1. Complete the table above and compute the sediment yield (a) for each subarea, and (b) for the entire site.
2. If the sediment is a clay/silt/sand mixture, what is the volume of the sediment lost?
3. Trace the path of sediment movement downstream from the site.
4. Assuming that 80 percent of the sediment load is sand and silt and 20 percent is clay, how much reservoir capacity will ultimately be lost? Let us assume that most of the clay will flow through the reservoir and that sediment contributions from other sources are negligible. Further, the reservoir has a surface area of 1.2 acres and an average depth of 4.0 feet.

II. Sediment yields, areas, and land use/cover for the seven sites shown in Fig. 12.10 are as follows:

Site	*Area*	*Land Use/Cover*	*Sediment Yield**
1	80 ac	cropland	7.5 tons
2	25 ac	pasture	1.5 tons
3	15 ac	construction	115 tons
4	30 ac	cropland	22 tons
5	32 ac	residential	2–4 tons
6	6 ac	abandoned cropland	24 tons
7	35 ac	construction	97 tons

*average annual per acre

1. Determine the total quantity of sediment released from each site per year.
2. Based on the location of each site, assign these quantities to the nearest stream. For those sites lying on a drainage divide, approximate the quantity going to each stream based on relative areas of the site lying on each side of the divide.

3. Trace the path of sediment movement downstream and bullet (■) those locations where coarse sediment will accumulate.
4. Assuming that 75 percent of the sediment reaching an impoundment will be trapped there, how many tons of the total annual sediment production from the watershed will actually leave the watershed?

SELECTED REFERENCES FOR FURTHER READING

Dissmeyer, G. E. "Erosion and Sediment From Forest Land Uses, Management Practices and Disturbances in the Southeastern United States." In *Proceedings of the Third Federal Inter-Agency Sedimentation Conference*, Water Resources Council, 1976, pp. 1-140–1-148.

Environmental Protection Agency. *Erosion and Sediment Control: Surface Mining in the Eastern United States*, U.S. EPA Technology Transfer Seminar Publication, 1976.

Ferguson, Bruce K. "Erosion and Sedimentation in Regional and Site Planning." *Journal of Soil and Water Conservation*, vol. 36, no. 4, 1981, pp. 199–204.

Heede, B. H. "Designing Gully Control Systems for Eroding Watersheds." *Environmental Management*, vol. 2, no. 6, 1978, pp. 509–522.

Hjulström, F. "Transport of Detritus by Moving Water." In *Recent Marine Sediments: A Symposium* (ed. P. Trask). Tulsa, Okla.: Amer. Assoc. Petroleum Geologists, 1939.

Soil Conservation Service. "Procedure for Computing Sheet and Rill Erosion on Project Areas." *Technical Release No. 51*, Washington, D.C.: Department of Agriculture, 1975, 15 pp.

Soil Conservation Service. *Standards and Specifications for Erosion and Sediment Control in Developing Areas.* College Park, Md.: U.S. Department of Agriculture, 1975.

Wolman, M. G. "A Cycle of Sedimentation and Erosion in Urban River Channels." *Geografiska Annaler* 49A, 1967, pp. 385–395.

13

SHORELINE PROCESSES AND COASTAL ZONE MANAGEMENT

INTRODUCTION

One of the nagging and costly problems in coastal zones is shore erosion. In the United States and Canada, this problem has grown significantly in the past two decades, not because the oceans and lakes are behaving differently than they did years ago, but because of increased development and use of the coast. This has given rise to heavy financial and emotional investment, and for many people, it has led to a bittersweet relationship with the sea.

The magnitude of this and related problems, such as flooding, destruction of wetlands, and commercialization of prized environments, has reached national proportions in the United States. In response, the federal government passed the Coastal Zone Management Act in 1972, which provides for the development of coastal planning and management programs at the state level. Among the responsibilities of the state programs is the classification of coastlines according to their relative stability, including the potential for erosion. To make such a determination, it is necessary to understand not only the physical makeup of the coast, but also the nature of the forces acting on it. The principal force in the coastal zone is wind waves. They are the source of most shore erosion and sediment transport, together which shape the shorelines, beaches, and related features.

WAVE ACTION AND NEARSHORE CIRCULATION

In deep water, waves cannot effect erosion because the motion of the wave does not reach bottom. In shallow water, waves not only touch bottom, but they can exert considerable force against it. Initially, this force is not very great, but as the wave nears shore, it increases rapidly. Where the shear stress of this force exceeds the resisting strength of the bottom material, displacement of particles takes place. Along coasts comprised of loose sediments, such as sand and silts, large quantities of particles are churned up, especially as the waves break near shore.

The water depth at which waves begin to move bottom particles is called *wave base.* This depth increases with wave size, and is roughly proportional to 1.0 to 2.0 times the wave height. Relative to wavelength, wave base falls at a depth between 0.04 and 0.5 wavelength. The shallow water

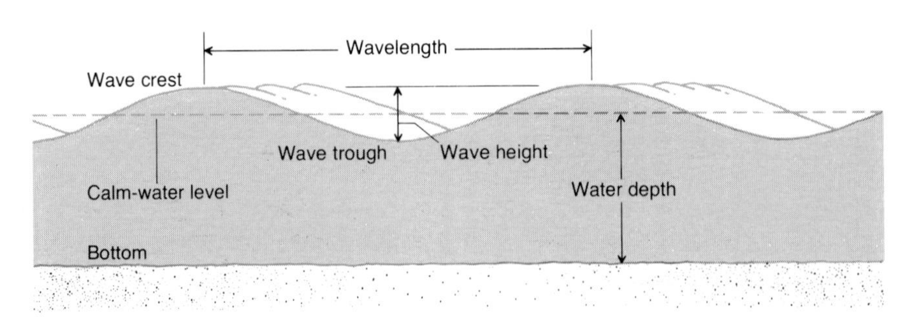

The principal dimensional properties of a wave. In shallow water the wavelength shortens and the height increases; when the height to length ratio exceeds 1:7, the wave becomes unstable and breaks.

zone begins at this depth, and since the wave base changes with wave size, the shallow zone actually fluctuates in width with different wave events. In the oceans, however, wave base for large waves is generally considered to average around 10 meters; in the Great Lakes it averages around 3 meters.

Once the sediment is waterborne, it is subject to transportation. On the outer edge of the shallow water zone, transport is slight, especially in sand-sized particles, with movement limited to a to-and-fro motion with the passage of each wave. Nearer shore, this motion is combined with a lifting action that carries the particle into the wave and on a turbulent ride before settling back to bottom. Although the individual movements of particles can be in any direction, the net direction of movement after the passage of many waves is usually parallel to the shoreline.

Two factors account for the parallel, or *longshore,* transport of sediment. One is that most waves approach and intercept the coast at an angle, therefore the direction of wave force is oblique to the shoreline. Although waves refract (bend) into a more direct approach angle near shore, most retain a distinct angle as they cross the shallow water zone. Thus, a large component of the energy for sediment transport is set up parallel to the shoreline.

The second factor takes the form of a current that flows along the coast. This current, called a *longshore current,* moves parallel to the shoreline in the direction of wave movement at rates generally less than 1 meter per second. Longshore currents are driven by wave energy, increasing in velocity and size (volume) with wave size and duration. When sediment is churned up by waves, the longshore currents transport it downshore a ways before it settles back toward bottom.

Other types of currents also operate along the shore, and one of the most prominent is *rip currents.* Flowing seaward from the shore across the lines of approaching waves, rip currents are narrow jets of water that intercept the longshore train of sediments carrying part of it into deeper water. Together, rip currents, waves, and longshore currents form near shore circulation cells, moving both water and sediments toward, away from, and along the shore. These cells are in turn nested in larger circulation systems that transport sediments from *source areas,* such as river deltas, to *sinks,* which are deposition areas such as embayments (Fig. 13.1).

WAVE REFRACTION AND ENERGY DISTRIBUTION

The ability of waves to erode and transport sediment is a function of the wave size and the size and availability of sediment. In deep water, wave size is a product of wind velocity, wind duration, and fetch. *Fetch* is the distance of open water over which a wind from a particular direction can blow across a water body. The greater the velocity, duration, and fetch, the larger the wave. Forecasts of the size of deep water waves can be made using the graph in Figure 13.2.

Upon entering shallow water, the rate of advance of the wave declines because of energy losses to friction and turbulence. If the water shoals gradually, the wave decelerates steadily toward shore. If the wave approaches the shore at an angle, the innermost segment or limb decelerates

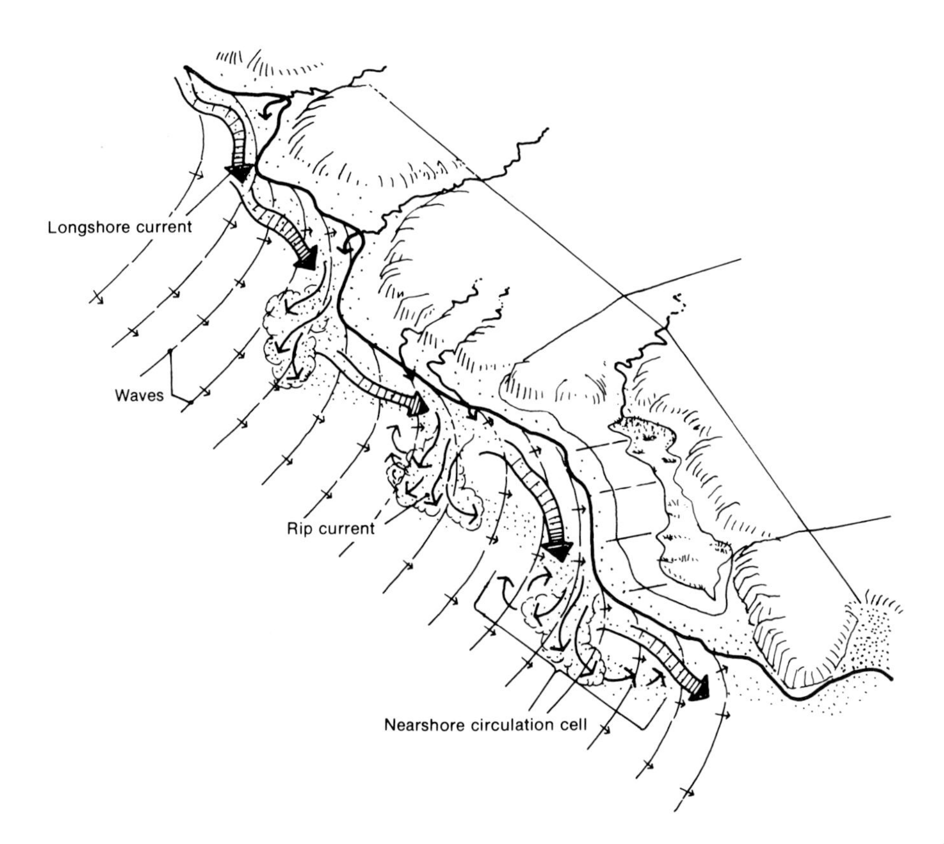

Fig. 13.1 Nearshore circulation showing waves, longshore currents, rip currents, and the transport of sediment from a source to a sink.

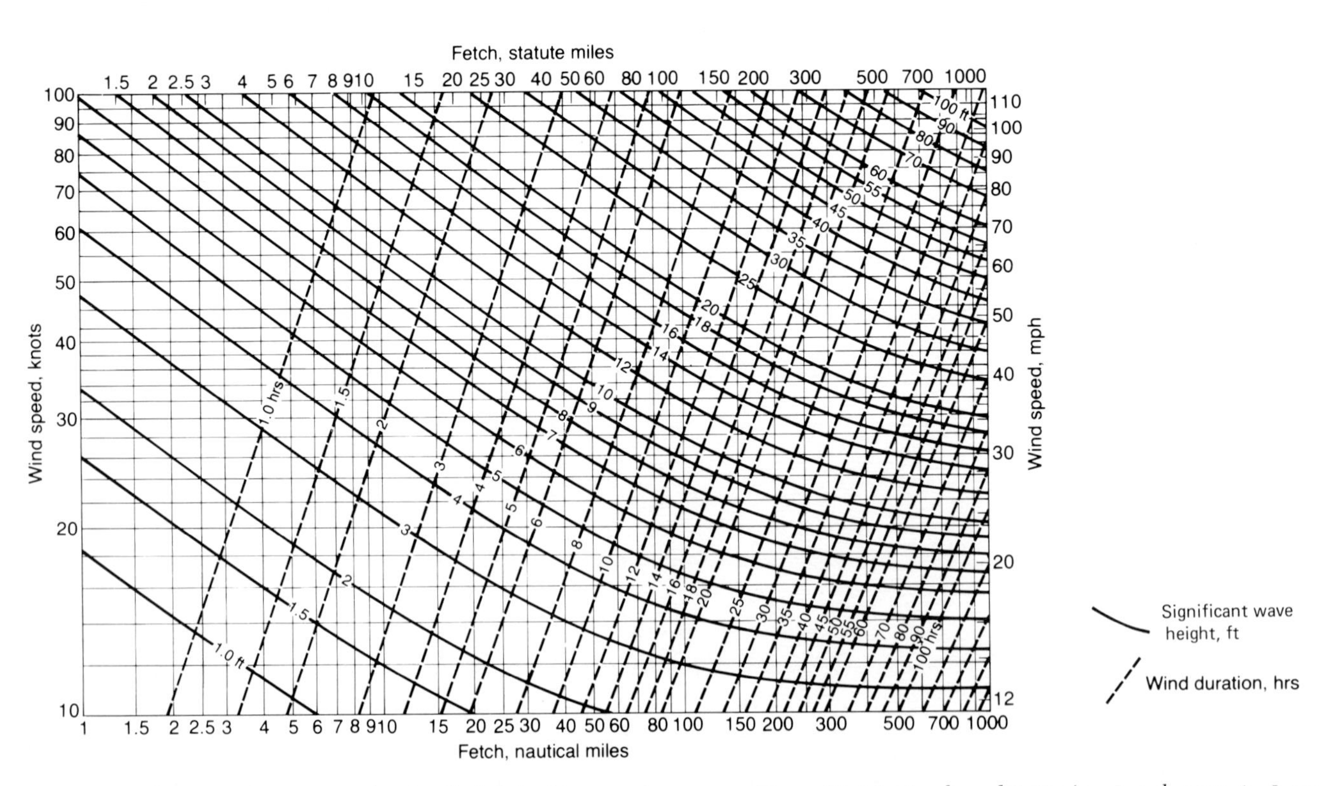

Fig. 13.2 Chart for forecasting the size (height) of deep water waves. (From W. M. Marsh and J. Dozier, *Landscape: An Introduction to Physical Geography*, © 1981, Addison-Wesley, Reading, Massachusetts. Fig. 28.3. Reprinted with permission. Adapted from Coastal Engineering Research Center, U.S. Army Corps of Engineers.)

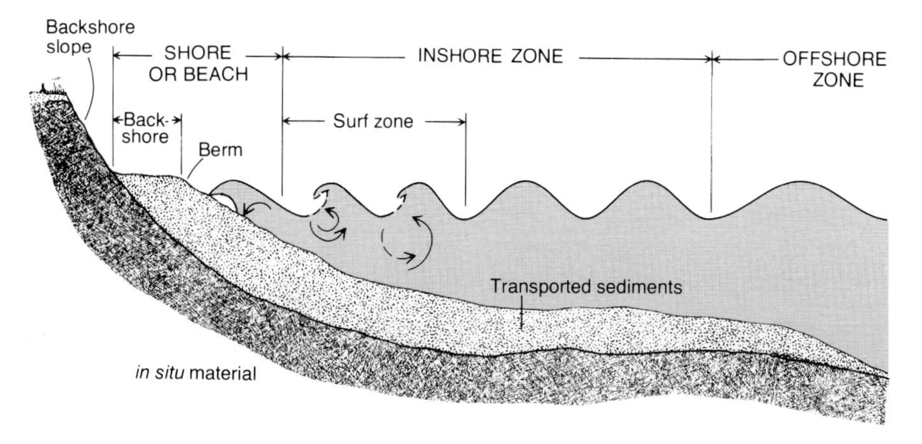

The principal zones and features of a coast comprised of transported beach sediments and *in situ* materials. (From W. M. Marsh and Jeff Dozier, *Landscape: An Introduction to Physical Geography*, © 1981, Addison-Wesley, Reading, Massachusetts. Fig. 28.4(a). Reprinted with permission.)

at the faster rate, causing the axis (crest line) of the wave to bend, or refract. Because refraction is controlled by the direction of wave approach and the bottom topography of the shallow water zone, refraction patterns can be estimated based on deep water wave direction and bathymetric maps. This can also be done by constructing lines called orthogonals perpendicular to the crest line of approaching waves as they appear on an aerial photograph. Assuming wave size (and therefore energy) is uniform along the entire wave crest in deep water, any convergence or divergence of the orthogonals in shallow water represents a change in the *relative distribution* and the orientation of wave energy. Normally, refraction causes wave energy to become focused on headlands and the seaward sides of islands, and diffused in embayments and the leeward sides of islands. These are often the sites of erosion and deposition respectively (Fig. 13.3).

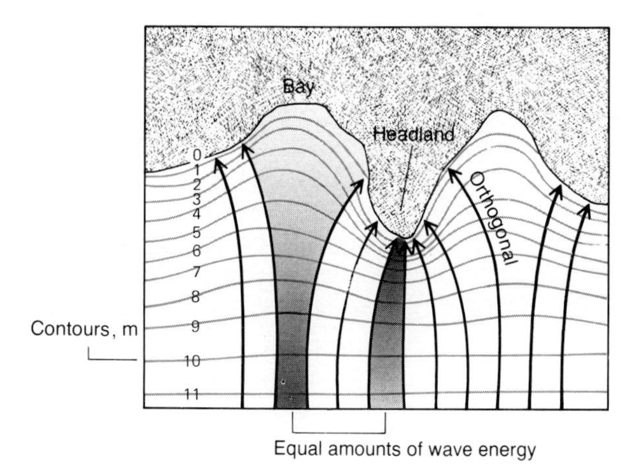

Fig. 13.3 Wave refraction around a headland and embayment. Notice that the distribution of energy wave is focused in one and diffused in the other. (From W. M. Marsh and J. Dozier, *Landscape: An Introduction to Physical Geography*, © 1981, Addison-Wesley, Reading, Massachusetts. Fig. 28.8(a). Reprinted with permission.)

NET SEDIMENT TRANSPORT AND SEDIMENT MASS BALANCE

On most shorelines, waves and currents may change direction with the passage of storms or the seasons, which in turn causes reversals in the longshore transport. The same sediments may be transported past one point on the shore many times in one year. A measure of this quantity, the total amount of sediment moved past a point on the coast, is called *gross sediment transport*. For obvious reasons, it is not a good indicator of the balance of sediment at a place on the coast, because it does not tell us whether the shore is losing or gaining sediment, nor whether the beach is growing or shrinking. Therefore, another measure is used, called *net sediment transport*.

$$\text{Net } Q_t = Q_p - Q_s$$

where

Net Q_t = net quantity of sediment/year
Q_p = longshore transport in the primary direction
Q_s = longshore transport in the secondary direction

Net sediment transport is the balance between the sediment moved one way and that moved the other way along the coast. If the longshore system is driven predominantly by waves from one quadrant of the compass, then net sediment transport can be large. By contrast, if the longshore system operates in both directions, then net transport may be small while gross transport may be large. Ultimately, it is the trend that we are concerned with, because it can tell us something about the development of the coastline.

At a local scale of observation, where only a short segment of beach is concerned, we would want to make a more detailed determination of the sediment balance (Fig. 13.4). This would include inputs from backshore slope erosion and runoff (R_i), losses (outputs) from wind erosion (W_o), onshore inputs from sand bar migration (O_i) in summer and offshore outputs from bar migration (O_o) in fall and winter, as well as longshore input (L_i) and output (L_o). The mass of sediment on the beach for any period of time is equal to inputs minus outputs:

$$\text{Sediment mass balance: } L_i - L_o + O_i - O_o + R_i - W_o = O$$

A value greater than zero means that the reservoir of beach sediment has gained mass; less than zero, that it has lost mass. Generally, these trends are manifested in larger- and smaller-sized beaches. In most places, detailed data are not available for computation of the sediment mass balance, nor is there adequate time or resources in most planning studies to acquire the necessary data. Therefore, in site planning one must usually resort to an interpretation of local records and features such as vegetation, old maps, and land surveys to gain an idea of local trends.

TRENDS IN SHORELINE CHANGE AND LAND USE

Rivers are the primary source of sediment for longshore transport, providing more than 90 percent of the total sediment supply yearly on the

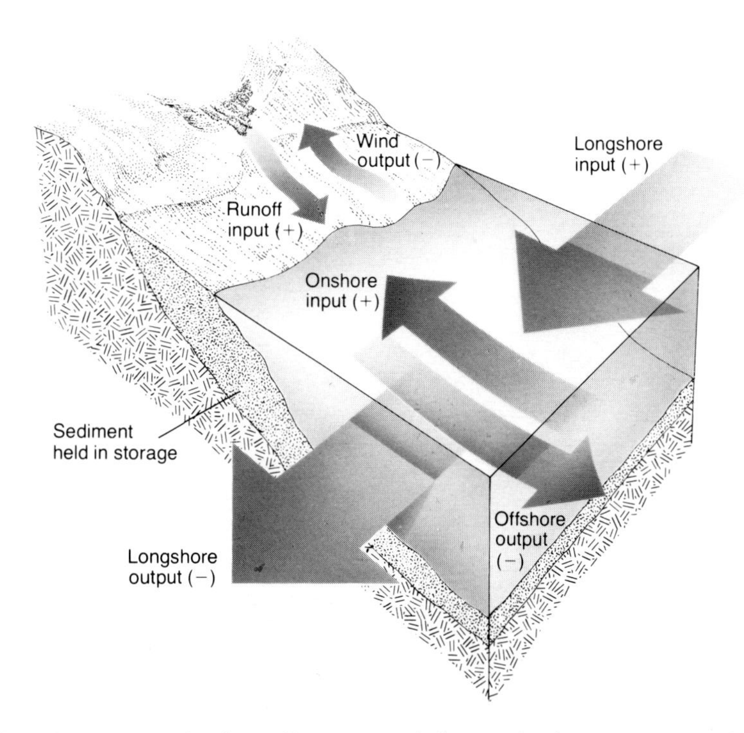

Fig. 13.4 Components in the sediment mass balance of a short segment of beach. (From W. M. Marsh and J. Dozier, *Landscape: An Introduction to Physical Geography*, © 1981, Addison-Wesley, Reading, Massachusetts. Fig. 28.18. Used by permission.)

world's coasts. Where river sediments are abundant, virtually all available wave energy may be expended in transporting the sediment load along the coast. Where river sediments are scarce and wave energy is abundant, however, the body of beach sediment is often small and only a fraction of available wave energy is expended in moving it. Moreover, by virtue of its small volume, the sediment mass offers little protection for the *in situ* material under and behind the beach. Under such circumstances, the backshore can be severely eroded and the resultant debris incorporated into the longshore sediment system. As erosion takes place, the shoreline retreats landward. Such coastlines are characterized by features such as sea cliffs, bluffs, or wave-cut banks, and are referred to as *retrogradational,* because over the long run they retreat landward as they give up sediment.

Retreat is defined as the landward displacement of the shoreline. It is usually caused by erosion; it may also be caused by a rise in water level, subsidence of coastal land, or any combination of the three. Where retreat is caused by erosion, the amount of *in situ* material actually lost can be computed by multiplying the retreat rate (R) times the backshore slope height (H) for a given length (L) of shoreline (Fig. 13.5):

$$\text{Erosion} = R \cdot H \cdot L$$

At the other extreme are those coastlines that collect sediments and build seaward, called *progradational* coastlines. These coasts are characterized by low relief terrain in the form of various depositional features. The most common of these are beach ridges, broad fillets of sand at the heads of bays, and spits and bars near the mouths of bays and behind islands.

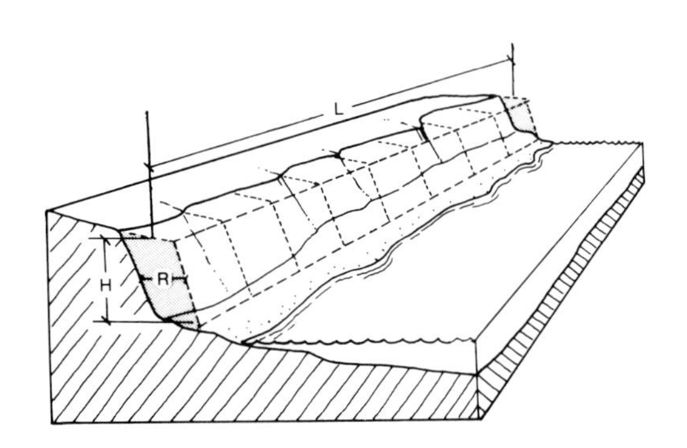

Fig. 13.5 The erosion resulting from shoreline retreat is computed from the retreat rate (R), the backshore slope height (H), and the length of the shoreline (L).

Most depositional features are prone to rapid changes in shape and volume; therefore, short-term trends should generally not be used as the basis for formulating land use plans in the coastal zone. In computing the volumetric change in a sand bar or a bayhead beach, for example, it is necessary to consider sediment added both above and below water level because the bulk of such features often develops under water. This is especially necessary if the rate of growth or decline of the feature is to be related to the net sediment transport rate over some time period (Fig. 13.6).

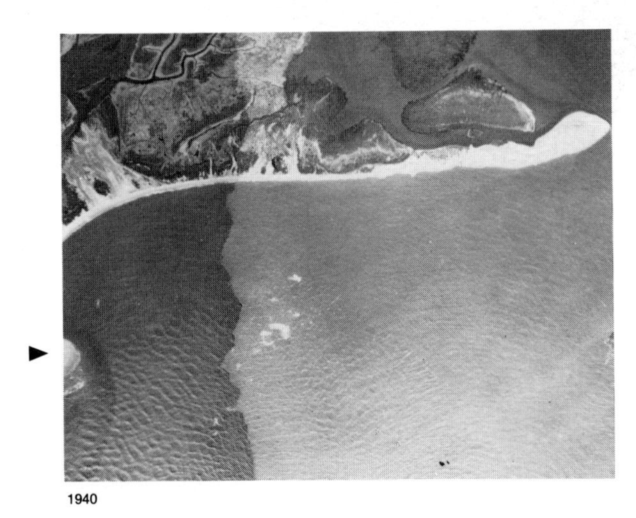

1940

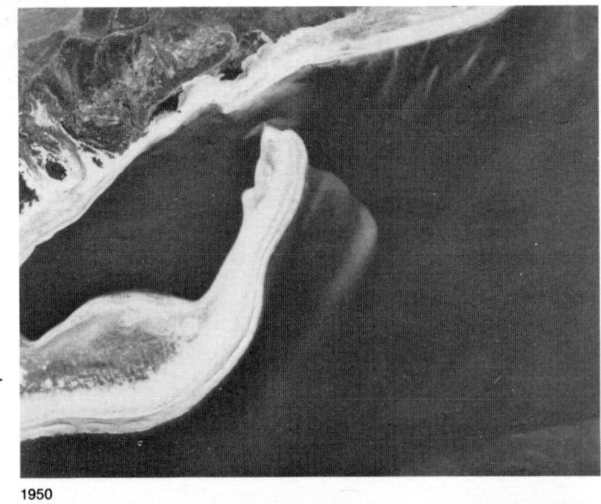

1950

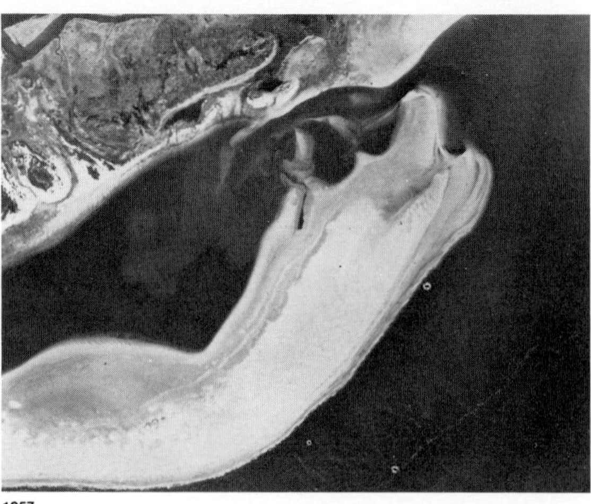

1957

Fig. 13.6 Changes in the configuration of Little Egg Harbor on the Atlantic Coast of New Jersey from 1940 to 1957. (Photographs from University of Illinois at Urbana—Champaign. Used by permission.)

APPLICATIONS TO COASTAL PLANNING AND MANAGEMENT

Proposals for development in the coastal zone today are subject to serious scrutiny in the United States and Canada. In order to evaluate a proposal it is first necessary to know the makeup and dynamics of the coastline involved. This usually involves an inventory of coastal landforms and lithology (composition), and an assessment of recent erosion and disposition trends (Fig. 13.7). Within this framework, more detailed studies can be carried out for the specific segment where the action is proposed. At the heart of such studies, especially when shoreline structures are involved, is an analysis of the sediment budget.

Two types of methods are available in net sediment transport analysis: shoreline change and wave energy flux. The most accurate one is based on measurements of volumetric changes in a shoreline. This usually involves comparing old charts and aerial photographs with newer ones, or making detailed measurements of sediment accumulation behind a shoreline bar-

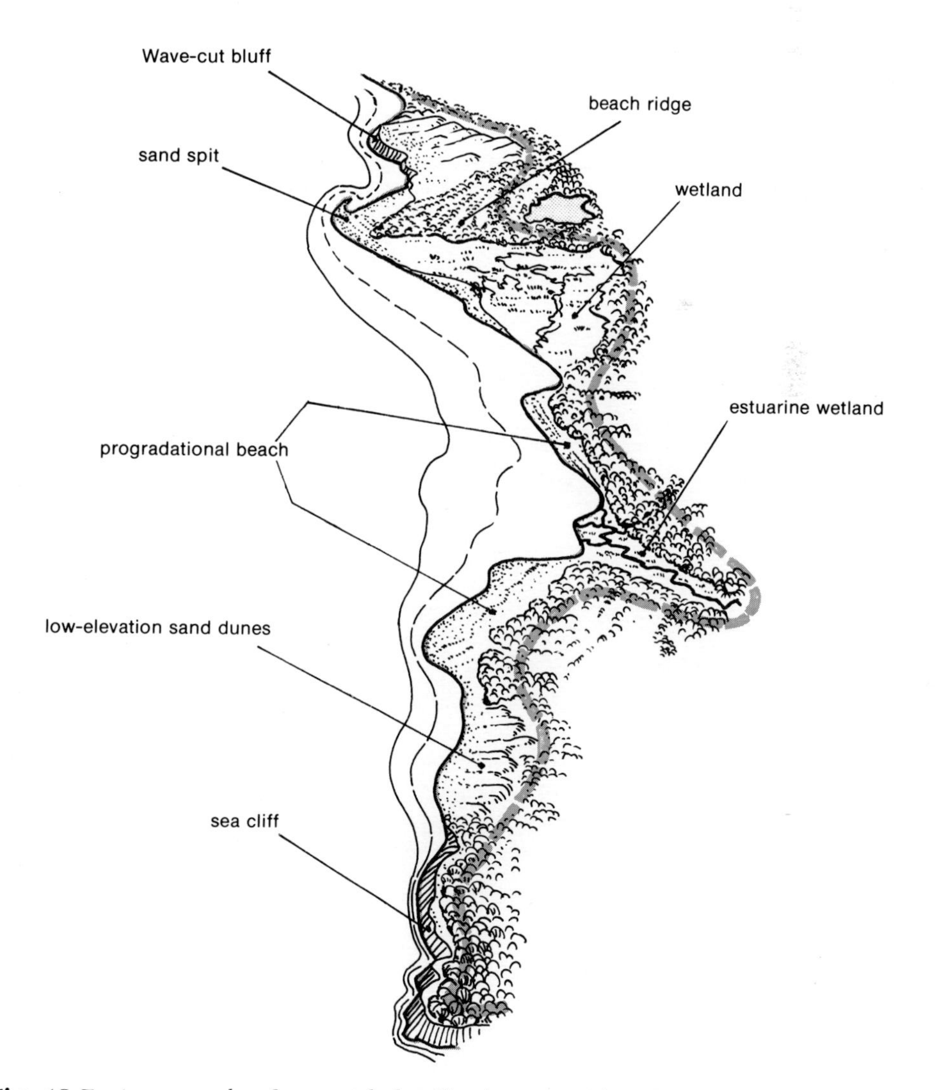

Fig. 13.7 An example of a coastal classification map, identifying critical features and trends of development in the coastal zone.

rier such as a breakwater. The latter necessitates making a topographic survey at the time of construction and one some years later and then measuring the net change between the two. Where harbor entrances are maintained by sand bypassing, the amount of sediment scheduled to be transferred from one side of a barrier to the other each year is determined by this method.

The second method is based on computations of wave energy and established correlations between wave energy flux and sediment transport rates. This method requires wave data for the location in question, and since such data are usually scarce, the wave energy computations are difficult to prepare.

Upon completion of the coastal inventory and estimates of the sediment budget, two important questions can be answered: (1) What is the nature of the system we are dealing with? and (2) What is the relationship among the features, processes, and trends of the coast? This information can then be used to test the feasibility of development schemes, evaluate community plans, or guide the formulation of coastal zone management plans.

CASE STUDY

Nantucket Shoreline Survey: The Planning Implications of Erosion Trends

Andrew Gutman

In a recent study it was found that since 1846 the vast majority of the Nantucket shoreline has been undergoing erosion. The highest rates were found on the south shore, where shoreline retreat averaged 6 to 10 feet per year. Erosion was much less severe on the east and north shorelines, with the section between Great Point and the East Jetty perhaps the most stable section of the entire Nantucket shoreline. Of the 41 miles of shoreline examined, less than 20,000 feet showed net accretion during the past 125 years. Characterized as a whole, the Nantucket shoreline has been, and remains, largely erosional but the rates varying appreciably from place to place.

Three methods were used to measure change in the position of the Nantucket shoreline. In 1961 cartographers of the U.S. Army Corps of Engineers Beach Erosion Board compiled coastal charts dating back a century

or more. From these charts shoreline changes were determined for the period 1846 to 1955. These data were supplemented with measurements made on aerial photographs, which in the Nantucket area were taken every thirteen years since 1938. Finally, field measurements were taken at some twenty sites on the island. This involved seasonal and annual measurements of the distance between the high water shoreline and fixed objects such as a nearby house or a telephone pole.

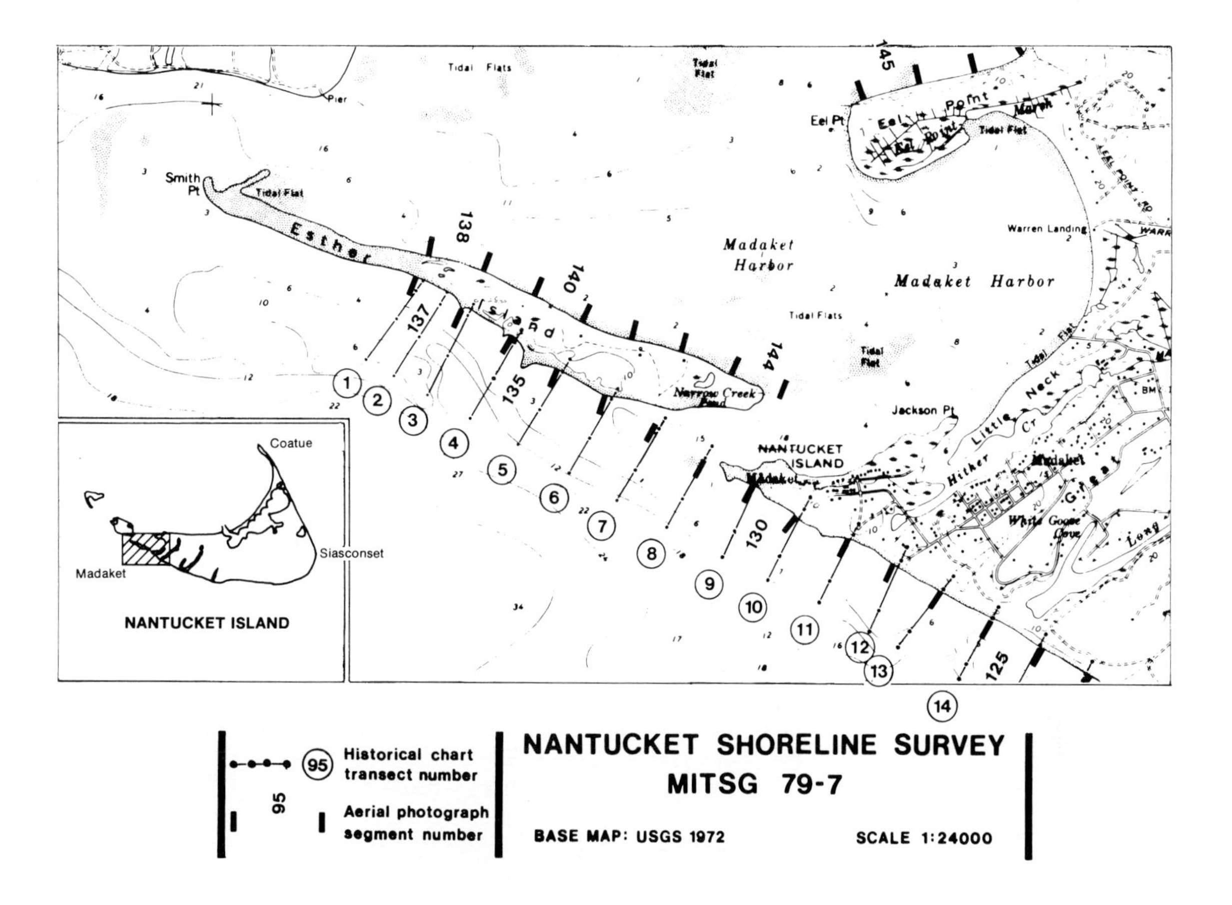

The shoreline trends documented by this study are intended to provide prospective purchasers of coastal property, real estate agents, banks, and regulatory officials with the information needed to properly manage development of Nantucket's coastal resources. Consider two hypothetical shoreline sites being considered for residential structures. Site A on the north shore has had an average erosion (retreat) rate of 3 feet per year and a range of 1 to 5 feet per year. Site B is on the south shore, and erosion here has averaged 11 feet per year with a range of 9 to 16 feet per year. Both sites have lots with the same dimensions and a maximum setback distance of 140 feet from the backshore bank. Both lots list for about the same price, and the buyer is trying to choose one, considering a thirty-year mortgage. With a setback of only 140 feet, a home built at Site B could fall into the ocean in fewer than thirteen years—hardly an attractive investment for a thirty-year mortgage! On the other hand, the property owner could be reasonably certain that the home built on Site A would be safe from damage for at least forty years.

(cont.)

CASE STUDY (cont.)

Aerial view looking over the Coatue toward Nantucket Harbor.

Vertical aerial of Siasconset.

(Photographs by Andrew Gutman. MIT Sea Grant Program, Massachusetts Institute of Technology.)

Data on shoreline trends can be useful not only in making decisions on individual sites but in developing land use plans. Large sections of coastline can be designated for certain uses or zoned against activities that are prone to costly damage. In addition, long-term trends need to be known in order for communities to make decisions on major expenditures such as harbor facilities and erosion defense structures. The limitations of this method should also be recognized, however. The rates and trends represent averages from charts and photographs covering intervals ranging from ten to sixty-eight years. At any particular site there may be no appreciable erosion for many years, then many feet of dune or bluff may disappear in a single severe storm. Many fall and winter storm seasons may pass with little hurricane or "noreaster" activity. For example, a severe hurricane has not struck Nantucket in many years. Then in one year several storms may strike the island. The trends and erosion rates presented here cannot be used to predict specific trends but are intended to be used as a framework to guide decision-making in land use and management of the coastal zone.

Andrew Gutman is a coastal engineer with the Massachusetts Institute of Technology Sea Grant Program. © Andrew L. Gutman et al. *Nantucket Shoreline Survey* published by the MIT Sea Grant College Program, 1979.

SITE PLANNING IN THE COASTAL ZONE

Because the coastal zone is so attractive for residential development, it is necessary to examine briefly some of the problems of site planning near shorelines. In evaluating a site for development, it is important first to examine the position of the site in the larger sediment system and determine whether it is in a retrograding, prograding, or stable part of the coastline (Fig. 13.7). Next, it is necessary to determine the composition of the site and its topographic configuration. If it is comprised of bedrock, the site is usually safe from serious erosion, though may pose significant problems in building construction, wastewater disposal, and stormwater drainage.

Shorelines composed of unconsolidated materials should be treated carefully where development calls for permanent structures because most are prone to seaward/landward fluctuations over various time intervals. Minor fluctuations occur seasonally with winter/summer changes in the sediment mass balance. Over larger time periods, the magnitude of the fluctuations can be expected to increase with the average return period; that is, big changes occur with the lowest frequency. Determination of the proper setback distances for buildings is difficult because the magnitudes of the fluctuations are often different for different segments of a continuous shoreline and few if any data are available to help in figuring out trends. In some instances, however, vegetation can be used to provide insight into trends and fluctuations. Exposed roots, tipped trees, and abundant driftwood are signs of shore erosion and retreat; whereas the establishment of numerous young plants on the backshore is a sign of progradation.

If tree stands of different age classes can be identified on the backshore they may reveal how far landward the sea or lake has advanced since the stands became established. In Figure 13.8, for example, the presence of

Signs of a retreating shoreline: tipped trees, driftwood, devegetated slope. (Photograph by Charles Schlinger.)

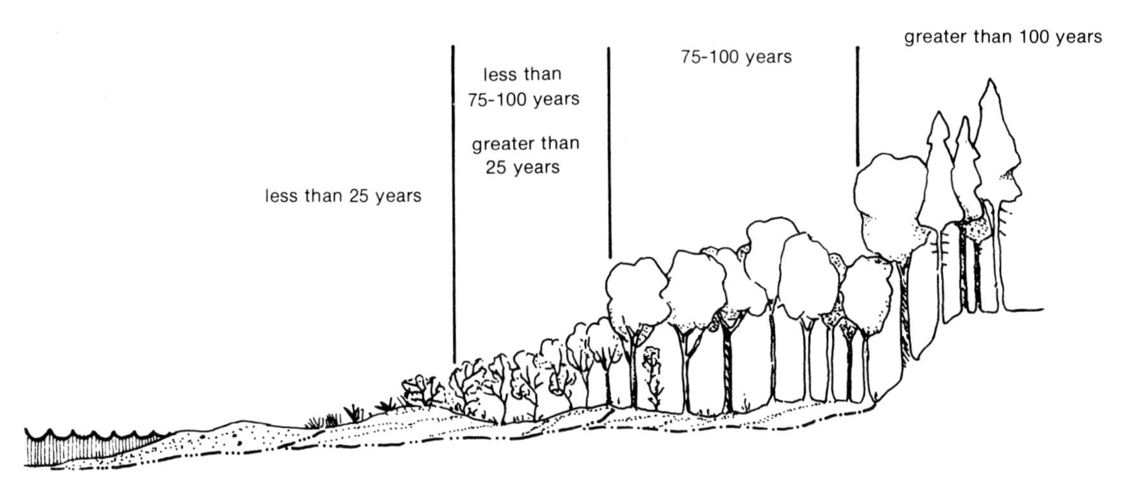

Fig. 13.8 Tree stands of different ages can sometimes be used as an indicator of the extent of past landward fluctuations in the shoreline.

seventy-five to one-hundred-year-old trees beginning at a distance of 100 m from the shore indicates that the area beyond this point has been free of transgression by waves for at least seventy-five to one hundred years, whereas at the 50 m distance the trees ages tell us that area has been eroded away in the last seventy-five to one hundred years and reformed in the past twenty-five to thirty years. This technique is best suited to areas where one stand gives way sharply to another, because the absence of a transition between neighboring stands is a sign that the older stand has been cut back by erosion and the younger stand subsequently established on newer deposits.

Erodible, low elevation coastal terrain generally requires large setbacks, not only because of the threat of wave erosion but also because of storm surges and flooding. The Atlantic and Gulf Coasts of southeastern United States are especially prone to hurricane surges, and development on islands and coastal plains are often severely damaged by large waves and high-water levels (Fig. 13.9). An important part of the federally supported coastal zone planning in such areas has been the development of evacuation plans that spell out a procedure for informing people of the danger level and providing directions for evacuation in the event of a hurricane.

For development sites situated near the crest of a sea cliff or backshore slope, the threat of wave erosion may not be great; however, the effects of wind and slope instability may be significant. The pattern of onshore airflow over a coastal slope produces a zone of low velocity on the lower slope and high velocity on the upper slope relative to wind velocities at comparable elevations over flat ground. The increase from the slope foot to the slope crest is about 4 times for slopes of 4:1 inclination or greater. Additionally, as the air crosses the brow of the slope it appears to cling to the ground (or tree canopy) for some distance inland and then the fast air separates from the surface.

Since the force of wind increases approximately with the cube of velocity, these zones are in reality more significant for site planning than raw wind data would lead one to believe. Sites situated near the brow of the slope are subject to severe wind stress during storms, and in cold climates, high rates of heat loss and deep penetration of ground frost. Downwind

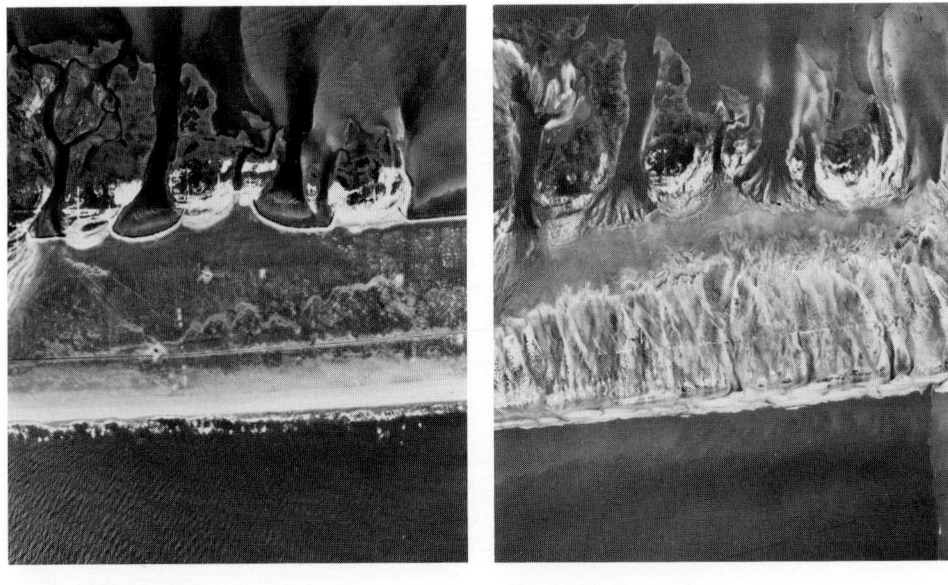

SUGAR POINT (1955)
Worcester County, Maryland
March 14, 1955
RF = 1:28,100 H = 10,000 feet

SUGAR POINT (1962)
Worcester County, Maryland
March 15, 1962
RF = 1:26,400 H = 7,500 feet

Fig. 13.9 Aerial photographs of a coastal island (Assateagne Island, Maryland) before and after it was overwashed by waves of the Great Atlantic Storm of March 1962. (Photographs from University of Illinois at Urbana—Champaign. Used by permission.)

from the brow, sand and snow accumulate where airflow separates, and this zone is often the site of sand dune and snowbank formation. If the brow of the slope is forested, sand and snow accumulation often take place along the brow (Fig. 13.10). Clearly, each of the zones presents limitations for development, and the nearer the crestslope, generally the more serious the limitations. Therefore, we can safely say that the placement of structures on or close to the brow of the slope is not advisable as a general practice; however, it is also necessary to add that development is feasible on stable sites where proper site analysis and appropriate engineering and architectural solutions have been worked out. This would necessitate slope analysis to determine stability against mass movement and erosion, soil assessment for stormwater and wastewater drainage, and wind velocity profiling to establish the aerodynamic conditions. Among many other things, the design of structures and landscaping on sites with exposures to strong winds should conform to the aerodynamic forces for best performance.

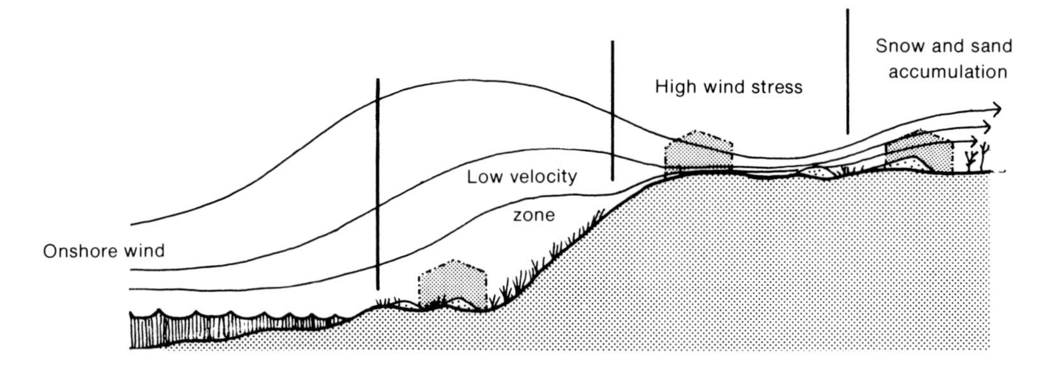

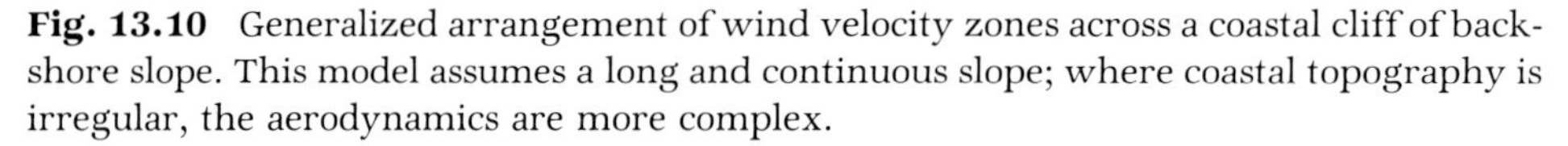

Fig. 13.10 Generalized arrangement of wind velocity zones across a coastal cliff of backshore slope. This model assumes a long and continuous slope; where coastal topography is irregular, the aerodynamics are more complex.

PROBLEM SET

I. Examine the accompanying map of a section of Pacific coast near Carmel, California, and classify the shoreline according to form and composition. The following classes and criteria are suggested:

Class	*Criteria*
Sea cliff	Bedrock, >20 feet high; >30 percent inclination
Shelf rock	Bedrock, <20 feet high and/or <10 percent inclination
Beach	Unconsolidated material (sand, pebbles, or cobbles) low elevation
Wetland	Unconsolidated material covered with vegetation at sea level
Artificial	Fill material or structure

1. For the Carmel Bay section approximate the refraction pattern and associated orthogonals for waves generated by a west wind. Assume that the wave base depth is 5 fathoms (30 feet). Assuming this

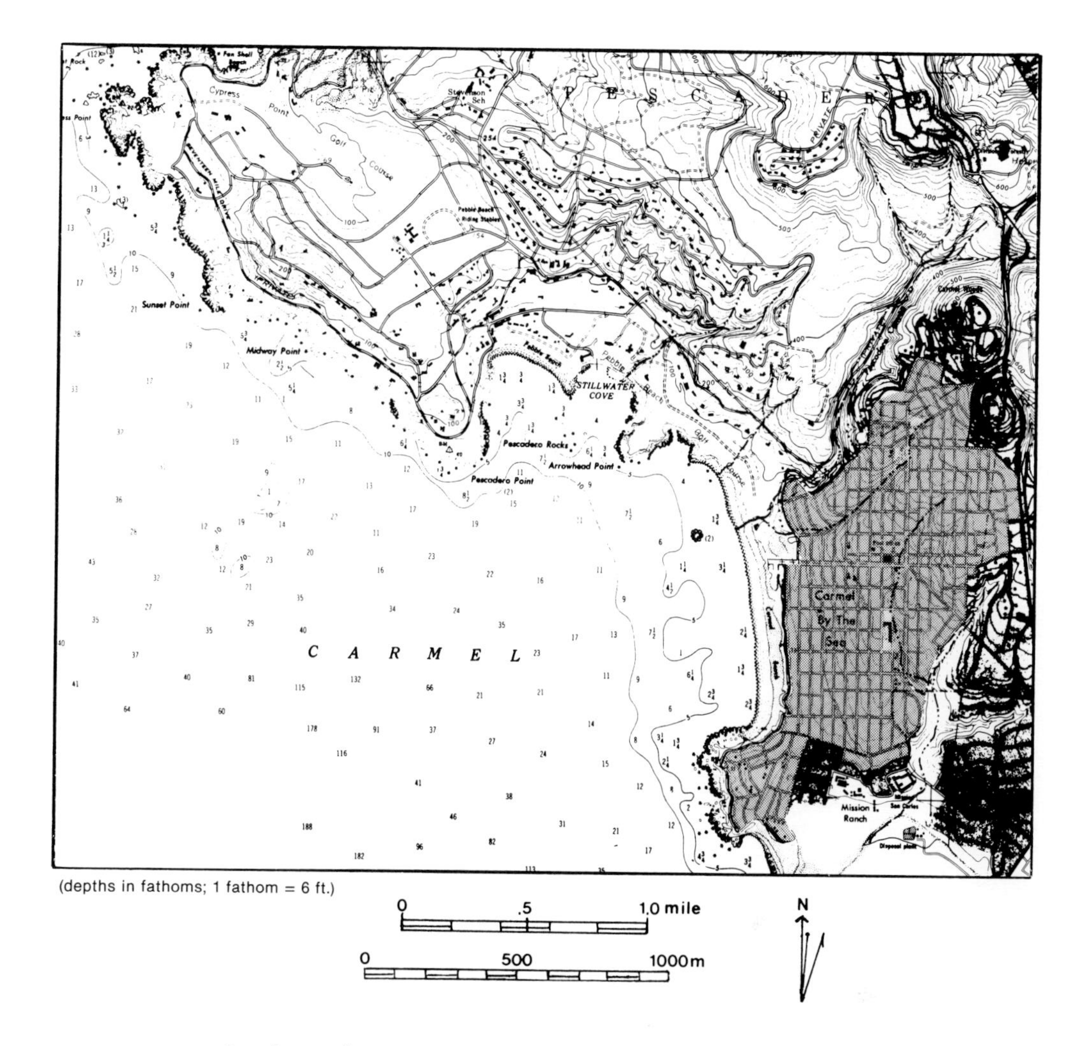

(depths in fathoms; 1 fathom = 6 ft.)

to be the only current-generating force in the bay, identify the direction(s) of the resultant longshore currents. On the basis of this pattern and the pattern of terrestrial drainage, identify the sources of sediment for this shoreline. Show the direction of sediment movement from each source.

2. If over the past thirty years the beach at the head of the bay has grown seaward by an average of 46 feet (14 m) above water and 66 feet (20 m) under water, what has been the approximate mean annual net sediment balance? Assume the average thickness of the added sediment to be 3.5 feet both above and below sea level and the length of the beach to reach from one border of the town to the other. Why is the word *approximate* used with the term *mean annual net sediment balance*? Would it be prudent to consider this to be a long-term trend? Why or why not? What does this imply for the development potential of such a beach for, say, a beach house or set of condominium units?

II. The segment of North Carolina coastline shown in the accompanying map was studied by the United States Army Corps of Engineers to evaluate the effects of a jetty (breakwater) and harbor opening on sediment transport. The jetty was constructed on the north side of the Mason-

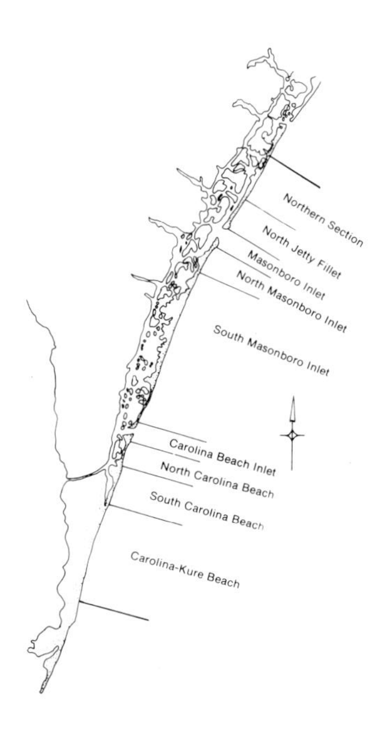

boro Inlet in 1966, and the harbor opening, called the Carolina Beach Inlet, was excavated in 1952. The shoreline is generally broad and sandy without bedrock control, although the Corps of Engineers has added sediment and hurricane protection features to the Carolina and Wrightsville beaches.

1. Examine the following wave energy data and determine the percentage of wave energy that is directed (a) northward, (b) southward, and (c) directly toward shore (within 5° of perpendicular to the shoreline). On the basis of these data, what would you expect the north-south ratio of longshore sediment transport to be?

Direction	*% Wave Energy*
NE	36.96
ENE	11.79
East	6.06
ESE	7.36
SE	11.42
SSE	10.89
South	15.52
TOTAL	100.00

2. The following volumetric change data for each littoral segment represent a preconstruction (1857–1933) and a postconstruction period (1966–1974). Construct a set of bar graphs showing the distribution of net change by segment for the two periods.

Littoral Segment	*Net Change, yds³/yr.*	
	(1857–1933)	(1966–1974)
Northern Section	+ 15,000	− 160,000
North Jetty Fillet	− 27,000	− 7,000

Masonboro Inlet	(not in existence)	+435,000
North Masonboro Inlet	+29,000	−155,000
South Masonboro Inlet	− 3,000	−310,000
Carolina Beach Inlet	(not in existence)	+163,000
North Carolina Beach	+ 9,000	+ 68,000
South Carolina Beach	−10,000	−160,000
Carolina-Kure Beach	−42,000	+ 28,000

3. What do you estimate has been the effects of construction of the two inlets on trends in beach change and why? What implications could this have for coastal zone management policy in the affected segments?

4. What options could be employed to correct the imbalances caused by the inlets?

5. What is the total annual net change considering all segments for the two periods? What explanations could be offered to account for any deviation from a net balance of zero?

SELECTED REFERENCES FOR FURTHER READING

Allen, James R. "Beach Erosion as a Function of Variations in the Sediment Budget, Sandy Hook, New Jersey, U.S.A." *Earth Surface Processes and Landforms,* vol. 6, 1981, pp. 139–150.

Bascom, W. N. *Waves and Beaches.* Garden City, N.Y.: Doubleday, 1964.

Bostwick, Ketchum N. *The Water's Edge: Critical Problems of the Coastal Zone.* Cambridge, Mass.: The MIT Press, 1972, 393 pp.

Coastal Engineering Research Center. *Shore Protection Manual.* Washington, D.C.: Government Printing Office, U.S. Army Coastal Engineering Research Center, 1973.

Davies, J. L. *Geographical Variation in Coastal Development.* London: Longman, 1977.

Gutman, A. L., et al. *Nantucket Shoreline Survey.* Cambridge, Mass.: MIT Sea Grant Program, 1979, 51 pp.

Inman, D. L., and Brush, B. M. "The Coastal Challenge." *Science,* vol. 181, 1973, pp. 20–32.

Jarrett, J. T. "Sediment Budget Analysis Wrightsville Beach to Kure Beach, North Carolina," Coastal Engineering Research Center Reprint 78-3, U.S. Army Corps of Engineers, 1978, pp. 986–1005.

Zenkovich, V. *Processes of Coastal Development,* trans. D. Fry. Edinburgh: Oliver and Boyd, 1967.

14

VEGETATION AND ENVIRONMENTAL ASSESSMENT

INTRODUCTION

Perhaps no component of the landscape is more directly related to land use and environmental change as vegetation. Besides being the most visible part of most landscapes, it is also a sensitive "thermometer" of conditions and trends in parts of the landscape that are otherwise not apparent without the aid of detailed observation and measurement. The loss of vigor in tree species near highways, for example, may be an indication of impaired drainage or heavy air pollution, thereby drawing attention to environmental impact problems that might otherwise be overlooked. In agricultural regions, changes in shrub and tree species in swales and floodplains may be a response to heavy sedimentation pointing up the need for erosion control.

Vegetation also plays a functional role in the landscape since it is an important control on runoff, soil erosion, slope stability, microclimate, and noise. In site planning, plants are used not only for environmental control, but to improve aesthetics, frame spaces, influence pedestrian behavior, and control boundaries. While other landscaping methods and materials can be used for the same purposes, few are as versatile and inexpensive as vegetation; not surprisingly, much of the work of the landscape architect involves planting plants.

Although it is not widely recognized, there are also some negative aspects to vegetation. The most serious are probably the noxious plants that inhabit most metropolitan and suburban landscapes today. These are mainly weed plants, such as poison ivy and ragweed, that are poisonous (usually in the form of allergic reactions) to certain people. The plants causing the greatest difficulty appear to be certain wind pollinators that cause or exacerbate respiratory disorders such as hay fever, bronchitis, asthma, and emphysema.

DESCRIPTION AND CLASSIFICATION OF VEGETATION

Most projects involving planning or monitoring of land use and environment call for a description of vegetation. This may be combined with land use inventories to produce land cover maps or treated as an independent task. In either case, the objective is to document the distribution and make-up of the vegetative cover, and this requires the use of an appropriate plant or vegetation classification scheme. Detailed descriptions of vegetation are virtually a universal requirement of environmental impact statements, and the conscientious investigator typically provides vast inventories and descriptions that draw on several classification schemes.

The schemes in greatest use today are: (1) the *floristic* (or Linnaean), which classifies individual plants according to species, genera, families, etc., using the universally recognized system of botanical names; (2) the *form and structure* (or physiognomic) schemes, which classify vegetation or large assemblages of plants according to overall form (for example, forest and grassland) with special attention to dominant plants (largest and/or most abundant); and (3) the *ecological* schemes, which classify plants according to their habitat (for example, sand dunes, wetlands, lake shores) or some critical parameter of the environment such as soil moisture or seasonal air temperatures.

Despite the conventions of the environmental impact methodology (that is, wholesale inventory and cross-classification), the type of description and classification used in a project should be governed by the nature of the problem and the form and variety of information that are called for. In many cases this requires using some mix of floristic, form and structure, and ecological schemes because in planning problems vegetation must not only be understood as a biological phenomenon tied to other biological phenomena such as animals and insects, but as a physical component of the landscape having height, volume, texture, color, and functional ties with soil, water, air, and land use.

The scheme given in Table 14.1 is organized into five levels, each addressing a different classification element. Level I is based on overall structure, level II on dominant plant types, level III on plant size and density, level IV on site and habitat, and level V on significant species. The latter is included to provide for rare, endangered, protected, and highly valued species, a standard requirement of environmental assessments and impact statements, as well as plants of value in landscaping for a proposed or existing land use.

Table 14.1 Vegetation Classification

Level I (vegetative structure)	*Level II (dominant plant types)*	*Level III (size and density)*	*Level IV (site and habitat or associated use)*	*Level V (special plant species)*
Forest (trees with average height greater than 15 ft with at least 60% canopy cover)	e.g., oak, hickory, willow, cottonwood, elm, basswood, maple, beach, ash	tree size (diameter at breast height) density (number of average stems per acre)	e.g., upland (i.e., well drained terrain), floodplain, slope face, woodlot, greenbelt, parkland, residential land	rare and endangered species; often ground plants associated with certain forest types
Woodland (trees with average height greater than 15 ft with 20–60% canopy cover)	e.g., pine, spruce, balsam fir, hemlock, douglas fir, cedar	size range (difference between largest and smallest stems)	e.g., upland (i.e., well drained terrain), floodplain, slope face, woodlot, greenbelt, parkland, residential land	rare and endangered species; often ground plants associated with certain forest types
Orchard or Plantation (same as woodland or forest but with regular spacing)	e.g., apple, peach, cherry, spruce, pine	tree size; density	e.g., active farmland, abandoned farmland	species with potential in landscaping for proposed development

(cont.)

Table 14.1 (continued)

Level I (vegetative structure)	*Level II (dominant plant types)*	*Level III (size and density)*	*Level IV (site and habitat or associated use)*	*Level V (special plant species)*
Brush (trees and shrubs generally less than 15 ft high with high density of stems, but variable canopy cover)	e.g., sumac, willow, lilac, hawthorn, tag alder, pin cherry, scrub oak, juniper	density	e.g., vacant farmland, landfill, disturbed terrain (e.g., former construction site)	species of significance to landscaping for proposed development
Fencerows (trees and shrubs of mixed forms along borders such as roads, fields, yards, playgrounds)	any trees or shrubs	tree size; density	e.g., active farmland, road right-of-way, yards, playgrounds	species of value as animal habitat and utility in screening
Wetland (generally low, dense plant covers in wet areas)	e.g., cattail, tag alder, cedar, cranberry, reeds	percent cover	e.g., floodplain, bog, tidal marsh, reservoir backwater, river delta	species and plant communities of special importance ecologically and hydrologically; rare and endangered species
Grassland (herbs, with grasses dominant)	e.g., big blue stem bunch grass, dune grass	percent cover	e.g., prairie, tundra, pasture, vacant farmland	species and communities of special ecological significance; rare and endangered species
Field (tilled or recently tilled farmland)	e.g., corn, soybeans, wheat; also weeds	field size	e.g., sloping or flat, ditched and drained, muckland, irrigated	special and unique crops; exceptional levels of productivity in standard crops

Adapted from W. M. Marsh, *Environmental Analysis for Land Use and Site Planning*, by W. M. Marsh. Copyright © 1978, McGraw-Hill, New York. Used with the permission of McGraw-Hill Book Company.

TRENDS IN VEGETATION CHANGE

Most of the major trends in vegetation change in North America are related to three land use activities: agriculture, lumbering, and urbanization. Forest clearing for agriculture over large parts of the Midwest, South, and East has resulted in the loss of virtually all the original forest cover over vast areas and in the formation of a landscape that is best described as agricultural parkland. Only in nonarable sites such as swamps and deep stream valleys has the forest escaped destruction. In many instances, however, it appears that such forests are quite different floristically than the original forest because increased runoff, sedimentation, and other sorts of disturbances from the surrounding lands have caused the elimination of certain tree species as well as changes in ground plants (Fig. 14.1).

Fig. 14.1 Vegetation on the floor of a small stream valley in Oklahoma. Farmfields on adjacent uplands have contributed large amounts of sediment to stream valleys, burying vegetation, and thereby eliminating plants not tolerant to this process.

A second trend has been the contraction of the agricultural landscape as the bulk of the rural population has shifted to the cities over the past five decades. With the abandonment of the small farms, much cultivated land has reverted back to natural, or unmanaged, vegetation. The second growth species, however, are often weedy plants; herbs, shrubs, and trees, such as thistles, sumac, and hawthorn, that are of limited aesthetic and economic value. On the other hand, these plants are effective ground stabilizers and, as such, have helped to reduce rates of runoff and soil erosion.

The trend toward urbanization after World War II has led to massive urban sprawl with the development of freeway systems, residential subdivisions, and shopping centers. Initially much of this growth was absorbed by abandoned farmland, but as the development rate accelerated and land values increased, active farmland was also absorbed (Fig. 14.2). Nearly everywhere that urban sprawl has taken place it has resulted in wholesale destruction of existing vegetation including the fencerows, woodlots, and orchards of active farmland as well as the second growth woodland of abandoned farmland. Only in the past decade or so have developers begun to seriously consider existing vegetation as an amenity and have incorporated it into their development plans.

Fig. 14.2 The loss of farmland to urban development is illustrated by these photographs of the Valley Stream area, Long Island, New York. Photographs taken in 1933 and 1959. (Used by permission of the Long Island State Park and Recreation Commission.)

As landscapes go, most of suburbia is new and the vegetative cover is still developing, meaning that it is undergoing comparatively rapid change as it adjusts to this new environment. In most areas, street trees are one or more decades from maturity, hedgerows are still being planted, and property owners are still in the process of making adjustments in yard plants by replacing exotic species with poor survival records with hardier species. One measure of the level of maturity of suburban vegetation is the diversity and abundance of wildlife such as squirrels, songbirds, and raccoons; generally, animal habitat improves with the density and diversity of the plant cover.

A fourth trend in vegetation is the establishment of noxious weed plants in disturbed areas. Though small by geographic standards, this trend represents a major health hazard. Again, there seems to be an association with urbanization, but not so much with sprawl as with decay. With the decline of inner cities, industrial areas, and old residential neighborhoods, the ground is taken over by weeds (Fig. 14.3). Many of these plants yield huge amounts of pollen to the atmosphere, and when inhaled, it causes allergic reactions in many people. A similar trend may be found on the suburban fringe where land is taken out of use and/or cleared and then held for several years before development.

Fig. 14.3 The introduction of vegetation to commercial districts has become a common practice in programs to renew the attractiveness of inner cities. (Photography courtesy of Korab Photography. Used by permission.)

Two additional trends in vegetation change are also noteworthy. One is the increased use and destruction of special plant environments, such as wetlands, sand dunes, mountain slopes, and shorelands. Wetlands have generally been treated badly by all varieties of land use, mainly because they were (and still are in many quarters) viewed as health hazards, barriers to development, or unproductive land. The second trend is the change in forest cover associated with lumbering. This has resulted not only in the harvesting of most of the original forests in the coterminous United States and southern Canada, but in the formation of new types of forests. Forests managed for a sustained yield of timber based on the selective cutting concept are usually comprised of multiple stories representing different tree ages. By contrast, planted forests that represent a complete replacement of some original cover consist of even-aged stands of floristically uniform trees (Fig. 14.4).

Fig. 14.4 Aerial photograph showing planted stands of conifers.

THE CONCEPT OF SENSITIVE ENVIRONMENTS

The concept of sensitive environments in planning has grown in part from a reaction to the wholesale mistreatment of what we might call minority environments, such as wetlands, stream valleys, and sand dunes, whose value cannot be measured accurately by standard economic criteria, that is, how much the land is worth on the open market. It has also grown from improved understanding of the role of such environments in the maintenance and quality of the larger landscape: Wetlands are often important groundwater recharge areas and small stream valleys are important in the maintenance of river water quality. Today it is not uncommon to find communities incorporating special provisions for sensitive environments into their master plans based on economic rationale (because it costs more to build in and manage such environments), social rationale (because these environments are valued by people for their scenic and general aesthetic value), and scientific rationale (because they are often necessary to the maintenance of larger environmental systems). Wetlands and shorelands are examples of landscapes that are widely recognized as sensitive environments.

Wetlands Prior to the mid-twentieth century, wetlands were generally viewed as third-rate landscapes of limited economic value. As a result, they were indiscriminately altered and destroyed on a wholesale basis to provide development land, improve agricultural production, improve navigation, control pests, and the like. Wetlands are now widely recognized for their roles as hydrologic, ecologic, recreational, forestry, and agricultural resources.

The definition and mapping of wetlands is generally based on two sets of criteria: vegetation and hydrology. Although hydrologic processes are usually the controlling force in the origin and formation of wetlands, the definition of wetlands based on hydrologic criteria has proven difficult for a variety of reasons. Vegetation, on the other hand, has provided a more useful set of criteria, especially indicator species.

Shorelands Coastal areas have long been attractive places to visit, but in the past fifty years development for residential and commercial purposes has increased substantially. This has been facilitated by the growth of highway and road systems, increased availability of land with the decline of agriculture, and the rising popularity of water-oriented living. The development has led to widespread alteration of coastal environments, especially the "softer" ones such as barrier beaches, beach ridges, sand dunes, and backshore slopes where prized plant communities are often found (Fig. 14.5). Because of the delicate balance that typically exists between plants, such as dune grasses, and environment, the imposition of roads and houses often leads to loss of entire plant communities and in turn the decline of an environment valued by many for ecological and aesthetic reasons.

VEGETATION AS A TOOL IN ENVIRONMENTAL PLANNING

The place of vegetation as a planning tool has improved significantly in the past several decades. Among other things, the costs of land clearing and landscaping alone have motivated developers to incorporate more existing (predevelopment) vegetation into site plans. In residential areas, a mature shade tree may have an estimated value of $500 to $1500 (1975 dollars), and the composite assemblage of plantings may improve the real estate value of an average residential lot by $3000 to $7000.

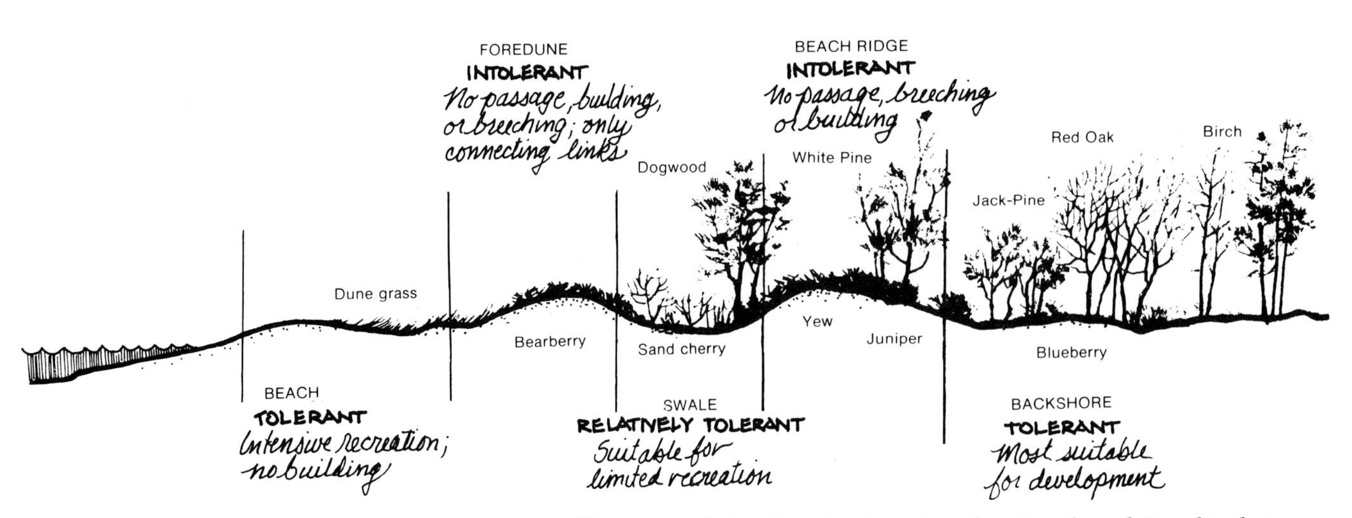

Fig. 14.5 A profile across a belt of sandy shoreline showing the relationship between plants and landforms including an assessment of the relative resistance of microenvironments to disturbance. (Illustration by Carl D. Johnson. Used by permission.)

Visual Beyond its direct economic value, vegetation is also recognized for its functional or environmental value in land use planning. In site planning, for example, it is regularly used to screen certain land use activities and features, abate noise, modify microclimate, and stabilize slopes. As a visual barrier, hedge rows and border trees can help separate conflicting land uses such as residential, commercial, industrial, and institutional. In this capacity, the density and the permanency of the foliage is critical because it controls the transmission of light (Fig. 14.5).

Noise Vegetation can also be used to help control noise. Under barrier-free conditions, the level (magnitude) of sound from a point source decreases at a rate of six decibels with each doubling of travel distance. (From a linear source such as a highway, the decay rate is nearer three decibels per doubling distance.) Placed in the path of sound, vegetation absorbs and diverts energy, and is somewhat more effective for sound in the high frequency bands (those above 1,000–2,000 hertz). In forests the litter layer (decaying leaves and woody materials) appears to be most effective in sound absorption. Table 14.2 gives an example of the reduction in locomotive noise associated with a 250-foot-wide belt of forest. For the most effective use of vegetation in noise reduction, it is best to combine plantings with a topographic barrier such as a berm or embankment.

Microclimate The influence of vegetation on ground-level climate can be very pronounced. A plant cover effectively displaces the lower boundary of the atmosphere upward from the ground onto the foliage. A microclimate of some depth is thus formed between the foliage (for example, under the tree canopy) and the ground where solar radiation, wind, and surface temperatures are lower than those over a nonvegetated surface. Heat exchange between the landscape and atmosphere is also influenced by vegetation; the Bowen ratio (sensible heat to latent heat flux) is lower because of transpiration, producing somewhat lower air temperatures over vegetated than nonvegetated surfaces (see Fig. 3.3).

Table 14.2 Locomotive Noise Reduction with and without a 200-Foot-Deep Forest

Frequency, Hertz	*Noise at 250 ft without Forest*	*Noise at 250 ft with Forest*
31.5 Hz	39 dB	39 dB
63	57	56
125	63	61
250	68	65
500	73	69
1,000	74	68
2,000	72	64
4,000	68	56
8,000	61	41
	79 dBA	73 dBA

dB = decibel; unit of measurement of sound magnitude based on pressure produced in air from a sound source
dBA = decibel scale adjusted for the sensitivity of the human ear; a correction factor applied to dB units that takes into account the pattern of sound frequencies perceived by the human ear
Frequency = the pitch of sound measured in cycles per second; higher pitches have higher frequencies (more cycles per second)
Hertz (Hz) = cycles per second

As a barrier to airflow, vegetation tends to force wind upward, thereby increasing the depth of the zone of calm air (called the boundary sublayer) over the ground. The taller and denser the vegetation, the thicker the sublayer; under a mature fir forest it is usually around 2.5 m deep, whereas in grass it is a hundred times less at 2.5 cm (Fig. 14.6). A related effect can be found on the downwind side of a vegetative barrier where a calm zone forms under the descending streamlines of wind. The breadth of this sheltered zone also varies with the height and density of the barrier; however, significant wind reduction can generally be expected over a distance of 10 to 20 times the height of the barrier (Fig. 14.7). In snowfall areas, this zone is subject to the formation of snow drifts, the length of which can be estimated using this formula:

$$L = \frac{36 + 5h}{K}$$

where:

L = snow drift length in feet
h = barrier height in feet
K = barrier density factor; 50 percent density is equal to 1.0 and 70 percent is equal to 1.28.

Air Pollution The influence of vegetation in reducing contaminants in polluted air is not well documented for urban areas, but the existing evidence suggests that it is relatively small. Plants are known to absorb certain gaseous pollutants, for example, carbon dioxide, ozone, and sulfur dioxide, but it is apparently limited to the air immediately around the leaf and thus has only miniscule effects on these pollutants in the larger urban atmosphere. Heavy herb covers and dense stands of shrub and tree-sized vegetation with full covers of foliage act as sinks (catchments) for airborne particulates, but their net effectiveness is questionable because a sizeable

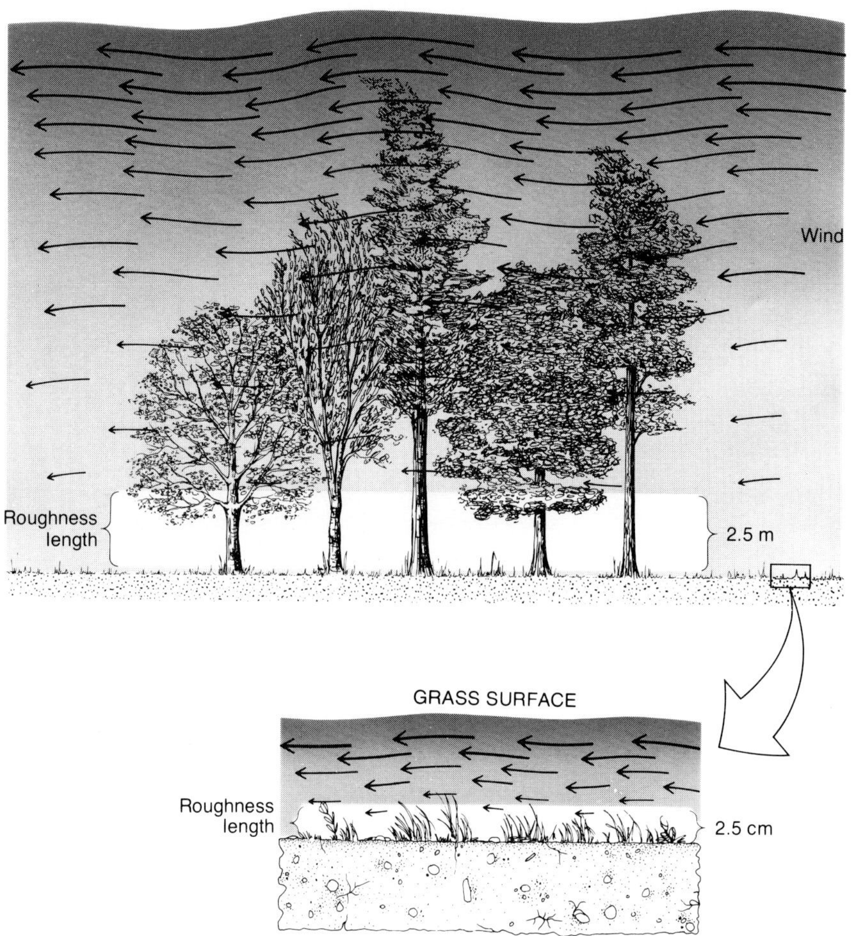

Fig. 14.6 The difference in the thickness of the boundary sublayer under coniferous trees and grass. This layer is important to the formation of ground-level microclimates. (From W. M. Marsh and J. Dozier, *Landscape: An Introduction to Physical Geography*, © 1981, Addison-Wesley, Reading, Massachusetts. Fig. 3.14. Reprinted with permission.)

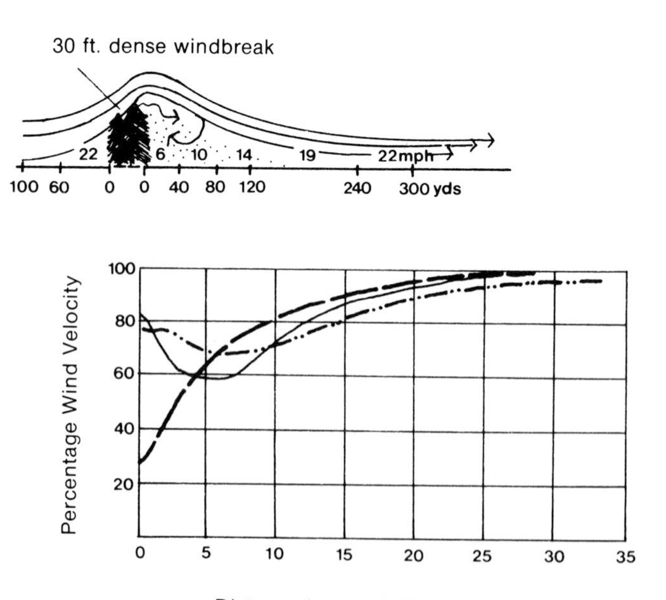

Fig. 14.7 Pattern of wind velocity leeward of a barrier. (Based on U.S. Department of Agriculture publications.)

percentage of particulates initially caught appears to reenter the atmosphere within hours or days. Overall, vegetation appears to be most effective in trapping large particles in air moving laterally within several meters of the ground.

In areas of heavy air pollution, a more pressing question may be that of the impact of pollutants on the health and survival of vegetation. Ozone and sulfur dioxide are the pollutants causing greatest damage to woody plants; other pollutants such as fluorides, dust, and chlorine are also known to cause damage, but it is usually localized around the point of emission.

Social Value People have long recognized the desirability of vegetation in neighborhoods, towns, and cities. This is related not only to the perceived role of plants in climate control, noise abatement, and the like, but also to socio-cultural norms that place value on living plants and the habitats they create. Residential preference surveys bear this out when people identify parks and green spaces as important reasons for choosing one neighborhood or community over another. Real estate data also support this because wooded and landscaped lots consistently bring higher prices than those without vegetation or with unkempt vegetation.

CASE STUDY

The Role of Vegetation in an Environmental Impact and Facility Siting Study

John M. Koerner

Despite its prominence in the landscape, vegetation rarely constitutes a major or controlling environmental determinant in most planning and design projects. Such significance tends to be relegated to special status species or to areas characterized by the presence of unique or relict stands of vegetation. Likewise, vegetation is occasionally given more significance because of its relationship to another environmental condition such as critical wildlife habitat, steep slopes, or watershed protection. Yet, because of this relationship with other environmental factors, vegetation is unmatched in its capacity to guide decision making and site planning by revealing how the environment functions. The structure and type of plant cover serve as indicators of important natural conditions such as floodplains, high groundwater tables, near-surface bedrock, and slope failure. Similarly, vegetative indicators allow various human influences including grazing, previous tillage, and forestry practices to be readily discerned and evaluated in terms of performance. Accordingly, vegetation can become the key to interpretation of the constraints and opportunities of a site, as well as the basis for forecasting the environmental impacts associated with a proposed project.

Vegetation played such a role in the siting and assessment of potential impacts associated with the proposed Rock Run Advanced Wastewater Treatment Plant, a 20-million gallon per day facility to be located in Potomac, Maryland. The facility is to be sited within the 1,000-acre Avenel Farm situated in the rolling landscape of the Piedmont section of the Appalachian Province. In this area, upland terrain is typically grazed or tilled while steep slopes and ravines are heavily wooded.

Numerous plant communities cover the bulk of the farm. Many of the communities are coincident with or reflective of particular environments. The riparian community is quite distinct along Rock Run, as is the wet

(cont.)

CASE STUDY (cont.)

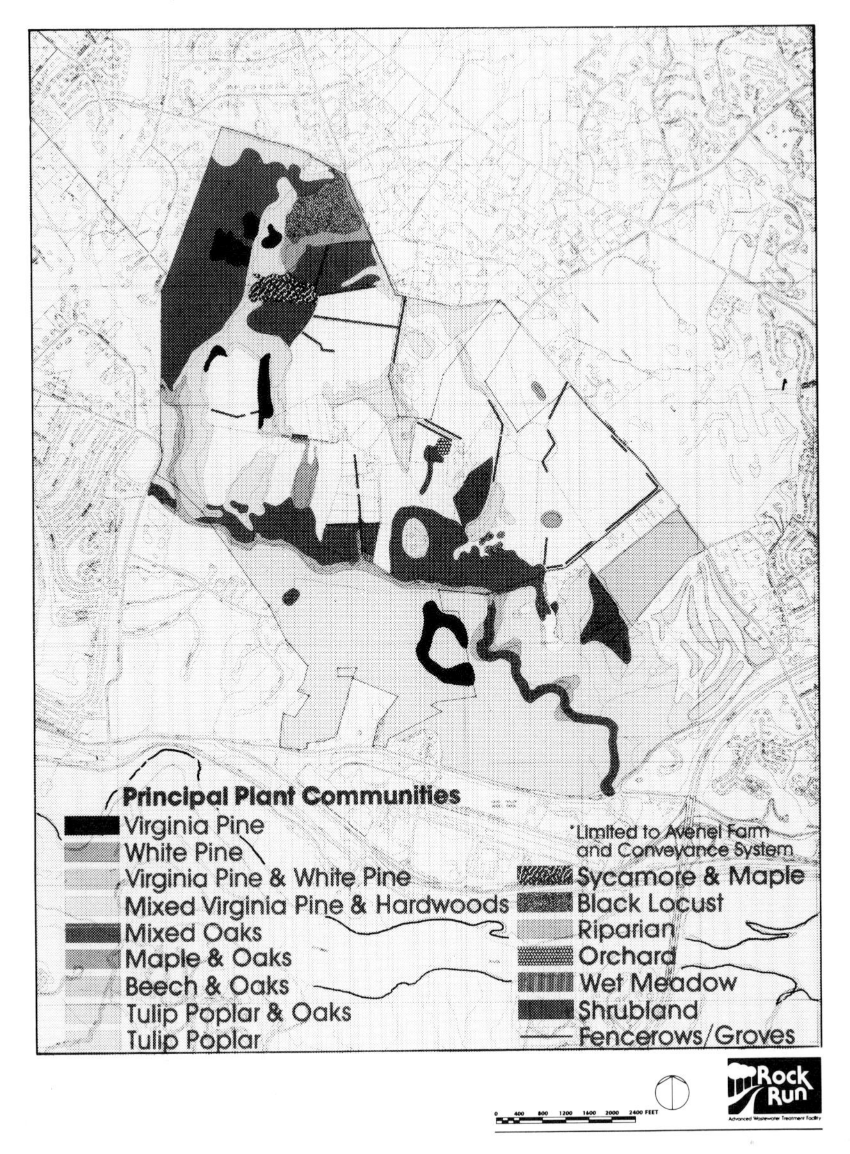

meadow and some of the shrubland along the stream valley. A large mixed oak community in the northwestern portion of the farm is coincident with an area of wet soils. Similarly, the sycamore/maple communities are found in zones of high groundwater and numerous springs. Curiously, the beech/oak community is found in a band with considerable near-surface bedrock, but that was also heavily disturbed during gold mining activities that occurred sporadically between 1865 and 1935.

Areas of pine cover on Avenel Farm reflect zones that were previously tilled or otherwise disturbed. Upland shrublands are similarly indicative. Most areas of the tulip poplar communities were previously disturbed to varying degrees by fire, grazing, or logging. Importantly, however, these areas were generally left in woodland because of their association with steep slopes and poor soils. Consequently, an unusually large expanse of mature forest, important to many forest species of wildlife, remains in this suburban area.

As the siting analysis was conducted, the environmental constraints coincident with the vegetation patterns were documented. Large areas of the farm were discarded as prospective facility sites because of these constraints and the potential for serious impacts to attendant environmental features. Conversely, existing fencerows and small groves of trees were utilized in the siting scheme as visual buffers between the main facility and the surrounding scenic roadways and neighborhoods. The potential for future plantings of vegetation was also considered in order to reinforce visual screening.

Vegetation played an additional role as part of the environmental planning for this facility by helping to determine the size and extent of a surrounding 500-acre buffer area. Some open pasture or shrubland areas were retained for future use as neighborhood recreational sites and to insure adequate wind flow and dispersion of potential odors. Vegetative screens were preserved and planned not only for visual screening, but also to aid in the capture of moisture droplets that contain odors and airborne particles of pathogenic fungi which may be associated with sewage sludge-composting operations.

John M. Koerner, a physical geographer by training, is manager of Environmental Programs for the NUS Corporation, an engineering and planning firm.

APPROACHES TO VEGETATION ANALYSIS

In most landscapes the distribution of plants can be highly variable, even at the local scale. The reasons for the variation are often complexly tied to existing environmental conditions, past events such as fires and land use, and the geographic availability of species to inhabit the area. Which of these myriad of variables exerts the greatest control is usually a difficult question to answer.

In searching for explanations for the distributions of plants, three basic types of studies or approaches are used. One approach is to map the distribution of plant types and environmental features and then examine the two distributions to see what correlations can be ascertained. For example, a comparison of topography and plant species may show that certain species consistently appear in stream valleys. What such a relationship means is not revealed by the correlation, but it may provide clues about what questions should be raised for analysis. The floors of stream valleys are usually wetter, subject to more flooding, and comprised of more diverse soils than other settings. Plants must spend much of the year under conditions of very wet soil and/or standing water. Moisture tolerance may thus prove to be a good candidate for detailed analysis of the plants that are found in stream valleys, especially if the other settings that support different vegetation in the study area are appreciably drier (Fig. 14.8).

A second approach begins with an examination of the environment in an effort to identify those features and processes that may influence plant distributions. The purpose here is to document the forces and processes that could control the plants and then propose where certain plants or groups of plants should and should not be found. The analytical part of this

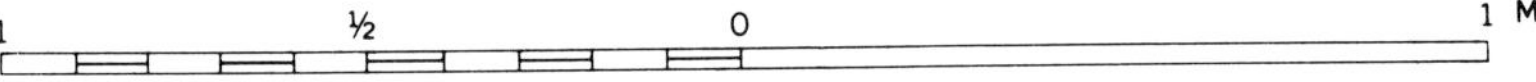

Chutes and channels created by the cloudburst flood of June 1949

Ground generally bare rock, soil, or gravel, cobbles, and boulders. Young trees abundant; especially black locust and, on flood plains, sycamore

Unit 1

NORTHERN HARDWOOD FOREST TYPE

Basswood (**Tilia americana**), *sugar-maple* (**Acer saccharum**), *and yellow birch* (**Betula lutea**), *or any one of the three, are present. Pitch-pine* (**Pinus rigida**) *and table-mountain pine* (**Pinus pungens**) *absent, or very rare. Usually contains red oak* (**Quercus rubra**) *and, in some places, chestnut-oak* (**Q. prinus**); *other species of oak absent; oaks generally few in number. Ground cover consists of ferns and thin-leaved herbaceous plants; climbing vines common. Characteristic of hollows, channelways, and flood plains*

Unit 2

YELLOW PINE FOREST TYPE

Pitch-pine (**Pinus rigida**) *and table-mountain pine* (**P. pungens**), *or either one of the two, are present. Basswood, sugar-maple, and yellow birch absent to very rare. Usually contains several species of oak (chestnut-oak, red oak, black oak—***Quercus velutina**, *and scarlet oak—***Q. coccinea**) *in the canopy layer and often scrub-oak* (**Q. ilicifolia**) *in the brushy understory layer. Ground cover brushy with heath plants* (**Ericaceae**) *abundant. Characteristic of noses*

Unit 3

OAK FOREST TYPE

*Pitch-pine, table-mountain pine, basswood, sugar-maple, and yellow birch absent or very rare. Forest consists largely of oaks (chestnut-oak, red oak, black oak, scarlet oak, and white oak—***Q. alba**). *Ground cover generally ericaceous. Characteristic of side slopes*

Fig. 14.8 The relationship of forest types to landforms in a section of the central Appalachian Mountains. Certain events (the cloudburst flood of 1949) and soil moisture conditions associated with these landforms are suggested by the authors as the basis for the distribution. (From J. T. Hack, and J. C. Goodlett, "Geomorphology and Forest Ecology of a Mountain Region in the Central Appalachians," *U.S. Geological Survey Professional Paper 347*, 1960.)

approach involves testing the proposed or expected distributions by making measurements in the field and then investigating the detailed relations between the affected plant(s) and the environmental variable (Fig. 14.9).

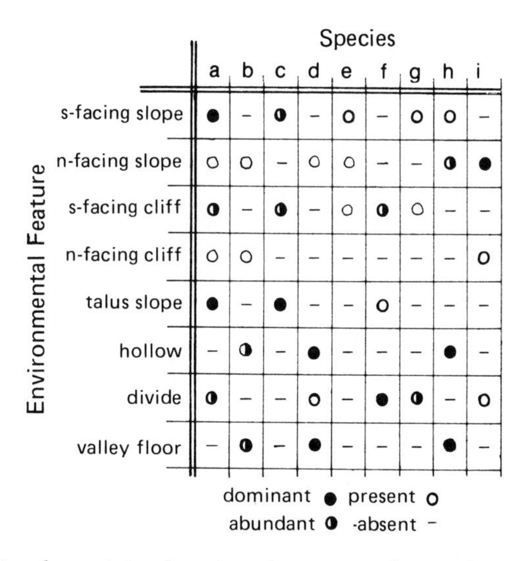

Fig. 14.9 A simple matrix showing the expected correlation between selected plant species and selected processes and features of the environment.

The third approach uses key plants as indicators of environmental conditions and events. Based on existing knowledge of the habits and tolerance levels of selected plants, the controlling forces in the environment can often be identified according to which plants are found in an area. At the simplest level, this entails determining the presence or absence of certain types of vegetation over an area. For instance, in a region where forests are the predominant natural vegetation, the absence of forest cover at a site is an indication that: (1) levels of stress (light, water, heat, and carbon dioxide) are too great for trees; (2) the resources of the site, for example, soil cover and water supply, are limited; and/or (3) a recent disturbance such as a tornado destroyed the forest at the site and it has not grown back yet. Table 14.3 lists a number of site conditions in different bioclimatic regions that can be interpreted from vegetation.

SAMPLING VEGETATION

Whether one's objective is to inventory and describe vegetation for the environmental impact and assessment studies, analyze vegetation for scientific purposes, or use vegetation as an indicator of site conditions and past events, it is usually necessary to sample vegetation in some way. Sampling is a means of selective observation that enables one to estimate various aspects of a plant population or vegetation community based on measurements of only a small portion of it. Sampling is attractive because it saves time and money; however, the proper use of sampling techniques can be difficult. Among the techniques used in vegetation studies are: quadrat

Table 14.3 Vegetation Indicators of Site Conditions

Climatic Region	*Absence of Plant Cover*	*Sparse Herb and Shrub Cover*	*Thick Herb and Shrub Cover*	*Brush and Small Trees*	*Blade and Reed Plants*	*Highly Localized Tree Cover*
Humid (Eastern North America, Pacific Northwest, South)	• bedrock at or very near surface • active dunes • recent human use, cultivation, etc. • recent fire • recent loss of water cover	• bedrock near surface • recent or sterile soils—dunes, fill • recently disturbed (fallow, fire, flood) • active slopes/erosion	• recently logged or burned • too wet for trees • managed grazing • organic soil • old field regrowth	• land-slide/fire, flashflood scars • old field or woodlot re-growth • shale/clay substrate • organic soil • moisture deficiency	• organic soil • standing water • high ground-water table • springs, seepage zones	• wet depression, organic soil, steep • slopes in agri-cultural areas • flood-prone areas
Semiarid (High plains, S. California) *Arid* (Southwest, Great Basin)	caliche or salt pan (playa) at or very near surface desert pavement rock surface unstable ground such as dunes or rockslides too dry	localized water sources eolian erosion overgrazing	overgrazing free from burning too dry for trees	channels of available moisture aquiferous substrate protected pockets favorable (moist) slopes logged/burned	(same as above)	aquiferous substrate seepage zone or spring stream valley (galleria) forest plantation
Arctic and Alpine (N. Canada, Alaska)	rock surface active slopes semipermanent ice or snow cover ponded water during growing season	above tree line semipermanent ice or snow cover active slopes periglacial processes active	above tree line ice, snow, and wind pruning mildly active slopes wet depressions	wind/ice pruning avalanche, landslide scars, fire recent logging near tree line permafrost near surface	(same as above)	protection pockets

Beyond this sort of exercise, the particular types of plants, their densities, and physiological conditions (health) can be examined to learn about the detailed nature of the environment and its forces.
From W. M. Marsh, *Environmental Analysis for Land Use and Site Planning*, New York: McGraw-Hill, 1978. Used by Permission.

sampling, stratified sampling, transect sampling, systematic sampling, and windshield-survey sampling.

In any sampling problem the first task is to define the relevant population. In the case of vegetation this is usually accomplished by defining the geographic area occupied by the vegetation under study. This area(s), which may be a development site or particular environmental zone, is outlined on a large-scale map; the vegetation within it can then be sampled, using either the quadrat method or the transect method.

Quadrats are small plots, the size of which varies with the type of vegetation being sampled. In quadrat sampling the first step involves subdividing the whole area into grid squares. For small study areas, individual squares may be used as a sample quadrat, whereas for large areas, it may be necessary to use some fraction of a square as a quadrat depending on the type of vegetation. Generally speaking, a one- to two-square meter quadrat would be used for grasslands and marshes, four to six square meters for low shrub covers, fifteen to thirty square meters for brushland, and thirty to one hundred square meters for woodland and forest.

Stratified sampling involves subdividing the population or study area into subareas or sets prior to drawing the sample. These are coherent subdivisions based on observable characteristics or prior knowledge of the area. A typical subdivision for many areas involves three strata: floodplains, valley slopes (walls), and uplands. Within these strata, quadrats are chosen on a random basis. Stratified sampling is widely used today with remote sensing imagery to define regional vegetation patterns (see Fig. 15.2 in Chapter 15). Indeed, aerial photographs are almost indispensable in modern vegetation studies, and more technically sophisticated remote sensing systems, such as line scanners, are showing promise for discriminating major types of vegetation. A form of stratified sampling is also used in soil mapping for land use projects where the strata are defined and sampling points assigned according to development and use zones.

In *random transect* sampling the study area is divided into a number of strips, called transects. The width of the transects should vary with the type of vegetation; five to ten meters would, for example, be appropriate for most forests. Of the transects selected for sampling, the entire transect may be sampled or individual quadrats may be selected for sampling within the transect.

Systematic sampling requires no prior knowledge of the population or area under consideration. A grid is drawn over the area and a sample is taken at each intersect in the grid. Quadrats can be used as the sample unit, and it is generally recommended that, together, the quadrats cover a minimum of 20 percent of the study area.

The *windshield survey* is the quickest and least expensive sampling technique. Though more commonly employed in land use surveys than in vegetation studies, it can be helpful in gaining an overview of vegetation types. One should be aware, however, that roadside vegetation may not be representative because it may be planted, cut back in road construction or maintenance, or atypical of the area owing to the establishment of second growth trees and weedy plants in the road right-of-way.

VEGETATION AND ENVIRONMENTAL ASSESSMENT

Finally, a note on vegetation as it relates to environmental assessment and impact analysis. As we mentioned earlier, few components of the landscape lend themselves to identification of environmental stress and change as does vegetation. At least five parameters or measures of impact related to vegetation can be highlighted in this context for evaluating a proposed action.

First, the sheer loss of cover, measured, for example, by the area of vegetation lost to development, is a very significant indication of impact because of its implications with respect to runoff, microclimate, and aesthetics, and so on. Second, the loss of valued species, communities, and habitats is a critical measure of environmental impact as mandated by law at various levels of government. Third, is the economic loss represented by the loss of merchantable vegetation such as timber and the longer term loss of profitable production areas. Fourth, vegetation is often an integral part of larger environmental systems such as microclimate, soils, and hydrology, and alteration or loss of plant cover can spell serious decline in these systems. And fifth, it is important not to lose sight of the fact that natural vegetation is adjusted to a certain set of environmental conditions, and changes in these conditions, even subtle ones, are often reflected in changes in the vigor, reproduction capacity, and make up of plant communities; therefore, plants serve as valuable "thermometers" of environmental performance, giving us warnings when things are not working well.

PROBLEM SET

I. A plan has been proposed for the development of a major highway near a residential area. The proposed alignment will bring the highway within 400 feet of the residences at the nearest point, and the highway planners would like to create a buffer of vegetation between the two.

1. Identify the particular types of disturbances (noise and others) that could be caused by the highway (from the residents' standpoint) and indicate the relative effectiveness of vegetation in mitigating them.

2. With regard to noise, what type of vegetation and planting plan should be recommended; that is, what types of vegetation and ground cover would be most effective in reducing noise?

3. If highway traffic at peak hour generates 80 dBA at a distance of 100 feet from the road, what will the estimated noise level be at the nearest residences without the benefit of a vegetative buffer? With a vegetative buffer?

II. The map in Fig. 15.2 of Chapter 15 was produced from imagery generated by a remote sensing satellite. Such maps are widely used today in vegetation studies.

1. What sort of sampling is actually represented by the map in Fig. 15.2?

2. If additional data were needed on one or two vegetation types, what sort of sampling technique would you recommend, considering the size of the area and assuming that you have limited time to complete the project?

SELECTED REFERENCES FOR FURTHER READING

Carpenter, Philip L., et al. *Plants in the Landscape.* San Francisco: W. H. Freeman, 1975, 491 pp.

Davis, Donald D. "The Role of Trees in Reducing Air Pollution." In *The Role of Trees in the South's Urban Environment* (Symposium Proceedings), University of Georgia, 1970.

Grey, Gene W., and Denecke, F. J. *Urban Forestry.* New York: Wiley, 1978, 279 pp.

International Union of Forestry Organizations. *Trees and Forests For Human Settlements.* Toronto: University of Toronto Centre for Urban Forestry Studies, 1976.

Luken, R. A. *Preservation Versus Development: An Economic Analysis of San Francisco Bay Wetlands.* New York: Praeger, 1976, 155 pp.

McBride, J. R. "Evaluation of Vegetation in Environmental Planning." *Landscape Planning,* vol. 4, 1977, pp. 291–312.

Robinette, Gary O. *Plants/People/and Environmental Quality.* Washington, D.C.: U.S. Park Service, Government Printing Office, 1972, 137 pp.

Thurow, Charles, et al. *Performance Controls for Sensitive Lands.* Washington, D.C.: American Society of Planning Officials, Reports 307 and 308, 1975, 156 pp.

U.S. Forest Service. *Better Trees For Metropolitan Landscapes.* Washington, D.C.: U.S. Government Printing Office, USDA Forest Service General Technical Report NE–22, 1976, 256 pp.

U.S. Forest Service. *National Forest Landscape Management* (Agricultural Handbook No. 478). Washington, D.C.: U.S. Government Printing Office, 1974.

III

TECHNIQUES AND TOOLS

15

REMOTE SENSING AND AERIAL PHOTOGRAPH INTERPRETATION*

*This chapter was prepared by **John M. Grossa,** Department of Geography, Central Michigan University.

INTRODUCTION

Aerial photographs and other remote sensing imagery are one of the most widely used sources of data and information in modern planning and geography. In fact, many of the topics addressed in this book rely directly on the use and application of remote sensory imagery. For example, land use inventories, vegetation classifications, and slope analyses are routinely done using remote sensing imagery. Such imagery is also used to document the magnitude of environmental events such as the extent of floods, tsunamis or storm surges, fires, or landslides. Physical processes including land use change, shoreline retreat, dune movement, stream channel shifts, and water conditions can be monitored over time with sequential remote sensing imagery. Indeed, remote sensing is so important a tool for environmental assessment that some basic skills of remote sensing image interpretation and analyses are essential for both students and practitioners in these fields. The objective of this chapter is to provide a primer for the development of these skills.

REMOTE SENSING SYSTEMS

Remote sensing is a means of detecting or imaging the environment from a remote and sometimes distant location. Although the format in which these images are displayed varies among remote sensing systems, aerial photography, line scanning, and side-looking radar are the systems most widely applied to environmental problems. Such systems selectively collect and record electromagnetic energy that has been reflected or emitted from the earth's surface.

Aerial photography, the most widely used remote sensing technique, normally focuses reflected sunlight from the earth surface through a camera lens onto photographic film. Although most photographic film is sensitive to ultraviolet and visible light of 0.3 to 0.7 micrometer (μm) wavelengths, specially prepared emulsions have extended sensitivity into the near infrared portion of the spectrum to 0.9 micrometer (μm). While this film *does* record infrared energy, it is reflected solar radiation in the near or photographic infrared wavelengths. This energy is *not* emitted by the earth and *does not* represent different surface temperatures.

Line scanning systems have a broader spectral range than photographic systems as they can be equipped to record energy over a range extending from 0.2 μm to 15 μm (Fig. 15.1). When selected infrared detectors are used, thermal infrared line scanners can record radiant energy being emitted from the earth in the middle (3 to 5.5μm) and far infrared (8 to 14 μm) wavelengths. Since this energy is being radiated by the earth at all times, thermal infrared systems can routinely produce imagery at night as well as day. Such imagery can accurately record small radiant temperature differences at the earth's surface. Multispectral line scanners can simultaneously record data in several wavelength bands ranging from ultraviolet through visible and into the infrared portions of the electromagnetic spectrum (Fig. 15.1). An advantage of all line-scanners data is that they are pro-

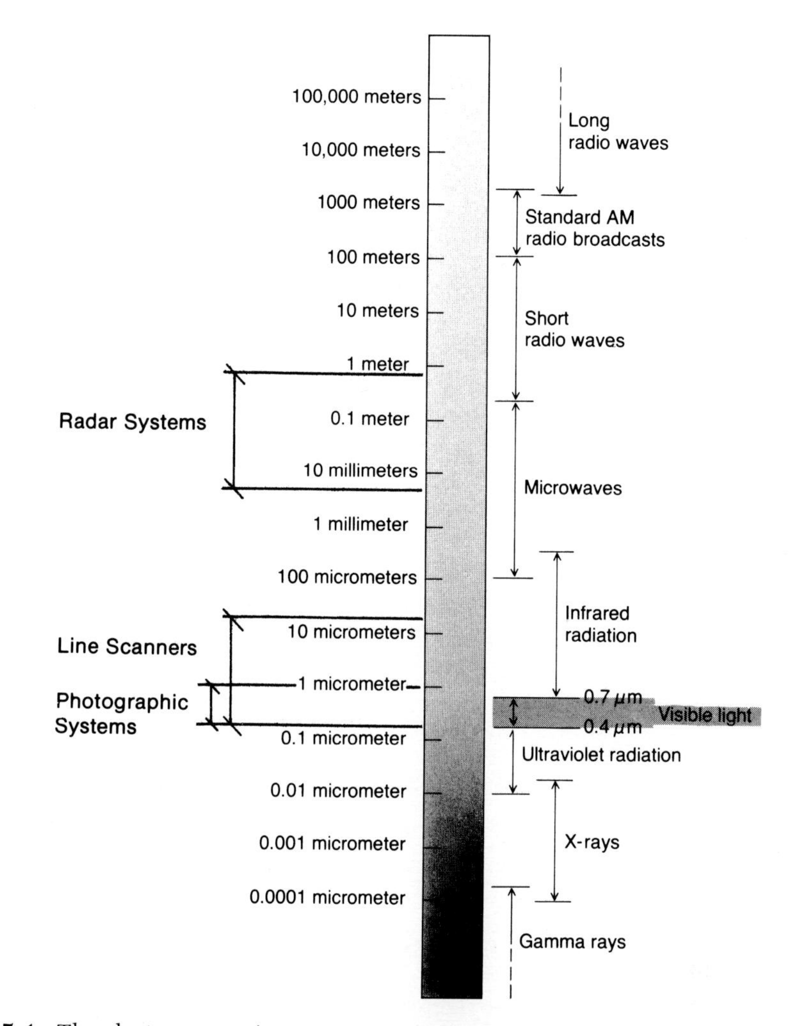

Fig. 15.1 The electromagnetic spectrum including the major categories of radiation and the wavelengths covered by various remote sensing systems. (From W. M. Marsh and J. Dozier, *Landscape: An Introduction to Physical Geography,* © 1981, Addison-Wesley, Reading, Massachusetts. Fig. 2.1. Reprinted with permission.)

duced as electronic impulses and can be stored on magnetic tape which can be transformed into computer-compatible formats as well as imagery formats. As a result, automated image interpretation techniques have been developed in which computers are used to develop land-use/cover and other thematic maps directly from line-scanner data (Fig. 15.2).

Side-looking airborne radar (SLAR) systems transmit pulses of microwave energy and then record the location of objects on the ground by systematically recording the time it takes for the pulsed energy reflected by that object to return to the radar antenna. Since SLAR systems generate and record their own "illumination" or signal, they can acquire imagery at any time of day. Furthermore, since this microwave energy has relatively long wavelengths (.3 to 20 + cm) which readily penetrate clouds, useful imagery of the ground can be acquired from above a layer of clouds. Imagery from side-looking airborne radar systems has been used to produce the first accurate maps of portions of the earth which are typically shrouded by clouds.

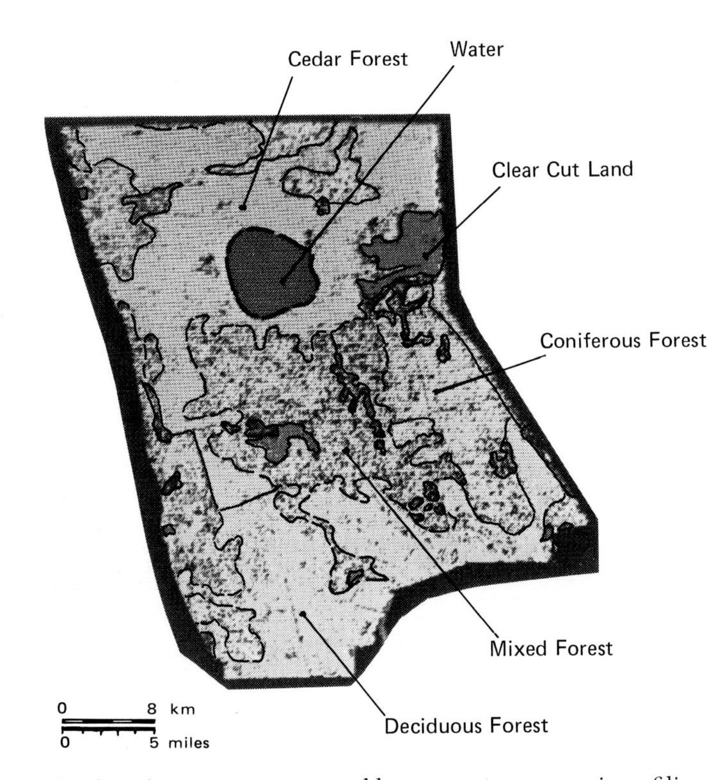

Fig. 15.2 A land use/cover map generated by computer processing of line scanner data. (Courtesy of NASA.)

SOME TECHNIQUES OF IMAGE INTERPRETATION

Aerial photographs and other remote sensing images usually provide a highly detailed record of some portion of the landscape. Such detail may evoke the question, "How do I begin?" or "What do I look for first?" While no single starting point or approach may be best for all image interpretation projects, one must be guided mainly by the objectives of the problem or project. Initially the environmental planner or landscape architect might find it helpful to scan the image to get a general overview of the landscape itself before more precise or site-specific analyses are undertaken.

Even before the image interpretation process begins, one should note the characteristics of the imagery such as the date and year, the scale, and the characteristics of the image system used. Remember, the way the landscape is portrayed in the image is determined mainly by these factors. Also, take advantage of any supplementary sources of information that may aid in the interpretation task. Topographic maps produced by the U.S. Geological Survey are available for most areas. They facilitate interpretation by providing baseline information such as road locations, stream courses, spot elevations, topographic contours, and significant land survey boundaries in a planimetrically correct format. Standard soil maps prepared by the Soil Conservation Service are also helpful. Other thematic maps that show land use, vegetation, or watershed information also may be available locally. Finally, other types of remote sensing imagery of the area may be used to augment or corroborate interpretation. Even imagery from very high altitudes or from space can support the interpretation effort by showing the study area in its surrounding regional context.

THE ELEMENTS OF INTERPRETATION

Aerial photographs and other remote sensing imagery are interpreted by examining and assessing the specific properties or elements of the features that make up the image. Although these elements may not always be consciously considered in the interpretation process, it is the collective assessment of them that leads to interpretation. What are some of the elements of interpretation and how are they used?

Tone or Color There must be an apparent difference in tone or color among objects in order to identify them as individual entities. Because the earth's surface is portrayed in gray tones on black and white imagery, such as panchromatic photographs, it is often difficult to distinguish between surfaces that have only slight reflective differences. For example, areas covered by herbaceous vegetation, barren soil, or concrete may appear in similar tones of gray on images using visible light but in distinctively different tones on images using reflected infrared wavelengths. On a color photo, however, these same surfaces would likely appear in different colors. With thermal infrared systems that record longwave energy emitted from the earth's surface, tone is a function of the radiant temperature of the object. In the preceding example, if the vegetation appeared dark in comparison to the concrete and soil surfaces, it would be radiating less energy because the radiant surface temperature would be cooler. Although tone conveys different information about objects in different remote sensing systems, it often provides the interpreter with more information than any other element.

Tonal variations in an aerial photograph related to soil types, soil moisture, crop types, and land use.

Shape Some features can be identified on aerial imagery by their characteristic shape. Circular fields clearly seen even on imagery obtained from orbiting satellites are diagnostic of center pivot irrigation systems, as are the cloverleaf patterns at some freeway interchanges. While features with geometric shapes are normally associated with human activity, there are unusual natural features that have linear, curvilinear, or polygonal shapes. Even when not diagnostic, shape often can be used along with other elements of interpretation to identify a landscape feature.

Size It is often helpful to compare the relative size (height, area, etc.) of a known feature that can be seen on the image with the size of those to be identified. Remember that some features have standard dimensions that can be used as a basis for size comparison. For example, on large-scale aerial photos, football or soccer fields or tennis courts can sometimes be seen as can railroad tracks which also have a standard width (4.71 feet or 1.46 m). On medium- and small-scale imagery, road systems that follow land survey section lines are one mile or 1.6 km apart. In other cases, the actual size of the feature may have to be calculated using techniques described later in this chapter.

Shadow The shadows cast by objects can be both an advantage and a disadvantage to the image interpreter. On a vertical aerial photograph the shape or silhouette of a vertical feature as seen in its shadow may often be diagnostic. Since shadows' lengths are proportional to the height of the objects casting them, the relative height of objects can quickly be determined. On the other hand, ground detail lying within the shadow may be lost completely or significantly reduced. Shadowed areas in panchromatic infrared aerial photographs normally appear completely black as the shorter wavelengths of diffuse light have been filtered out. On imagery produced by side-looking radar, featureless shadows result when the transmitted radar beam is blocked by buildings or terrain features. Such shadow areas can be quite extensive due to the relatively low angle at which the radar beam strikes the earth's surface and the geometry of radar image formation.

Pattern The repetition of certain features over the landscape may produce a characteristic pattern on remote sensing images. Quite often they reflect human use of the land such as the patterned arrangement of orchards, contour plowing, and strip cropping which typify modern agricultural land use practices. The patchwork of rectangular fields and road networks oriented in the cardinal directions produces a landscape pattern that reflects the grid-like township and range survey system used in many parts of the United States. Drainage lines may reveal distinctive patterns related to geologic controls. On large-scale aerial photographs, gullies and stream channels, the smallest components of drainage networks, are usually detectable, whereas on images acquired from space entire drainage networks may be evident. Similarly, longshore currents can be plotted from satellite images that reveal plume-like patterns of sediment transported along coastlines or into sediment sinks.

Patterns in an agricultural landscape: strip-cropped fields, orchards, tree plantations, and natural forest.

Texture Recognizable patterns apparent on large-scale imagery may dissolve to textural differences at medium- or small-image scales. For example, the row pattern in corn, soybean, or cotton fields may merge to a coarse-textured appearance on a medium-scale aerial image and to a smooth texture on a satellite image. The visual impression of texture is produced when individual features or objects are too small to be clearly discerned on the image.

Like pattern, texture will often provide information on land use, vegetative cover, and plant conditions. On aerial photographs planted vegetation will often appear to have smoother texture than natural vegetation. Actively cultivated farmland will normally have a finer, and more uniform, texture than abandoned fields with their diverse mixture of herbaceous and shrub vegetation. Mature stands of deciduous trees will exhibit a rough, mottled texture while dense stands of younger trees will appear fine-textured. Woodlots in which trees have been selectively cut may also display a very coarse, cobbled appearance due to the fuller crowns of the remaining trees and by their shadows cast on the forest floor or understory.

Site and Association Stereoscopic analysis of medium- to large-scale aerial photographs can provide information that is not detectable in the two-dimensional image of a single photograph. In river valleys, the boundary between valley walls and floodplain can usually be located in a stereomodel, as can the relief and slope characteristics in hilly terrain (see Fig. 8.8 in Chapter 8). As sites vary from uplands to bottomlands, from hilltops to swales, and from gentle to steep slopes, the soil and vegetation assemblages found at these sites will likely change as well.

Many features of the cultural landscape are also site-specific. Certain manufacturing industries and electricity-generating plants are commonly located along navigable waterways, while agricultural practices such as contour plowing and terracing are indicative of sloping terrain. In addition, certain objects are normally found in association with others. An aerial photograph of a college campus would show the complex of facilities that are integral components of this type of institution. Large parking areas and close proximity to major streets are requisites for retail shopping centers.

Resolution In order for a feature to be considered in the interpretation process it must be discernible on the image. The ability of the remote sensing system to resolve fine image detail is referred to as its *resolving power* and it varies considerably between systems. The resolving power in photographic systems depends on the characteristics of the lens, the type of film and filter, the exposure time, and film processing. With cameras and films designed for high-resolution photography very small objects may be seen when examined under a microscope. Although most line-scanning systems in theory have somewhat lower resolving powers than cameras, the electronic output used to produce the imagery can be further processed to improve resolution. Most available SLAR imagery has poor resolution characteristics when compared with photograph and line-scanner imagery, however. Synthetic aperture radar (SAR) systems coupled with sophisticated processing techniques can, however, produce imagery with acceptable ground details.

High-resolution infrared photograph showing the texture contrasts between fields and forest and the sharp boundary along a water feature (upper right).

In order for even a sizeable feature to be seen on a remote sensing image it must appear in a different tone or color than its background or adjacent objects. On a normal panchromatic or color photograph the exact shoreline of a lake might not be apparent due to a lack of tonal difference between the shallow water and beach. A similar lack of resolution may exist between coniferous and deciduous trees growing in a mixed stand because of their similar reflectance characteristics. However, both the shoreline and the different tree species would be apparent on either a color infrared or panchromatic photograph because of sharp tonal differences between the water and land and the coniferous and deciduous trees.

The solution of most interpretation problems relies on the use of several elements of interpretation with each providing evidence required to solve the problem. Remember that accurate interpretation involves the synthesis of evidence drawn from the careful and systematic study of the remote sensing image itself, coupled with the use of supplementary information sources such as topographic maps and soils data.

STEREOSCOPIC VIEWING

When a pair of overlapping vertical aerial photographs is viewed stereoscopically, the landscape appears in a three-dimensional perspective. Such a view often facilitates more complete and accurate interpretation than can be made from individual photographs of the same area. For example, the subtle undulation of gently rolling topography not apparent on single photographs becomes obvious when viewed stereoscopically. Uplands are readily distinguished from lowlands, as are tall buildings from low and trees from shrubs on the stereoscopic model.

Such a perspective is achieved by simultaneously viewing two photographs taken from slightly different camera positions but which show the same ground features. As a result, each eye is viewing the same scene from the position of the aerial camera when each of the two photographs making up the stereo model was acquired. When this model is viewed through a simple instrument, called a *stereoscope* (which allows each eye to focus independently on each component photograph), a three-dimensional perspective of the terrain results.

CONSTRUCTION OF THE STEREOSCOPIC MODEL

Because aerial photographs are usually acquired so that each photo overlaps the next by 60 to 70 percent, a stereoscopic model can be produced from successive photographs in a flight line. This is accomplished by locating the center of each photograph, called the *principal point.* The principal point (PP) is located at the intersection of straight lines drawn from the fiducial marks on opposite sides of the photo. The fiducial marks are printed on the margins of the photograph (Fig. 15.3).

After locating and marking the principal points on both photos (PP_1, PP_2) with a pin prick, transfer and mark the position of each principal

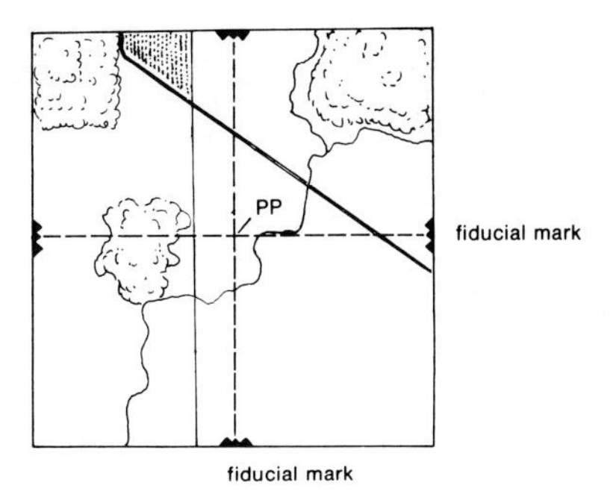

Fig. 15.3 The features of a standard aerial photo necessary for building a stereoscopic model; the principal point (PP) is located between the fiducial marks.

point on the other photographs. These are called the *conjugate principal points* (CPP). Next, draw a straight line between the CPP and PP on each photograph. Each line represents the aircraft flight line between photo exposures, and is known as the *air base.*

Having located the principal and conjugate principal points, the two photographs can be arranged for stereoscopic viewing (Fig. 15.4).

1. Orient the two photographs on the viewing surface so that the aircraft flight line and the long axis of the stereoscope are parallel and so shadows on the photographs fall toward the viewer.
2. After taping down the edges of one photograph, place the second photograph so that the corresponding CPP is about 60 mm (2.3 in) from the PP of the first photo and secure it.
3. With the lenses of the stereoscope about 60 mm apart, place the stereoscope so that the left lens is over the left photo and the right lens is over the right photo, ensuring that the corresponding images on each photo will be viewed separately.
4. Looking through the stereoscope, slightly adjust its position until the stereoscopic effect is achieved. Remember that the stereoscope must be oriented so that its long axis is parallel to the flight line in order to see in stereo.

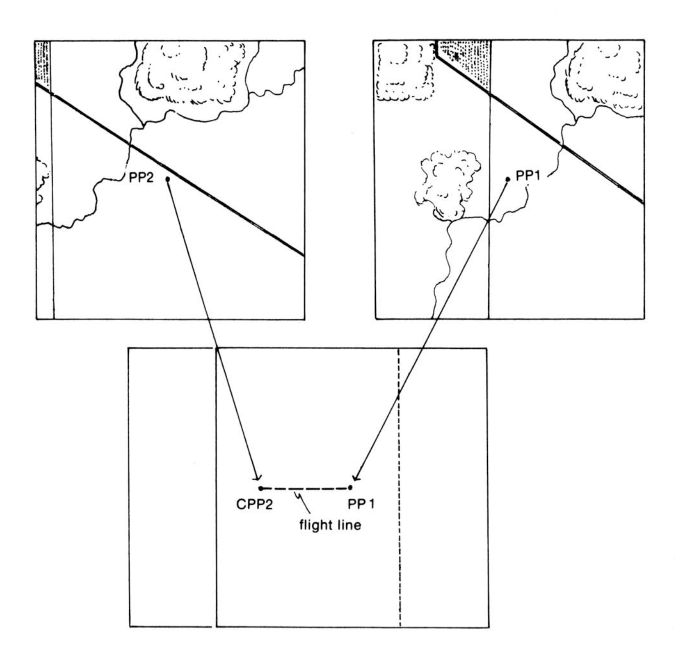

Fig. 15.4 The basic stereoscopic model in which two successive photographs in the flight line are arranged with the principal points about 60 mm apart.

The perspective of the stereo model may seem somewhat abnormal as elevation changes and objects' height appear to be exaggerated. While at first this characteristic may be somewhat misleading, it soon becomes useful not only in discerning slight variations in height or elevation, but in the overall interpretation process. The stereo model is also used to make accurate elevation measurements and is the basic tool for the construction of large- and medium-scale topographic contour maps.

MAKING MEASUREMENTS ON AERIAL PHOTOGRAPHS

It is relatively easy to obtain rather accurate measurements of landscape features and other objects from vertical aerial photographs. Indeed in the photo interpretation process it is usually essential that quantitative data are generated as a part of the description or analysis.

Scale The scale of an aerial photograph must be known if measurements such as distance, area, size, volume, or height are to be computed. Scale is the relationship between distance on a photograph and the corresponding distance on the earth's surface. Because it is so important, a scale may be stamped on the photograph. However, when this scale is not adequate or not known, one may calculate a scale based on a known ground distance that is visible on the photo. This scale, termed a *representative fraction*, is determined by using the relationship:

$$RF = \frac{PD}{GD}$$

where:

RF = representative fraction
PD = photo distance
GD = ground distance

To compute the representative fraction, a feature of a known size (ground distance) must be identified on the photograph and the corresponding photo distance measured on the photo. For example, if the ground distance between two points was known to be 500 meters and the photo distance was 5 centimeters, the photo scale would be:

$$RF = \frac{5 \text{ cm}}{500 \text{ m}} = \frac{5 \text{ cm}}{50{,}000 \text{ cm}} = \frac{1}{10{,}000}$$

This scale indicates that 1 unit of distance (usually centimeters or inches) on the photograph represents 10,000 of the same units on the ground.

In the preceding example, a ground distance was provided. However, if no actual ground distance were known, how could scale be determined? Two methods can often be used. In many urban and suburban areas there are sports facilities, such as football and soccer fields, tennis courts, and baseball diamonds, that have at least some standard dimensions. Often these fields can be seen on aerial photographs and if the standard ground distance can be measured accurately, photo scale can be determined. When no such feature-known dimensions can be seen, a topographic map of the area can be used to calculate a ground distance. The distances between two points that show on both the map and the photograph can be measured.

Since the map scale (RF_{map}) is given, the map distance is converted into the ground distance ($GD = MD \div RF_{map}$) and the photo scale can be obtained.

There is still another method to determine the scale of a vertical aerial photograph that depends on the geometric relationship between the sides of a triangle formed by the earth's surface within the camera's field of view, the height of the aircraft above the ground, and a similar triangle formed by the image plane of the camera and its focal length (Fig. 15.5).

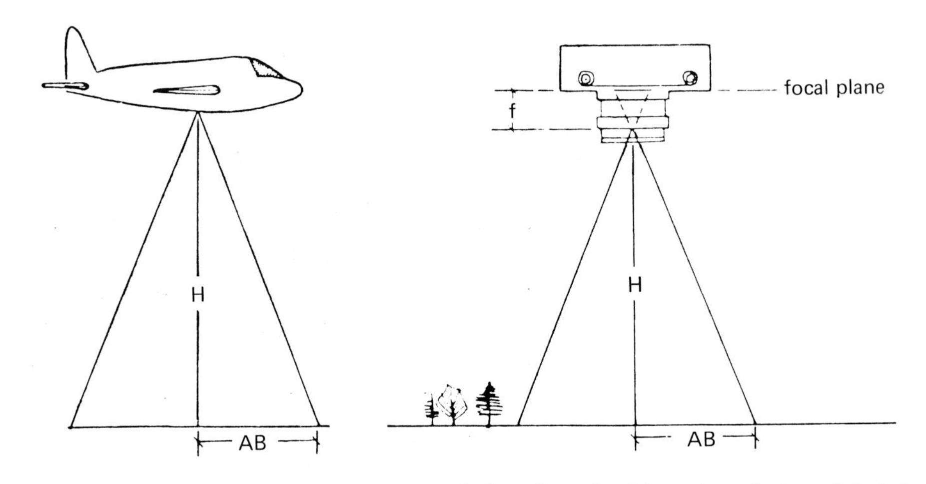

Fig. 15.5 Model for determining photo scale based on focal length and aircraft height.

Because sides of similar triangles are proportional, if the focal length (f) of the camera and the height of the camera in the photo aircraft (H) are known, then a similar proportional relationship would apply for given photo distance (ab) and ground distance (AB).

$$\text{Since } \frac{f}{H} = \frac{ab}{AB} \text{ and } \frac{ab}{AB} = \frac{PD}{GD} \text{ then } RF = \frac{f}{H}$$

Because focal length and altitude data are often provided with the aerial photography, this method affords a simple way to calculate the nominal scale of an aerial photograph. For example, if the focal length of a camera lens was 150 mm and the height of the photo aircraft was 1000 meters above the ground, the photo scale would be:

$$RF = \frac{150\text{ mm}}{1500\text{ m}} = \frac{0.15\text{ m}}{1000\text{ m}} = \frac{1}{6667};$$

that is, 1 m on the photograph is proportional to 6667 m on the ground. It should be noted that this method of scale determination assumes that the ground surface is flat. Hilly terrain would result in scale variation over the photograph. On photographs taken at relatively low altitudes significant scale differences can occur due to local relief. If the local topographic relief varied by 100 meters on the photo in the example used previously, the photo aircraft would be 950 meters to 1050 meters above the ground. This would result in scales ranging from 1/6333 to 1/7000.

Distance Distance calculations can easily be made on vertical aerial photographs once the photo scale is known. If the unknown distance can be measured on the aerial photograph, it can be converted into a ground distance as $GD = PD \div RF$ (photo). This approach yields accurate results where the terrain is relatively flat, but it becomes less accurate when measured on photographs where significant elevation differences occur.

Area The area occupied by landscape features or any other objects with regular geometric shapes can be easily calculated on vertical aerial photographs by measuring the dimensions on the photograph, applying the appropriate mathematical formula, and converting the photo area to ground area. The area of rectangular woodlot measuring 5 cm by 10 cm on a 1:5000 photograph would cover 50 cm^2 or 0.005 m^2 on the photograph; this would represent a ground area:

$$GA = PA \div (RF)^2$$

$$= 0.005 \div \left(\frac{1}{5000}\right)^2$$

$$= 125{,}000 \text{ m}^2 \text{ or } 12.5 \text{ hectares}$$

An effective method for determining the area of irregularly shaped features employs a grid of uniformly spaced dots on a transparent overlay. The area represented by each dot is determined by the density of the dots and the scale of the photo. On a grid with 25 dots/cm^2, each dot would be spaced 0.2 cm apart. At a photo scale of 1:5000, the area represented by each dot would be:

$$\frac{PA}{\text{dot}} = (0.2 \text{ cm})^2 = 0.04 \text{ cm}^2 \text{ or } 0.000004 \text{ m}^2$$

$$\frac{GA}{\text{dot}} = PA \div (RF)^2 = 0.04 \text{ cm}^2 \times (2.5 \times 10^6) = \underline{1{,}000{,}000} \text{ cm}^2 \text{ or } \underline{100} \text{ m}^2 \text{ or } 0.1 \text{ ha}$$

By placing the dot grid over the feature, the number of dots lying entirely within the feature and one-half of those that fall on the boundary are counted. The ground area is computed by multiplying the total number of dots by the area value of each dot.

This method is attractive because it is simple, quick, and requires no specialized equipment. The accuracy of this method increases with larger areas and with higher dot densities. As with scale and distance measurements, errors will develop when this method is used on photographs with significant local relief.

Height Determination On medium- and large-scale vertical aerial photographs the height of a building, tree, or other vertical feature can be determined using several different methods. One method is based on the characteristic photo displacement of vertical objects because of their height. The geometric relationships between the height of the object and the amount of displacement on a vertical aerial photograph can be seen in

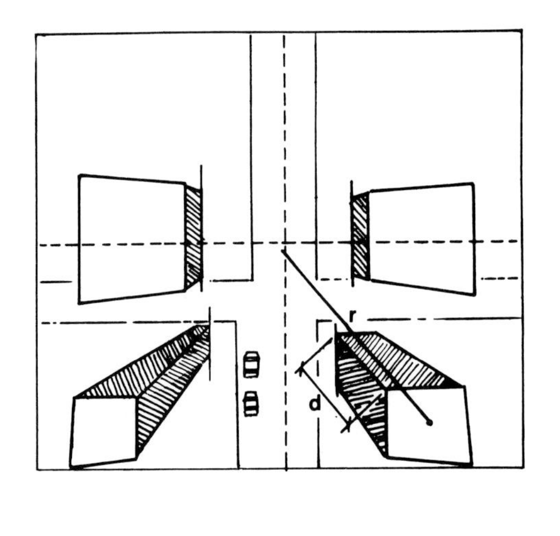

d = vertical displacement
r = radial distance from top of object to principal point

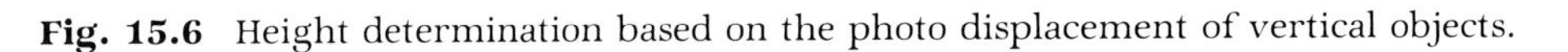

Fig. 15.6 Height determination based on the photo displacement of vertical objects.

Fig. 15.6. Notice that photo displacement is proportional to height of the object and the distance of the object from the center of the photograph (the principal point). Furthermore, notice the direction of displacement is radial from the principal point. As a result, the height of the object can be calculated with the following information:

1. By measuring on the photograph the displacement of the vertical feature from top to base along a line radial from the principal point. (d)
2. By measuring the radial distance from the principal point to the top of the object. (r)
3. By determining the height of the camera above the base of the object. (H)

Using the following formula, the height of the object (h) can be computed:

$$h = H\frac{d}{r}$$

Other methods for determining the height of objects involve the use of a stereometer. For a description of these techniques, refer to a standard book on aerial photograph interpretation, such as Avery (1977).

CASE STUDY

Documenting Landscape Change Using Aerial Photographs

Roy Klopcic

Documenting change in the landscape has been facilitated greatly by the use of vertical aerial photographs showing the same area at different time periods. In some instances, it is actually possible to make various measurements on the amount and rate of change related, for example, to shore erosion, sand dune development, and damage caused by storms. However, the vertical aerial photograph is not without its problems; among them is an inherent geometric distortion caused by the photograph's single point perspective. Additionally, every aerial photograph, due to a variety of factors, has variation in scale. Failure to recognize and accommodate these factors when making measurements on aerial photographs can lead to erroneous answers and misleading conclusions.

One type of scale problem occurs in aerial photographs of terrain comprised of surfaces at different elevations. Those surfaces closer to the camera appear at larger scales, whereas those farther away appear at smaller scales. This is often encountered in measuring shorelines because of the elevation differences between the top of a backshore slope or bluff and the beach near the toe of the slope. A simple solution to this problem involves making two sets of measurements: one for the lower datum and one for the upper datum. This requires that the photo scale used be specific to the datum where measurements are to be made.

To illustrate this measurement technique, a location was chosen on the Lake Michigan shoreline in an area that has experienced severe shore erosion with considerable loss of and damage to property. The actual measurements are applied to three specific sites identified on aerial photographs flown on four separate dates; April 1962, April 1969, November 1972, and March 1977. The coastal zone in this area is comprised of two distinct topographic levels: one represented by the lake and the active shore and the other by an upper, terrace-like surface 40 to 60 feet above lake level. The surfaces are separated by a steep bluff which is retreating because of wave erosion at its toe.

Because land use development is limited to the upper surface, measurements could be made by scaling the distance between the location of buildings and the edge of the bluff. The position of the shoreline could be extrapolated by constructing a model (profile) of the bluff at the appropriate angle and then measuring the distance to the shoreline on the horizontal axis of the model.

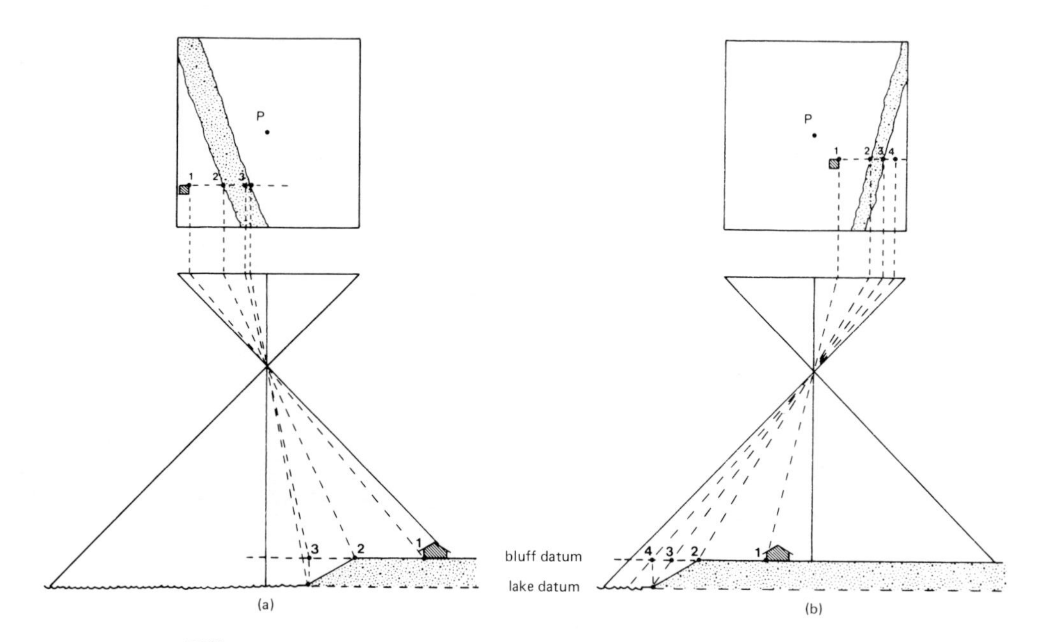

Differences in the apparent locations of the shoreline and bluff using (a) lake datum and (b) bluff datum elevations.

(cont.)

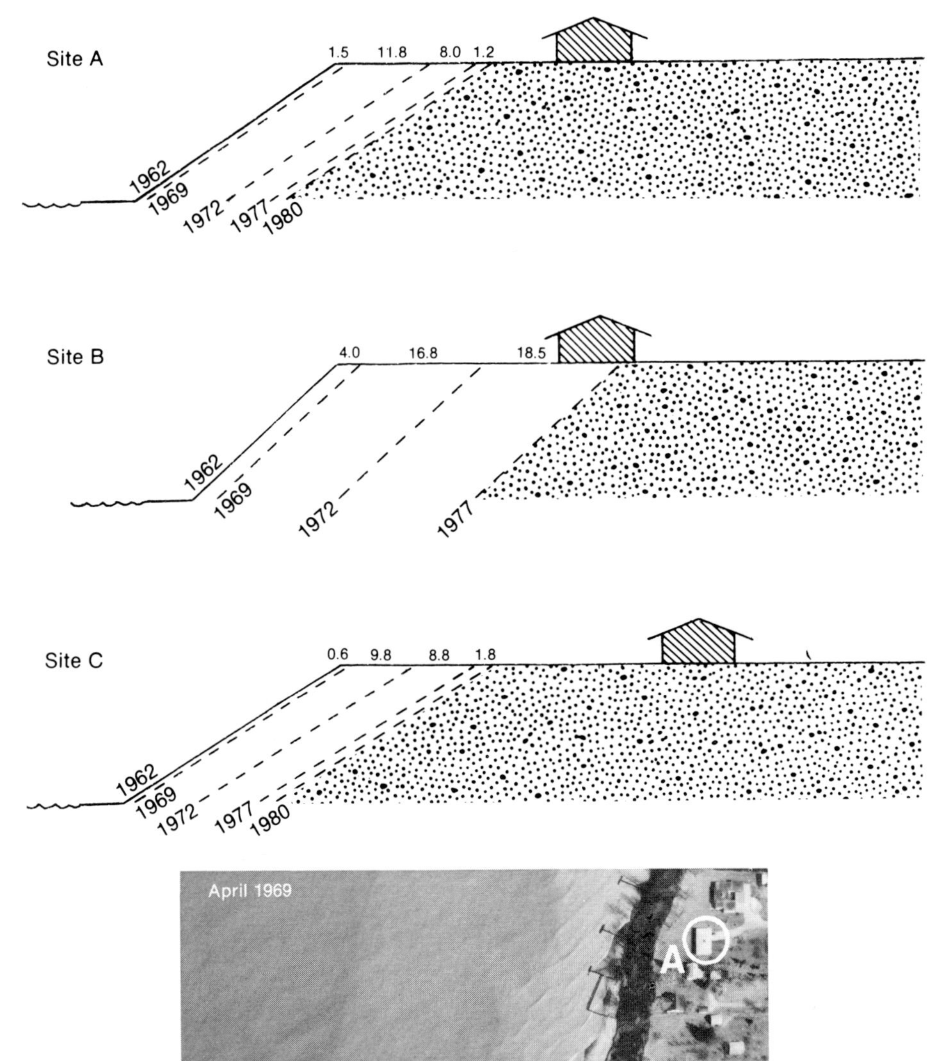

Pattern of bluff retreat from 1962 to 1977 (Site B) and 1980 (Sites A and C) based on measurements from aerial photographs. The accompanying aerial photograph shows the three sites (houses circled) in 1969.

For major mapping projects, such as those undertaken by the U.S. Geological Survey or by private aerial mapping firms under contract to a planning agency, special machines are used to overcome distortion problems and improve the efficiency of interpretation. Stereoplotters and transfer scopes are commonly employed, not only to correct the distortion but remove scale variations by adjusting each photograph to fit an accurate planimetric base such as a topographic map. Consequently, it is a relatively routine matter to reconstruct whole coastal environments and make the desired measurements of the changes in selected features as they appear among photographs of different dates. The instruments involved in accomplishing this task are, however, expensive and do require special training if accurate results are expected.

Finally, it is necessary when using aerial photographs, or any sort of landscape sampling or measurement procedure for that matter, to realize

An example of the rate of change that is possible in coastal environments. The March 1977 photograph shows sand being placed on the shore in an attempt to rebuild the beach; by April 1978 most of the sand has been washed away. Such change is not detectable unless monitoring is done frequently.

(cont.)

CASE STUDY (cont.)

that environmental change may be episodic, rather than continuous, and may thus escape detection because it falls between the sampling (overflight) dates. This is most likely to occur in river valleys and coastal zones where each event in a series of events produces change opposite that of its predecessor; for example, erosion, deposition, and erosion. The net change may be negligible while the gross change is great. If the frequency of overflights is such that photographs are taken only after the first and third events, then not only does it *appear* that no or little change has taken place, but an entire event or series of events crucial to understanding the behavior of the landscape has been masked out.

Roy Klopcic is a geographer at Central Michigan University and a specialist in remote sensing and natural resources.

SOME CHARACTERISTICS OF AERIAL FILMS AND FILTERS

Several types of photographic film have been produced specifically for aerial photography. As these films have been designed for certain applications, it should be useful to briefly mention some of the characteristics of each. Panchromatic films, which have been available for aerial photography for decades, continue to be widely used in photo interpretation and mapping work. Most of these films are sensitive to both ultraviolet and visible light (0.3 to 0.7 μm), and panchromatic infrared films have extended sensitivity into the reflected infrared portion of the spectrum (0.3 to 0.9 μm) (see Fig. 15.1). These films are sometimes referred to as black-and-white films because they depict reflected light in gray tones.

For panchromatic film sensitive to reflected infrared wavelengths, a filter is used to block visible light while transmitting the infrared to the film. The resultant infrared photographs often appear quite different than photographs taken with visible light. Water features appear black on a true infrared photo because of the absorption of incoming infrared radiation at the water's surface, whereas water may appear in a wide range of gray tones on a normal panchromatic photo. Healthy, leaf-covered vegetation, which appears dark gray on normal panchromatic photos, is light-toned on the infrared photograph because it reflects a relatively high percentage of near infrared wavelengths. Although both deciduous and coniferous species are good reflectors in the infrared, the foliage of deciduous trees is somewhat more reflective than that of conifers. As a result, the deciduous trees usually appear in lighter tones than the coniferous species on an infrared photograph (Fig. 15.7).

In recent years color aerial films have been more widely accepted and used for aerial photography. Normal color film is sensitive to nearly the same spectral range as panchromatic film; however, color film contains three emulsion layers, each of which records only one primary color. When the film is developed, dyes within the emulsion layers combine to replicate the original colors reflected by the photographed object. Al-

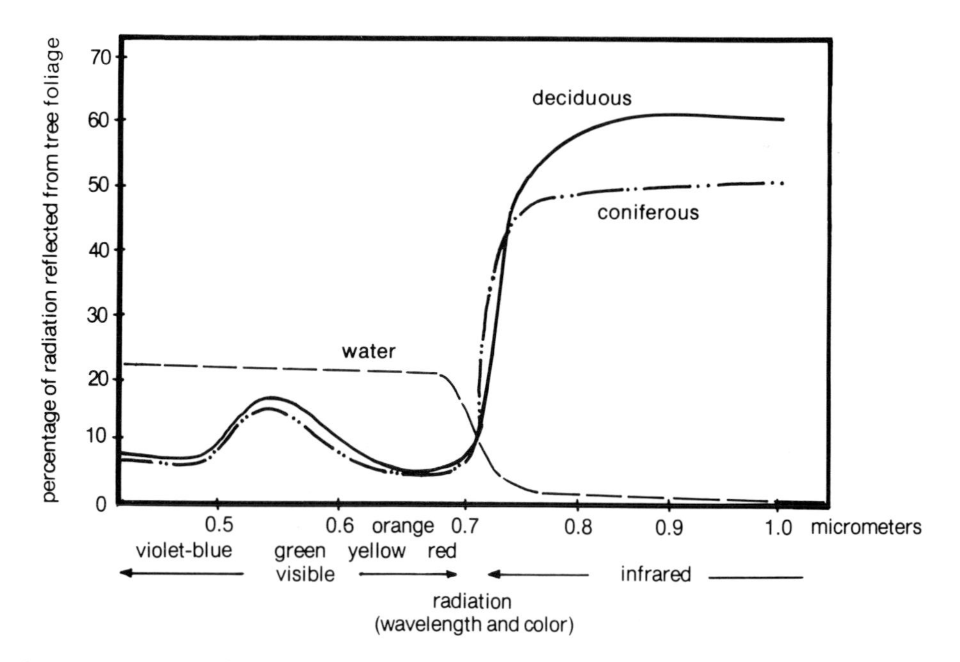

Fig. 15.7 Some of the differences in panchromatic infrared photographs with respect to water and vegetation.

though color films can produce excellent quality aerial photographs, they have lower resolution than panchromatic films because they are sensitive to scattered skylight and haze. Scattered light reduces both photo resolution and contrast; therefore, the utility of color film may be somewhat limited, particularly for space and high-altitude photography.

In color infrared film the spectral sensitivity of the film can be extended to record reflected infrared up to 0.9 μm (see Fig. 15.1). This is done in film manufacturing by replacing the blue-sensitive layer in normal color film with infrared sensitive materials. Since infrared radiation is invisible to our eyes, it must be assigned a color by the manufacturer. As a result, color infrared film portrays the environment in false colors so that objects reflecting a high proportion of infrared radiation appear in a wide range of unfamiliar hues. Although the blue-sensitive emulsion layer has been removed, the other layers (green, red, and infrared) all have some sensitivity to blue light. For this reason a "minus-blue" filter is normally used with this film to eliminate all blue light from reacting with the other emulsion layers. Although a color infrared photograph renders the earth in strange colors, it does significantly reduce the problems with scattered light that are inherent in normal color photography.

Color infrared photography is used for similar applications as panchromatic infrared photography such as forest classification and surface water and wetland detection. In addition, color infrared photography is used to monitor the outbreak and spread of disease in agricultural crops and forests. Plant stress induced by disease manifests itself quickly through reduced infrared reflectance which can be detected on color infrared film before any visual signs of stress are apparent. Such pre-visual detection of plant stress may sometimes point up conditions that can be corrected or contained before large losses occur.

LINE-SCANNER IMAGERY

Among the most important technological advances in remote sensing have been the development of image-forming line-scanning systems. Not only can multispectral scanners simultaneously produce images in several discrete bands of the electromagnetic spectrum, but thermal infrared scanners can produce images by recording longwave infrared radiation from the earth's surface. Furthermore, as the scanner systematically collects energy from the ground an image can be displayed instantaneously aboard the aircraft or spacecraft and can be transmitted by microwave (radio) signal to receiving stations on the ground, or it can be stored on magnetic tape for later use. This flexibility in data storage, transmission, and image display is further enhanced because the scanner data can be processed by computer. Because of this characteristic, a computer can be programmed to interpret, classify, and display certain landscape features directly from the energy recorded and transformed by the scanning system. For example, the land cover map in Fig. 15.2 was produced from line-scanner data.

Although line-scanner imagery may appear to look something like an aerial photograph, it is produced in a much different manner. Figure 15.8 is a schematic diagram of a typical optical-mechanical line-scanning system. A small rotating mirror systematically scans the ground perpendicular to the heading of the aircraft. Energy from the ground that strikes the scanning mirror passes through an optical system and is focused on one or more detectors, which instantaneously converts the received energy into an electrical impulse. The signals generated by the detectors are electronically amplified and recorded on magnetic tape. Each time the mirror scans across the landscape, a narrow swath or scan line of data is recorded. When the tape is played through a cathode ray tube (television picture tube) or recorded on photographic film from a glow tube, each scan line represents one line on the resulting image. A photographic image is produced by exposing film that is synchronized with each scan line on the glow tube.

THERMAL INFRARED SYSTEMS

Line-scanning systems equipped with special infrared detectors are usually used to record thermal infrared energy at wavelengths from 3 to 5 μm and 8 to 14 μm. Because this energy is being emitted by the earth's surface at all times, thermal infrared imagery can be acquired day or night. The amount of energy radiated by an object in the landscape is a function of its surface temperature and a constant physical property of the object called *emissivity*. As a result, ground features with warmer radiant temperatures emit more infrared energy than those with cooler radiant temperatures. On thermal infrared (TIR) prints, these variations in radiation intensity from different surfaces are usually portrayed so that the surfaces having relatively warm radiant temperatures appear in lighter tones while the cooler

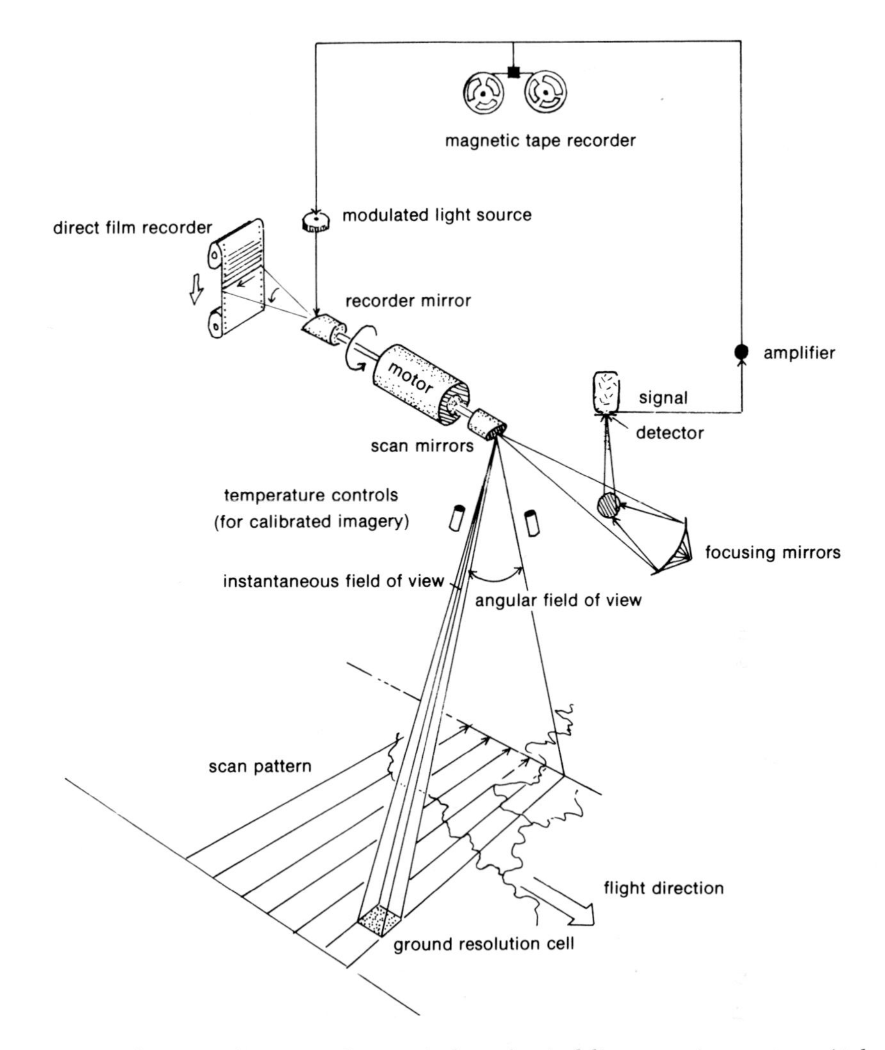

Fig. 15.8 Schematic diagram of an optical-mechanical line-scanning system. (Adapted from *Remote Sensing: Principles and Interpretations* by Floyd F. Sabins, Jr. W. H. Freeman and Company. Copyright © 1978.)

surfaces appear darker. However, on TIR imagery from the meteorology satellites, a negative format is used that shows the cooler surfaces such as high clouds in light tones, in contrast to the warmer surface features that appear dark.

Because the surface temperature of landscape features usually changes considerably during the course of a day, features will often appear in different tones on night imagery than they do on day imagery. For example, the water in Fig. 15.9b appears dark on the day-time TIR image. At night, however, it appears light, indicating that it has a relatively warmer radiant temperature than many of the other surfaces (Fig. 15.9a). This common phenomenon of diurnal tonal change occurs because water tends to heat and cool more slowly than other surface materials. On most TIR imagery the variations in tone indicate only relative radiant temperature differences. However, some TIR line scanners are designed to display actual radiant surface temperatures by assigning various gray tones or colors to

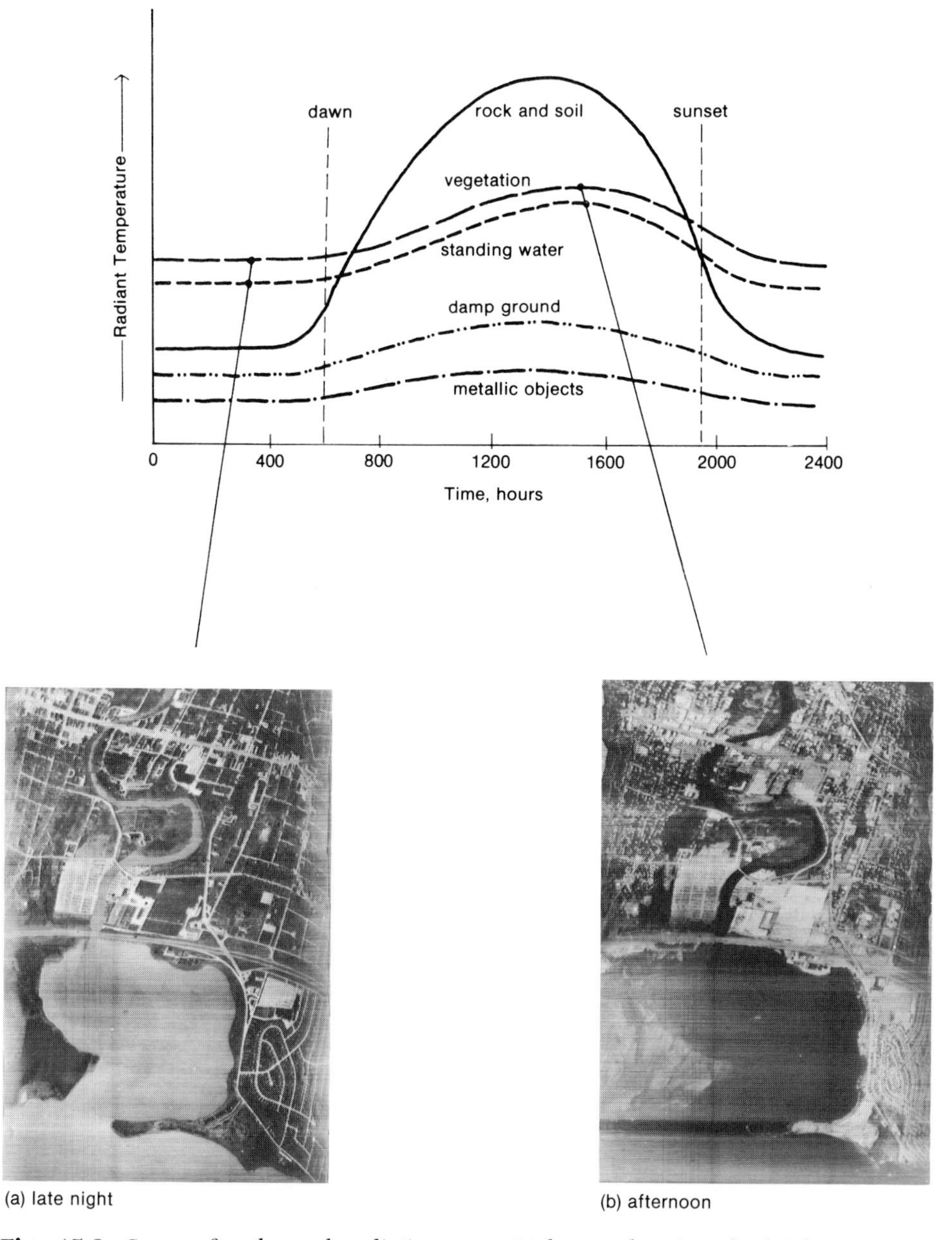

Fig. 15.9 Curves for thermal radiation over 24 hours showing day/night contrasts. (Adapted from *Remote Sensing: Principles and Interpretations* by Floyd F. Sabins, Jr. W. H. Freeman and Company. Copyright © 1978.)

specified ranges of signal intensities that correspond to certain temperatures, which can then be translated into an isothermal map (Fig. 15.10).

Today, thermal infrared line-scanning systems are used in a wide variety of applications. For example, airborne and ground systems are routinely used in water pollution programs and in energy conservation programs to monitor and detect heat loss from buildings. Meteorological satellites have used these systems to provide day and night imagery and other information about world-wide weather phenomena on a continuous basis since the 1960s.

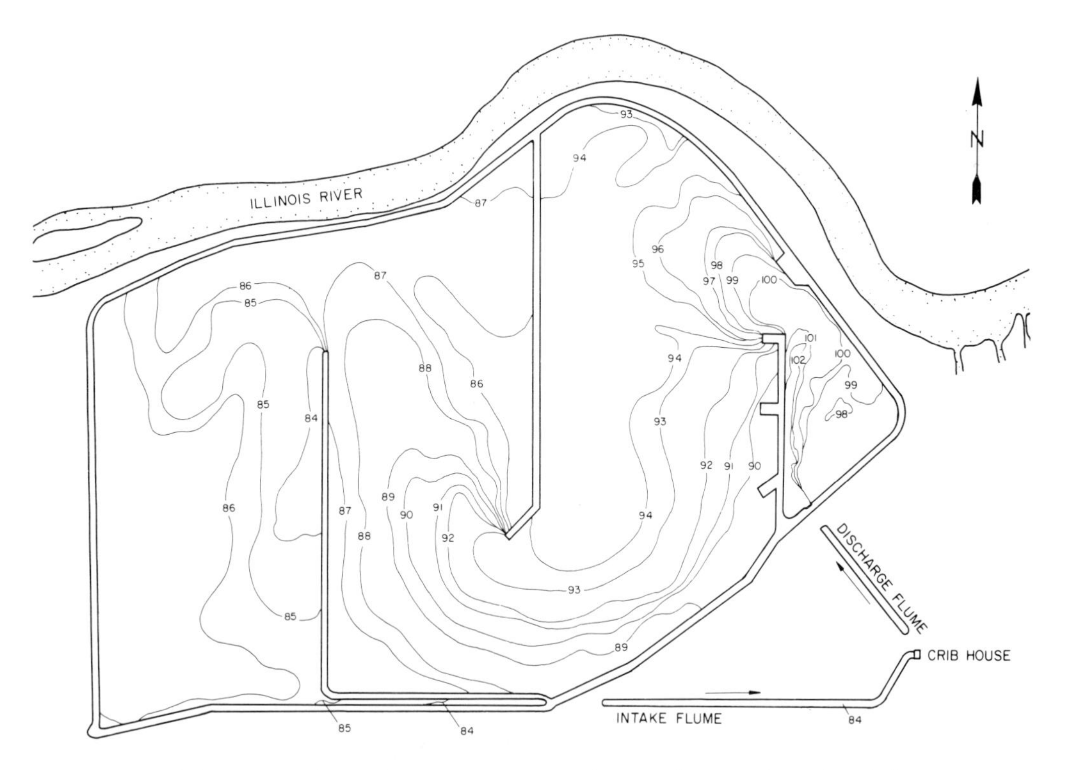

Fig. 15.10 An isothermal map of a power plant cooling pond produced from temperature specific thermal infrared imagery. (Powerton Cooling Pond study performed for Sargent & Lundy Engineers and Commonwealth Edison of Illinois by Daedalus Enterprises, Inc.)

MULTISPECTRAL SCANNING SYSTEMS

As previously mentioned, some line-scanning systems can simultaneously produce images in two or more discrete bands of electromagnetic energy. These systems, called *multispectral scanners* (MSS), can be designed to image from the ultraviolet wavelengths, through the visible and reflected infrared, and on into the thermal infrared part of the spectrum. Figure 15.11 illustrates three bands of imagery obtained from an airborne MSS.

One of the most familiar sources of MSS imagery is produced by the Landsat (formerly ERTS) series of satellites. The MSS system onboard Landsat 1 and 2 received reflected energy from the earth's surface in four wavelength bands. Two bands were in the visible spectrum, green (0.5 to 0.6 μm) and red (0.6 to 0.7 μm), and two recorded reflected infrared energy at 0.7 to 0.8 μm and 0.8 to 1.1 μm. In addition to these four bands, the MSS aboard Landsat 3 launched in 1978 had a thermal infrared channel (10.4 to 12.6 μm). Unfortunately, the thermal channel failed shortly after launch; however, the remaining four channels have continued to operate normally.

Black and white imagery from Landsat, as from any MSS system, can be produced from single bands, or bands can be combined to produce a composite image. Landsat imagery is frequently produced as a false color composite of all four MSS bands which gives the appearance of a color in-

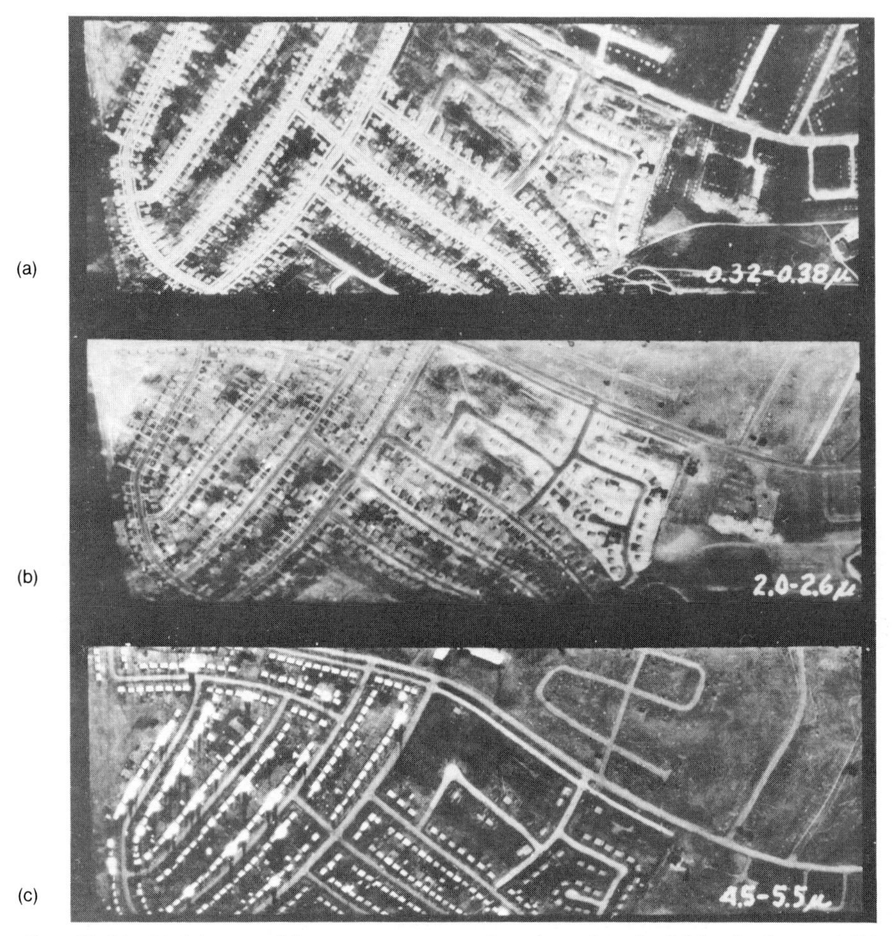

Fig. 15.11 Multispectral imagery representing three bands: (a) in the long visible wavelengths, (b) in the near infrared, and (c) in the thermal infrared. (Courtesy of the Environmental Research Institute of Michigan.)

frared photograph. Figure 15.12a, b, and c shows three bands of a full Landsat scene, which covers an area of more than 34,000 square kilometers (185 km or 115 miles on a side), and Fig. 15.12d is a black and white rendition of a color composite of all four bands.

Although it has relatively low contrast, the green band image usually provides more information about water conditions than the other bands. Notice how longshore currents are manifested by plumes of turbulent water in the coastal waters. The red band is often useful for land use classification because the shapes, tones, textures, and patterns of agricultural land use stand out clearly in contrast to the relatively dark toned, irregularly shaped forest areas. Major transportation arteries can be seen on both the green and red bands. While no details of water features can be seen, the sharp contrast between land and water on the infrared bands facilitates the precise location of the shorelines and the identification of smaller streams, lakes, ponds, and wetlands. Differentiation between deciduous and coniferous forest stands is often possible on the infrared bands due to differences in infrared reflectance characteristics between the two tree types. On Fig. 15.12c the light-toned amorphous areas are predominantly

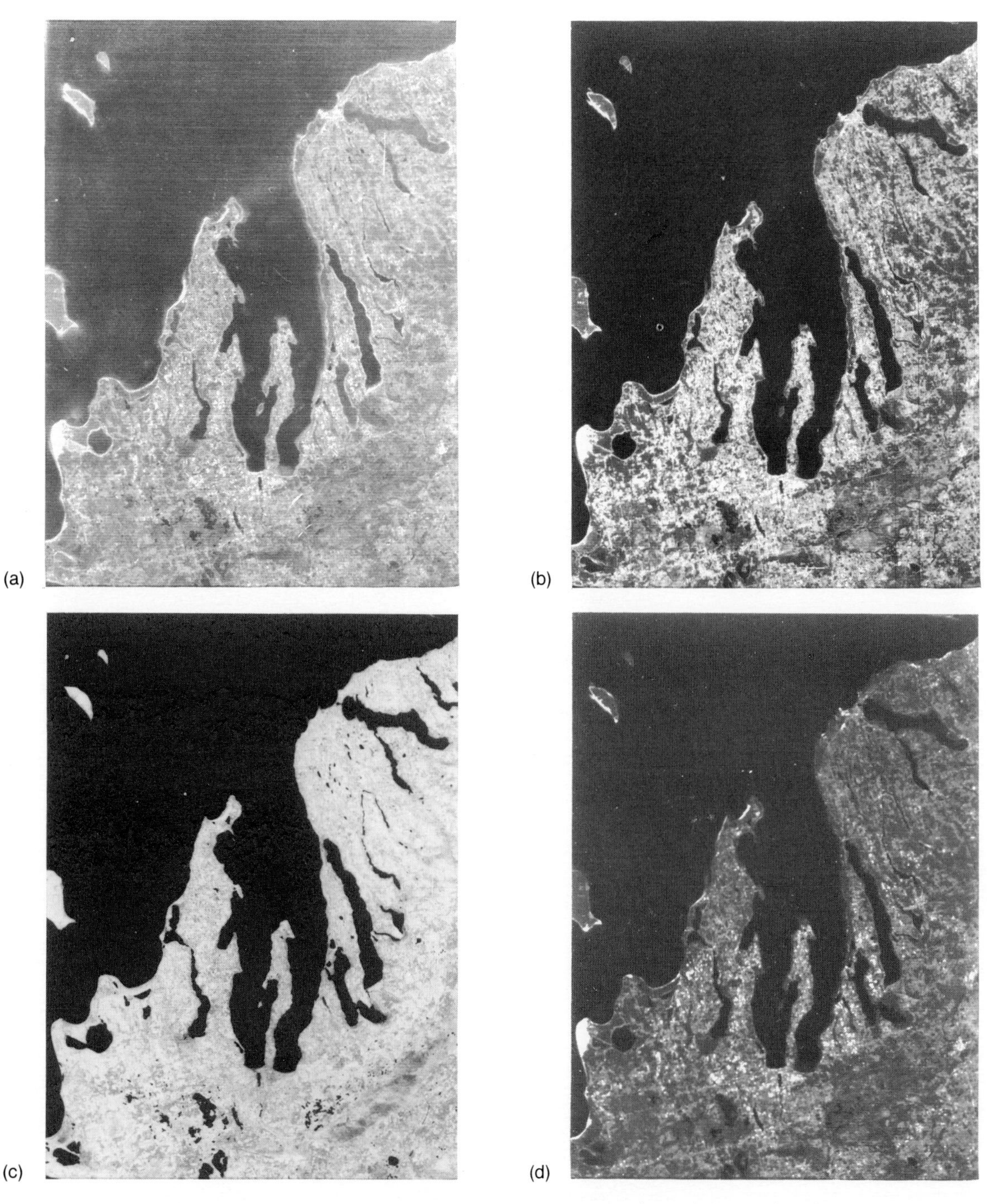

Fig. 15.12 Four LANDSAT images representing three different bands in green (a), red (b), infrared (c), and a composite in (d). (Courtesy of NASA.)

deciduous forests while coniferous stands in the center left of the image appear relatively dark toned.

Line-scanner systems are well suited for use in satellite remote sensing because data gathered by them can be transmitted to receiving stations on the ground. Furthermore, these electronic signals can be readily converted into digital data which make them suitable for computer assisted analysis and interfacing with geographic data from other sources that have been adapted to a computer format.

SIDE-LOOKING AIRBORNE RADAR (SLAR) SYSTEMS

Remote sensing images produced by radar are different in several respects from photographic and line-scanner systems discussed previously. To begin with, radar systems are "active" sensors in that they generate and transmit energy in pulses or bursts to the earth's surface. The portion of this energy that is reflected by landscape features directly back to the radar antenna provides the electronic record from which the imagery is produced. Furthermore, SLAR systems transmit electromagnetic energy in the microwave portion of the spectrum at various frequencies or wavelength bands. Microwave signals are used for radar as well as other communication systems since the earth radiates very little energy at these wavelengths, and because they are transmitted through the atmosphere with minimal interference. Unlike visible light and infrared radiation, radar energy is not significantly scattered, reflected, or absorbed by atmospheric haze, clouds, or even precipitation. As a result, SLAR systems can generate earth imagery day or night in nearly any weather conditions. Finally, the perspective of SLAR imagery is very much different from that of other remote sensing systems as it is acquired from highly oblique angles rather than from the near-vertical perspectives typically afforded by photographic and line-scanner systems.

In order to image the landscape, remote sensing radar systems must "see" the landscape bit by bit at different ranges or distances. This is best done by "looking" down at the landscape from an oblique position. The imagery sequence begins when the SLAR antenna transmits a focused beam of pulsed energy toward a narrow strip of ground. As the speed of this radar pulse is constant, ground features at different ranges or distances from the radar intercept this beam at slightly different times. As a result, the energy reflected back by objects at closer ranges reaches the radar antenna slightly before the signal returns from features at more distant ranges. Because radar systems can precisely measure these tiny time differences in signal returns, the range position can be accurately determined. The energy reflected from each narrow strip of ground is recorded on magnetic tape and the sequence begins again. The radar image is formed as data from adjacent strips (which have seen systematically recorded) are transformed into a video signal through a cathode ray tube and on to photographic film. Since SLAR systems record only that energy that is reflected directly back to the radar antenna, any characteristics of landscape features that influence the quantity and direction of reflected energy can be significant in radar image formation and interpretation. The orientation of the surface with respect to the radar antenna may significantly affect the way a feature appears on the radar image or whether it even appears at all. In hilly mountainous terrain the slopes that face toward the radar antenna will intercept more of the radar beam than those oriented in other directions. In fact, some slopes facing away from the radar may receive no energy at all as they lie in the "radar shadow" of the hill. Not only does an orientation toward the radar make it likely that energy will be reflected back to the radar, but the geometry of radar image formation is such that the foreslopes are compressed and the backslopes are expanded on the radar image.

In landscapes where slopes are not significant features, other surface characteristics affect radar returns. Surfaces such as water that are relatively smooth will result in a specular or mirror-like reflection of the radar energy. Point A in Figure 15.13 shows that these surfaces will not normally return any energy back to the radar. Conversely, rough surfaces produce a diffuse reflection which diverts some energy toward the radar antenna (point B in Fig. 15.13). As a result, surfaces with rough textures usually appear in lighter tones on the radar image than smooth surfaces. One notable exception occurs when two smooth surfaces oriented toward the radar intersect at right angles. Such "corner reflections" result in a specular reflection which is redirected back to the radar (point C in Fig. 15.13).

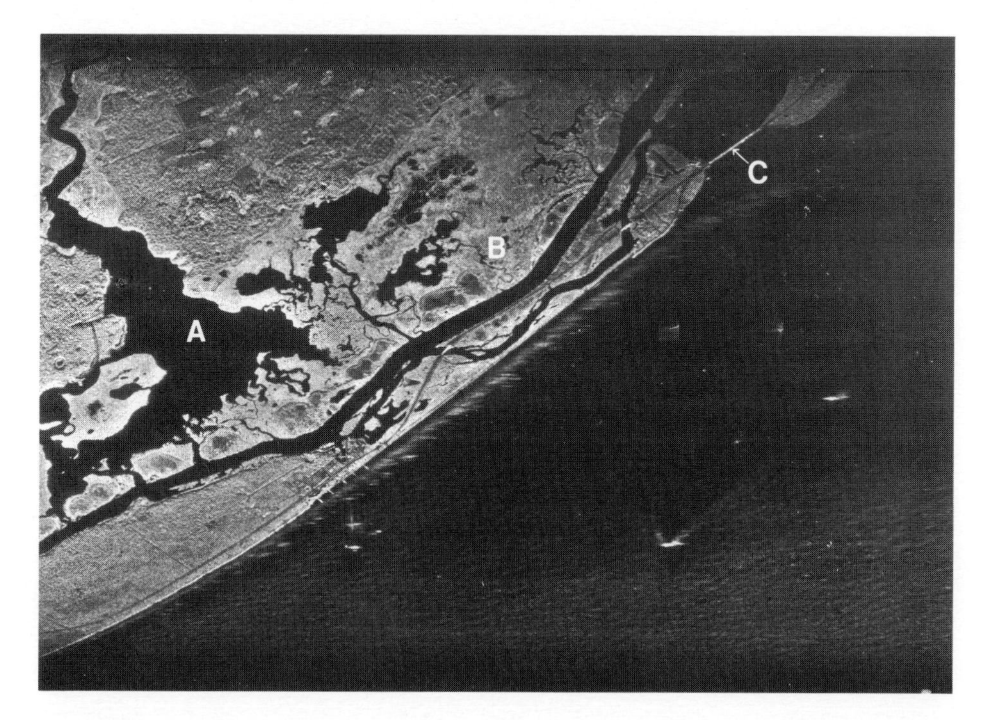

Fig. 15.13 A radar image of a coastal environment showing a water surface with low return (A), a vegetated surface with moderate return (B), and corner surfaces with high returns (C). (Courtesy of the Environmental Research Laboratory of Michigan.)

It should be noted that most surfaces that produce specular reflections on radar imagery produce diffuse reflections when imaged by photographic systems. This occurs because the relatively long radar wavelengths are not affected by the very small surface irregularities which diffusely reflect the much shorter wavelengths of visible light. Furthermore, surfaces that are rough for shorter radar wavelengths, such as X-band, may appear smooth at longer L-band wavelengths (Fig. 15.14a and b).

Yet another characteristic that affects the intensity of return from a surface feature relates to its reflectance properties of incident radar energy. For example, high moisture content in soils and vegetation will normally increase their radar reflectivity. Metal objects such as bridges, buildings, poles, and rails are excellent reflectors, so they will typically appear as bright spots on SLAR imagery. In addition, controlled characteristics of the radar signal itself can affect the appearance of landscape features, includ-

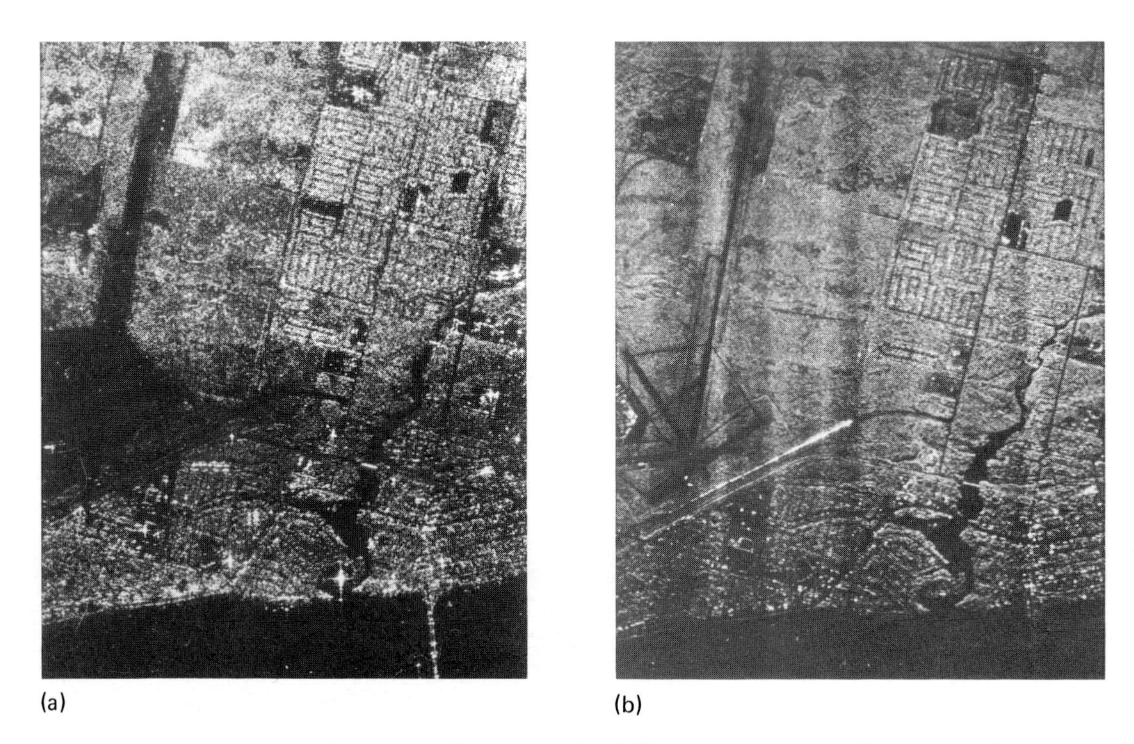

Fig. 15.14 Radar images illustrating the difference in energy reflection as a function of wavelength, (a) being shorter, and (b) being longer. (Courtesy of the Environmental Research Laboratory of Michigan.)

ing the wavelength and the polarization of the electromagnetic energy that the radar transmits and receives. Images that are produced from radar signals that transmit and receive in the same polarization mode such as the H-H (horizontal transmit and horizontal receive) image in Fig. 15.15a bring out certain features while the H-V (horizontal transmit and vertical receive) cross-polarized image accentuates others (Fig. 15.15b).

Relatively small-scale SLAR imagery has been used experimentally to map vegetation and crop types, geologic features (including faults, rock units, and surficial materials), as well as surface drainage patterns. Radar image mosaics have been made for the cloud-shrouded Darien Province in Panama and for much of the Amazon Basin, providing the first detailed look at areas that have never been photographed or mapped before. Although SLAR imagery can be acquired at relatively large scale (Fig. 15.13), it typically is used at small scale (Fig. 15.15) so that extensive areas can be analyzed relatively quickly since it does not have the high resolution characteristics of photographic or line-scanner imagery. Furthermore, SLAR imagery is not routinely collected as part of operational programs of civilian government agencies so that imagery is not widely available. As a result, it is not currently used as much in environmental analysis as other types of remote sensing imagery.

For most applications remote sensing using SLAR systems is still in the experimental stage, although it appears to have great potential for regional scale environmental inventories, preliminary planning, and mapping projects. Not only can large areas be imaged quickly with radar systems, but missions can be flown on a timely basis nearly independent of weather conditions.

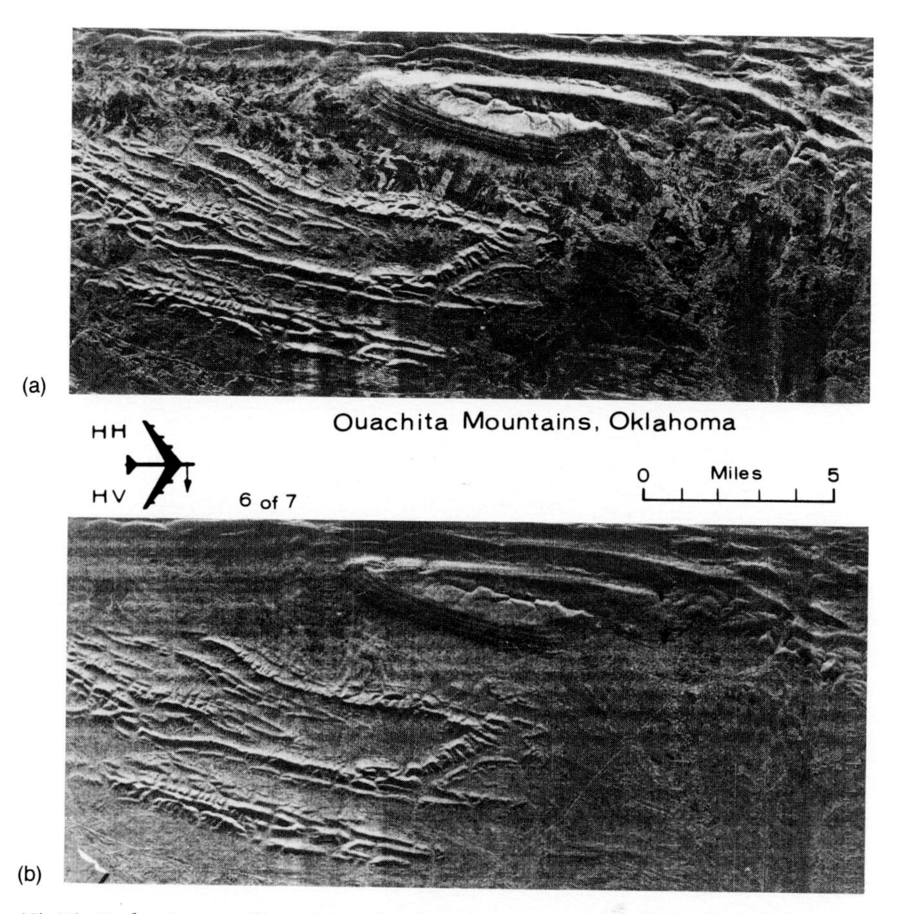

Fig. 15.15 Radar images illustrating the difference in energy reflection as a function of polarization: (a) is H-H (horizontal transmitted and received), and (b) is H-V (horizontal transmitted and vertical received).

SELECTED REFERENCES FOR FURTHER READING

Avery, T. E. *Interpretation of Aerial Photographs.* 3rd ed. Minneapolis: Burgess Publishing Co., 1977. 392 pp.

Bryan, M. L. *Remote Sensing of Earth Resources: A Guide to Information Sources.* Detroit: Gale Information Services, 1979.

Lillesand, T. M., and Kiefer, R. W. *Remote Sensing and Image Interpretation.* New York: John Wiley and Sons, Inc., 1979. 612 pp.

Reeves, R. G., ed. *Manual of Remote Sensing.* Falls Church, Va.: American Society of Photogrammetry, 1975, vol. 1, no. 2. 2144 pp.

Sabins, F. F., Jr. *Remote Sensing Principles and Interpretation.* San Francisco: W. H. Freeman and Co., 1978. 426 pp.

Short, N. M.; Lowman, P. D., Jr.; Freden, S. C.; and Fince, W. A. *Mission to Earth: Landsat Views the World.* Washington, D.C.: National Aeronautics and Space Administration, 1976. 459 pp.

Siegal, B. S., and Gillespie, A. R., eds. *Remote Sensing in Geology.* New York: John Wiley and Sons, Inc., 1980. 702 pp.

Smith, J. T., ed. *Manual of Color Aerial Photography.* Falls Church, Va.: American Society of Photogrammetry, 1968. 550 pp.

Townshend, J. R. G., ed. *Terrain Analysis and Remote Sensing.* Winchester, Ma.: Allen and Urwin, Inc., 1981. 232 pp.

Williams, R. S., and Carter, W. D., eds. *ERTS–1 A New Window on Our Planet.* Geologic Survey Professional Paper 929. Washington, D.C.: U.S. Government Printing Office, 1976. 326 pp.

16

MAPS AND MAP READING

INTRODUCTION

Maps are models, or miniature replicas, of the landscape. As models, they are abstractions of reality and highly selective in the phenomena they portray or represent. They are the most widely used form of graphics in geography and planning, both for recording spatial phenomena and displaying the result of analysis. In order to use maps effectively, one must be familiar with some of their basic properties, such as direction, location, and scale.

DIRECTION: VARIATIONS ON NORTH

In order to be read, a map must be oriented; that is, placed in its correct relation to the earth. This is a simple matter; in essentially all maps, north is at the top of the sheet; south, at the bottom; east, at right; west, at left. Practically all of the maps in this book are oriented with north at the top. The orientation is shown by an arrow or similar symbol pointing north. On some maps two arrows are shown—one pointing to true north, and one to magnetic north; the map should be oriented to true north.

True north represents the straight line direction to the North Pole, whereas magnetic north is the direction of the compass needle as determined by the earth's magnetic field. The locations of the North Pole and the magnetic pole do not agree, the latter being situated in northern Canada; thus there is a deviation between north on a compass and north on a map. This deviation is known as *magnetic declination*, and it is read as degrees east or west of the 0 degree declination, which is the meridian where true north and magnetic north coincide. For the coterminous United States and southern Canada, magnetic declinations range from 0 degrees to 25 degrees east or west. Because the earth's magnetic field shifts somewhat from year to year, accurate surveying and detailed field mapping require that the most recent magnetic declination readings be used. In the United States these are prepared by the National Ocean Survey.

A third arrow, representing *grid north*, will also be shown on certain maps. This arrow defines north according to the grid lines used in different mapping systems. Each mapping system has its own geometric bias that yields a north that is rarely the same as magnetic north or true north (Fig. 16.1).

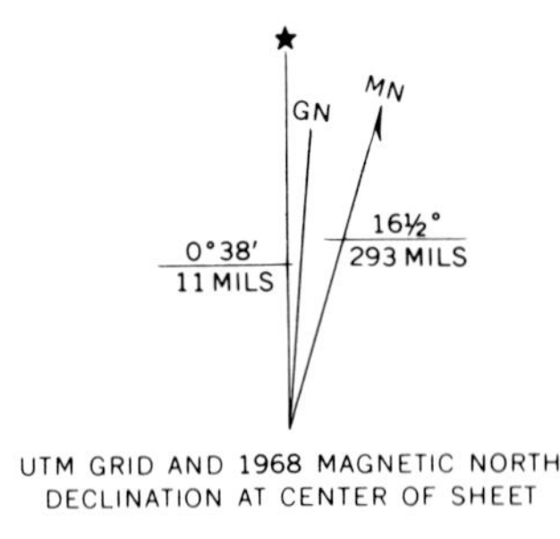

Fig. 16.1 North arrows that appear on U.S. Geological Survey topographic maps; magnetic north, true north, grid north.

LOCATION: THE GLOBAL COORDINATE SYSTEM

Location is a second important consideration in map reading. Several standard location and grid systems are used for the identification of an area covered by a map. The principal geographic coordinate system consists of a rectilinear network of orthogonally intersecting lines. At the global scale this network is comprised of parallels and meridians, which are designated in degrees, minutes, and seconds. Each parallel encircles part or all of the earth, running parallel to the equator. Only one parallel, the equator, traces the full circumference of the earth. Therefore, it qualifies as a *great circle*, defined as the perimeter of any plane that passes through the center of the earth. All of the other parallels are *small circles*, because the planes they define do not pass through the center of the earth, and hence their perimeters represent less than the earth's full circumference.

To construct parallels, the first step is to bisect the globe from pole to pole. Next, a protractor is placed on the plane of bisection, with the base of the protractor aligned with the equator. Starting at the equator, angles northward to the pole are ticked off; the procedure is repeated toward the South Pole (Fig. 16.2a). The angles are then numbered, beginning with 0 degrees at the equator and ending with 90 degrees at each of the poles. Finally, the parallels themselves are drawn by rotating each angle entirely around the earth so that the ray of the angle inscribes a line into the surface of the globe. Thus the latitude of any location on the earth's surface represents an angle between the equatorial plane and a line drawn from the center of the earth to the location (Fig. 16.2b).

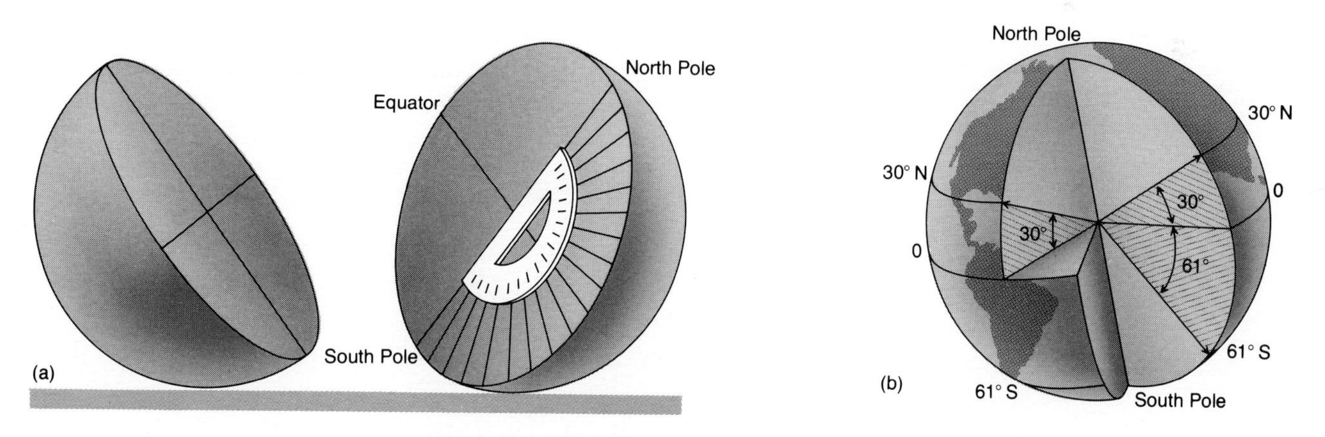

Fig. 16.2 Latitude is measured (and parallels drawn) according to the angle formed between the plane of the equator and any location on the earth's surface; the reading is referenced north or south of the equator. (From W. M. Marsh and J. Dozier, *Landscape: An Introduction to Physical Geography*, © 1981, Addison-Wesley, Reading, Massachusetts. Fig. 7.2. Reprinted with permission.)

The north-south lines, called *meridians*, are constructed in the same fashion as the parallels (Fig. 16.3). The earth is bisected along the equatorial plane, and angles are measured around the perimeter. However, there is a problem as to where to start the system because there is no convenient point at which to place zero. International agreement has specified an arbitrary starting point that coincides with the Royal Observatory at Greenwich, England (Fig. 16.3a). A north-south line drawn through this point to the North Pole and the South Pole is the Greenwich (or Prime) Meridian, and it is labeled 0 degrees longitude. From the Greenwich Meridian all meridians westward to 180 degrees are designated *west longitude*, and those eastward to 180 degrees are designated *east longitude.*

As half of a great circle, every meridian connects the North and South Poles. Since the 0° meridian does not completely encircle the earth, but only half of it, longitudes can vary from 0° to 180° east and west. The line at 180°E coincides with the one at 180°W, both being half-way around the globe, in opposite directions, from the Greenwich Meridian; the direction indication for 180° is thus omitted (Fig. 16.3b).

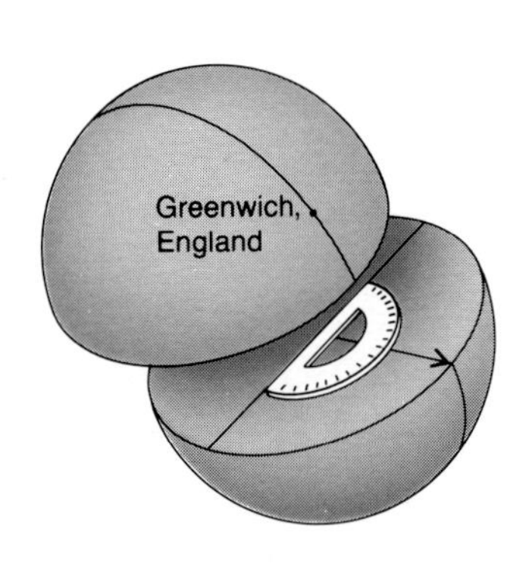

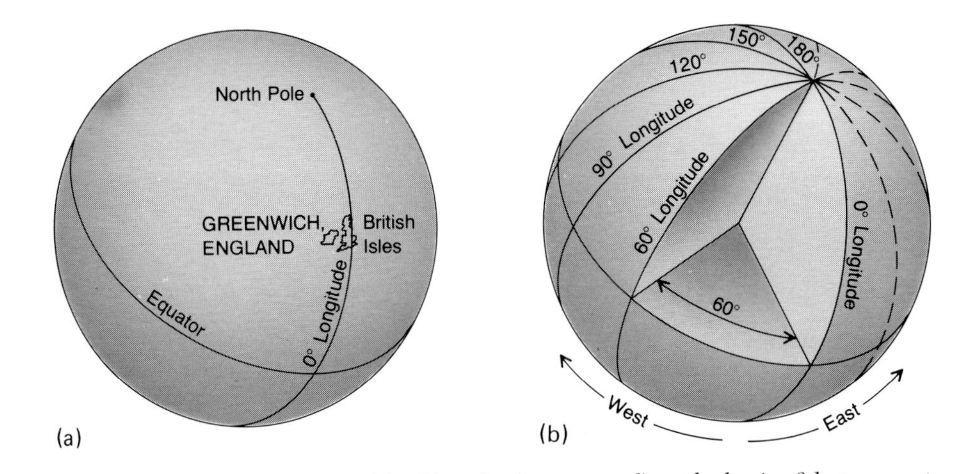

Fig. 16.3 Meridians are constructed (and longitude measured) on the basis of degrees east or west of the Prime Meridian, which itself is referenced as 0° longitude. All meridians converge at the north and south poles and thus follow a true north-south alignment. (From W. M. Marsh and J. Dozier, *Landscape: An Introduction to Physical Geography*, © 1981, Addison-Wesley, Reading, Massachusetts. Fig. 7.3. Reprinted with permission.)

Locations for both longitude and latitude are always given as portions of a circle and are usually measured in degrees, minutes, and seconds (symbolized °, ′, and ″). There are 360 degrees in a complete circle; each degree is divided into 60 minutes; and each minute is divided into 60 seconds. In this book our considerations of angles are generally not precise enough to justify their measurement to the second, so we have usually used just degrees and minutes.

The distance represented on a sphere by a degree of latitude is always the same. On the earth this distance is 111 km (69 miles), although if we consider the true, ellipsoidal shape of the earth, there is some variation—about 1 km between the equator and the poles. All meridians, on the other hand, converge at the poles; so the length (distance) represented by a degree of longitude varies from 111 km at the equator, to 96 km at 30° latitude, to 56 km at 60° latitude, to 0 km at 90° latitude.

THE TOWNSHIP AND RANGE SYSTEM

Throughout much of North America, the grid system referenced on maps is the township and range system. Originally devised for the division of the landscape in the old Northwest Territories of the United States, the township and range system has since been extended to the majority of the United States as well as to northern Ontario and the western provinces of Canada. This modified rectangular grid is based on a set of selected meridians, termed *principal meridians*, and parallels, called *baselines*, which intersect at an initial point (Fig. 16.4).

Distances are measured in the four cardinal directions from the initial point, and locations are identified at twenty-four-mile intervals along the baseline and principal meridian. However, due to the earth's shape, meridians converge toward the poles, making it impossible to fit an exact square to the earth's surface. Consequently, the ideal planimetric grid that forms the basis of the township and range system is intentionally and necessarily distorted at certain points in order to conform to the earth's curvature.

Within each set of twenty-four-mile-wide strips, six-mile strips are defined. The strips oriented east-west are defined by the parallels and are termed *townships;* those oriented north-south and bounded by the principal guide meridians are termed *ranges.* Each township and range strip is assigned a number to indicate its position vis-à-vis the initial point. Thus each small square, generally referred to simply as a township and measuring six miles on a side, is easily identified by a notation such as T4N, R3W ("township 4 north, range 3 west"). This notation identifies the township that is formed by the convergence of the fourth township strip north of the baseline and the third range strip west of the principal meridian (Fig. 16.4a).

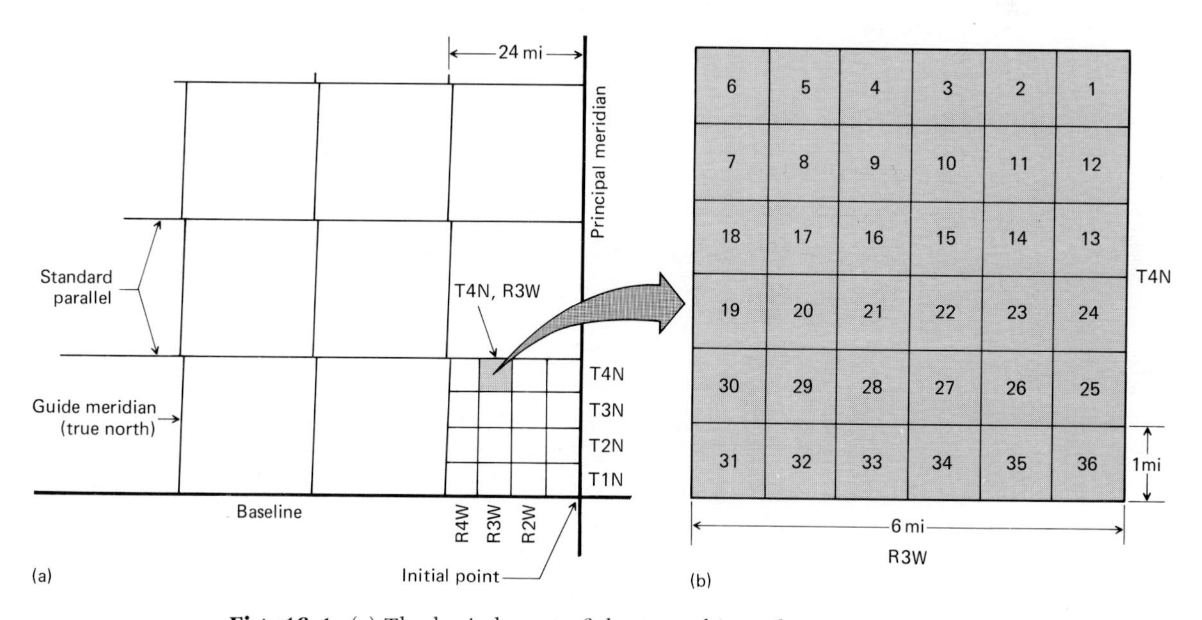

Fig. 16.4 (a) The basic layout of the township and range system; (b) each township is subdivided into thirty-six sections and each section is one square mile (640 acres) in area. (From W. M. Marsh and J. Dozier, *Landscape: An Introduction to Physical Geography*, © 1981, Addison-Wesley, Reading, Massachusetts. Fig. A3.1(a). Reprinted with permission.)

Every township is subdivided into thirty-six units, termed *sections*, each measuring one mile on a side. Each section within a township is given a number designation, beginning with section 1 in the northeast corner and proceeding sequentially westward to section 6, then dropping down to the next tier and proceeding back to the east, and so forth, as shown in Fig. 16.4b.

The errors due to the convergence of the meridians toward the poles are accumulated along the eastern and northern column and row in each township. Thus sections, 1, 2, 3, 4, 5, 6, 7, 18, 19, 30, and 31 are often a fraction less than 640 acres (one square mile) in area.

OTHER LOCATIONAL SYSTEMS

On many maps reference is made to additional grid coordinate systems. In the United States, the National Ocean Survey has designed a grid system for each state called the Plane Coordinate System. The basic unit or cell of this system is a square measuring 10,000 feet on a side. The topographic maps prepared by the United States Geological Survey give state plane coordinates, township and range (where applicable), longitude and latitude, as well as reference to UTM coordinates.

The letters UTM stand for the Universal Transverse Mercator grid, a system based on square grids with 100,000 meter spacing arranged within 6 degree sections of longitude extending between latitudes 80 degrees south and 80 degrees north. The UTM grid is drawn on topographic maps prepared by the United States Army Topographic Command (formerly the Army Map Service), whereas longitude and latitude are referenced by tick marks on the map margins.

SCALE

Scale is defined as the relationship between distance on the map and the corresponding distance on the earth's surface. Maps of small areas, such as planning sites, are called *large-scale maps,* whereas those of large areas are called *small-scale maps.* The level of detail on a map varies with scale; the smaller the scale, the less detail possible.

Scale is generally indicated on a map in either a graphic or an arithmetic form, and occasionally both are included as part of the map legend. The simplest scale indicator employed is the graphic, or bar, scale. This consists of an actual line or bar calibrated to indicate a precise map distance and labeled to indicate the corresponding ground distance (Fig. 16.5). Any linear measurement on the map can be compared directly to the bar scale to determine the actual ground distance.

The arithmetic scale represents a ratio of units on the map to like units on the ground and is called a *representative fraction.* (See the discussion on aerial photo scale in Chapter 15.) A representative fraction of 1:50,000, or 1/50,000, indicates that 1 unit on the map is equivalent to 50,000 of the same

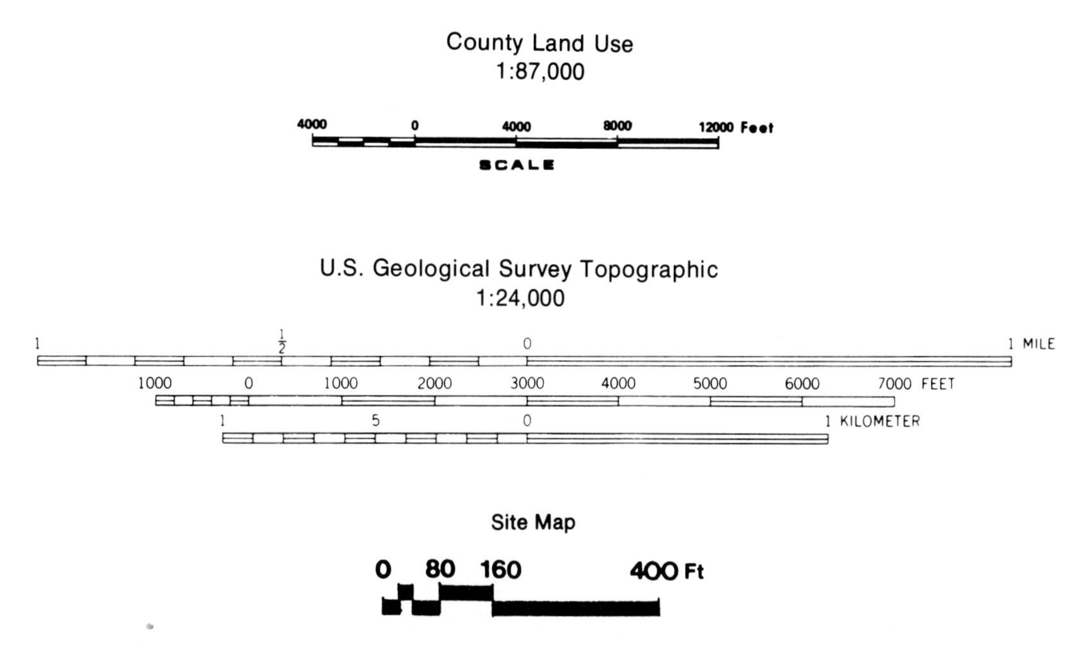

Fig. 16.5 Examples of graphic (bar) scales employed in modern maps.

units on the earth's surface. Since the scale is expressed in terms of a ratio, the proportion between the two distances (map and ground) is constant. Thus the representative fraction is applicable to all systems and all units of measurement simultaneously. Hence 1:50,000 can be read as "1 map inch to 50,000 ground inches" or "1 map centimeter to 50,000 ground centimeters." Similarly, any other unit of measurement can be substituted. The need for conversion factors between measurement systems, for example, U.S. Customary and metric, is thus avoided.

MAP TYPES

The symbols you see on maps depend on the type of map and the phenomena portrayed by the map. There are three basic types of maps used in geography and planning: choropleth, dot, and isopleth. A *choropleth map* may be used to portray either numerical or nonnumerical phenomena. The key feature of this map is its patch-like appearance. Each patch (area) represents a different class or category, and any class may abut against any other. Sections or counties are often used as the mapping units and both natural and human features can be portrayed with choropleth maps (Fig. 16.6).

A *dot map* is usually used to portray numerical phenomena such as population or crop production. The placement of the dots is usually intended to be representative of the location of the phenomena being portrayed. Each dot may represent a fixed value, or the dot may be sized in proportion to different values. Dot maps are useful in portraying water use by state in the United States or irrigated acreage by county (Fig. 16.7).

An *isopleth map* is designed to show the pattern or trend of numerical values over an area. This type of map utilizes lines, called *isolines,* to con-

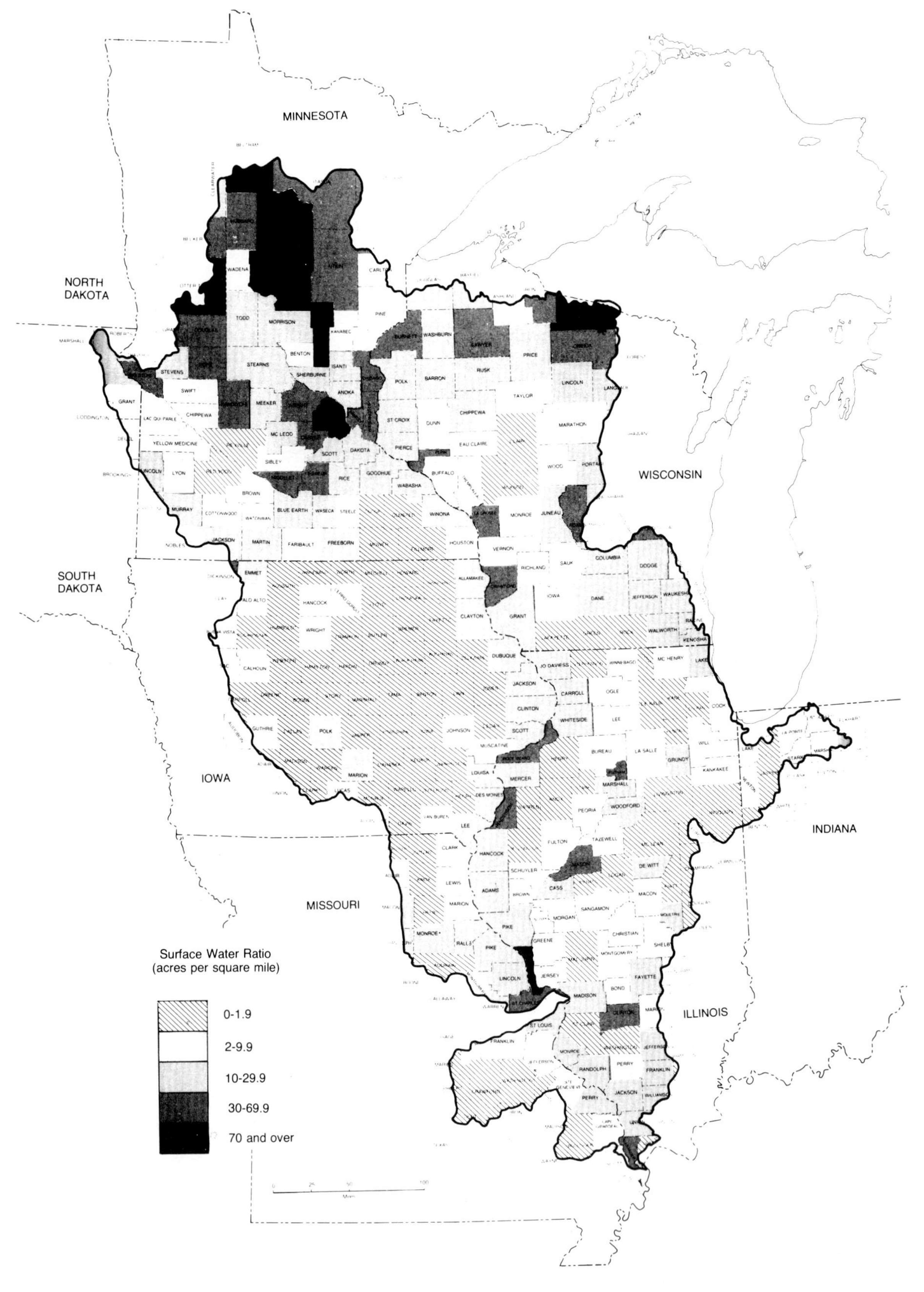

Fig. 16.6 An example of a choropleth map showing relative area cover by water, using the county as the mapping unit.

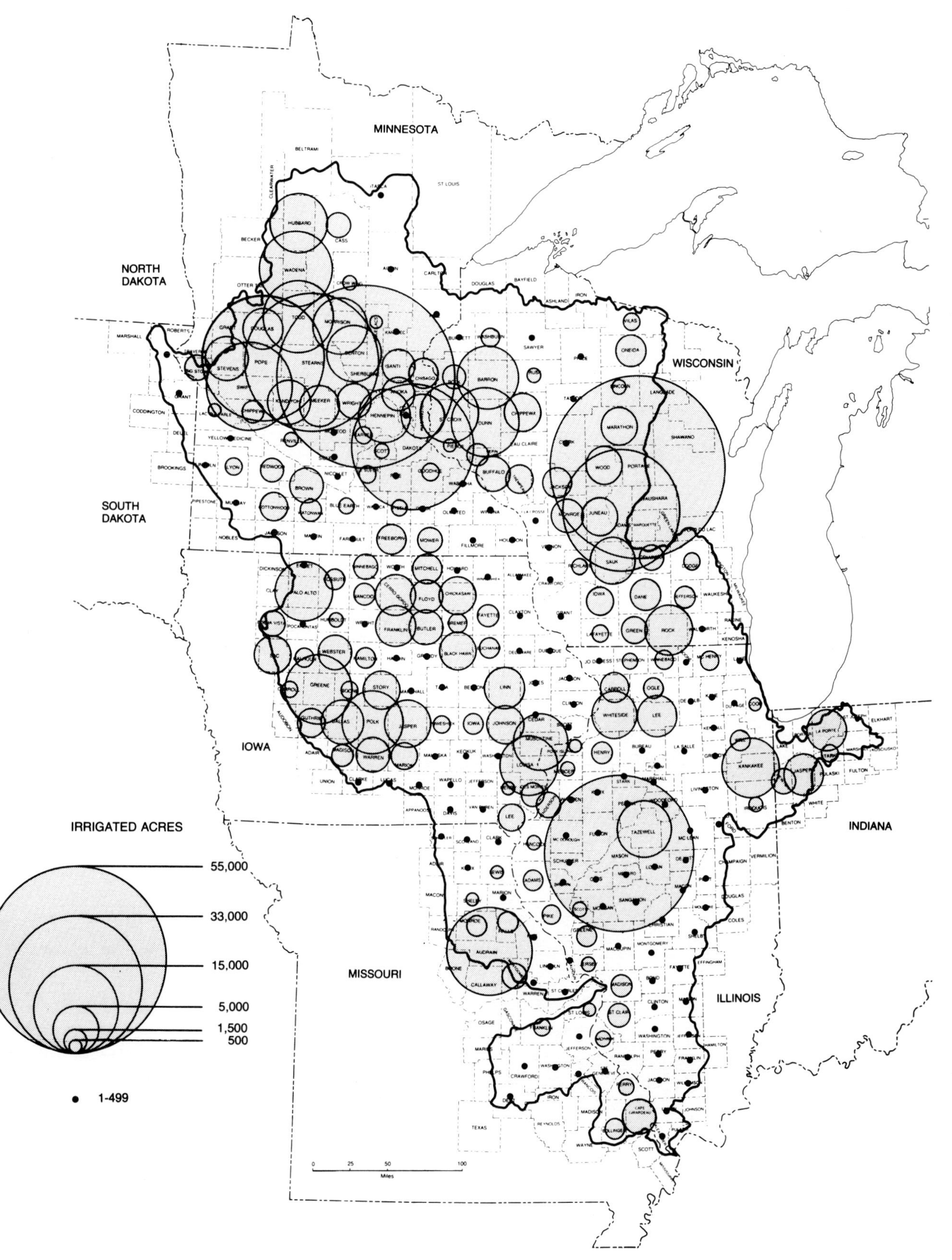

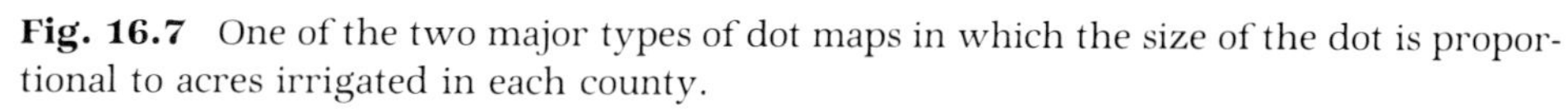
Fig. 16.7 One of the two major types of dot maps in which the size of the dot is proportional to acres irrigated in each county.

nect points (places) of equal value. If a value is not known, the location of the line is interpolated on the basis of the nearest known values. Among the rules governing isopleth maps are: (1) a given isoline must have the same value over the entire map; (2) isolines cannot cross each other; and (3) the change in value from one line to the next must not exceed the iso-interval; that is, the specified difference in value of adjacent lines in a sequence.

Isopleth maps are used extensively in physical geography and planning, especially for regional phenomena such as solar radiation, precipitation, and soil erosion rates. One of their most common uses is in portraying the topography of the earth (Fig. 16.8). (See pages 203–206 for a description of slope measurement and mapping from contour maps.)

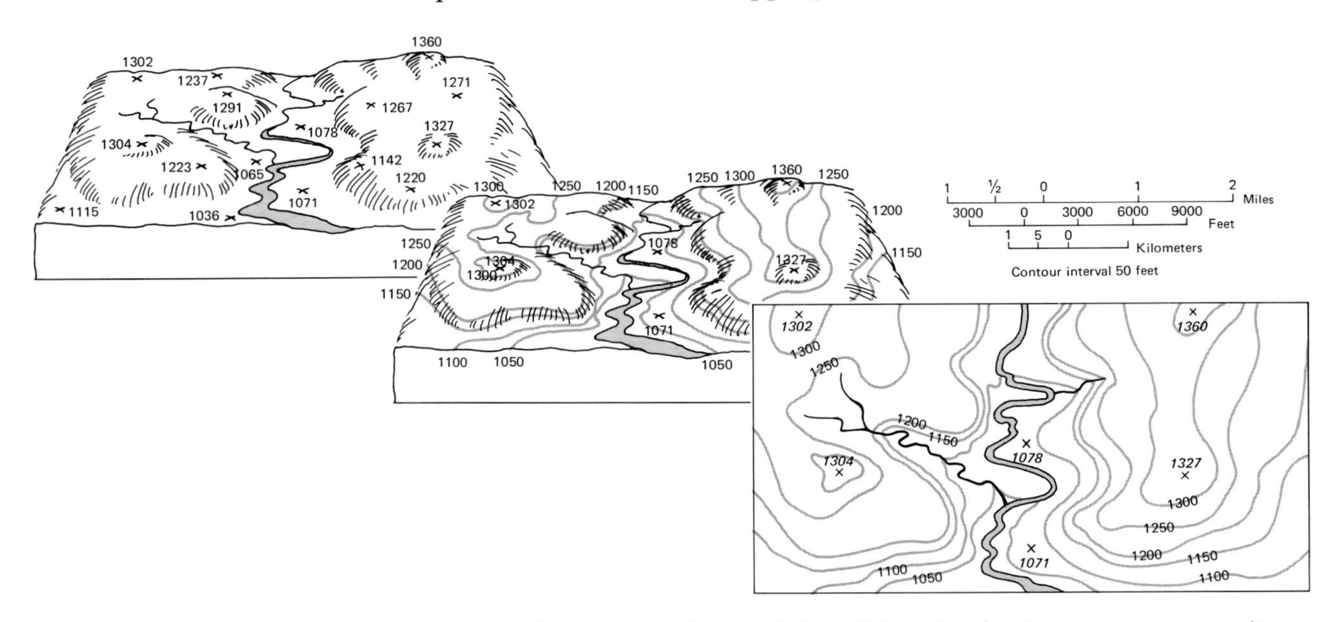

Fig. 16.8 An illustration of the translation of elevation data into a contour map. (From W. M. Marsh and J. Dozier, *Landscape: An Introduction to Physical Geography*, © 1981, Addison-Wesley, Reading, Massachusetts. Appendix 3. Reprinted with permission.)

TOPOGRAPHIC CONTOUR MAPS

Topographic contour maps are perhaps the most widely used maps in the world today because they are so valuable in terrain analysis, planning, and development. In the United States the U.S. Geological Survey is charged with the task of preparing topographic maps for the nation. These maps, called *topographic quadrangles,* are prepared at a variety of scales, for example, 1:24,000, 1:62,500, and 1:250,000, and are available to anyone at a relatively low cost. In Canada, topographic maps are prepared by the Department of Energy, Mines and Resources under a program called the National Topographic System.

In addition to contours and elevation data, the U.S. Geological Survey quadrangles provide a great deal of other information about the land. This includes drainage features, forested areas, wetlands, roads, highways, urbanized areas, and even individual structures such as homes and schools in rural areas (Fig. 16.9).

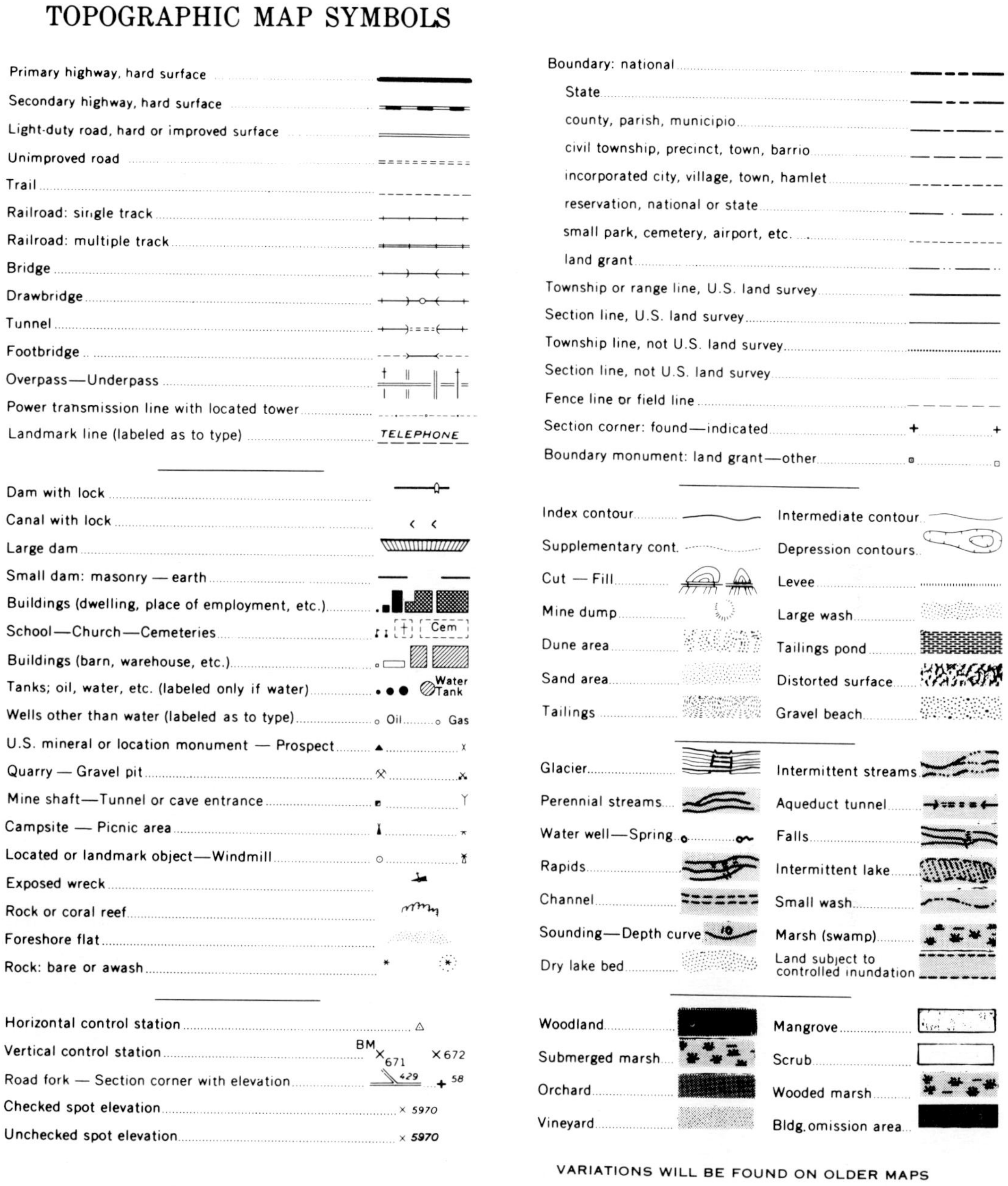

TOPOGRAPHIC MAP SYMBOLS

Primary highway, hard surface
Secondary highway, hard surface
Light-duty road, hard or improved surface
Unimproved road
Trail
Railroad: single track
Railroad: multiple track
Bridge
Drawbridge
Tunnel
Footbridge
Overpass—Underpass
Power transmission line with located tower
Landmark line (labeled as to type) TELEPHONE

Dam with lock
Canal with lock
Large dam
Small dam: masonry — earth
Buildings (dwelling, place of employment, etc.)
School—Church—Cemeteries Cem
Buildings (barn, warehouse, etc.)
Tanks; oil, water, etc. (labeled only if water) Water Tank
Wells other than water (labeled as to type) Oil Gas
U.S. mineral or location monument — Prospect
Quarry — Gravel pit
Mine shaft—Tunnel or cave entrance
Campsite — Picnic area
Located or landmark object—Windmill
Exposed wreck
Rock or coral reef
Foreshore flat
Rock: bare or awash

Horizontal control station
Vertical control station BM ×671 ×672
Road fork — Section corner with elevation 429 58
Checked spot elevation × 5970
Unchecked spot elevation × 5970

Boundary: national
State
county, parish, municipio
civil township, precinct, town, barrio
incorporated city, village, town, hamlet
reservation, national or state
small park, cemetery, airport, etc.
land grant
Township or range line, U.S. land survey
Section line, U.S. land survey
Township line, not U.S. land survey
Section line, not U.S. land survey
Fence line or field line
Section corner: found—indicated
Boundary monument: land grant—other

Index contour
Intermediate contour
Supplementary cont.
Depression contours
Cut — Fill
Levee
Mine dump
Large wash
Dune area
Tailings pond
Sand area
Distorted surface
Tailings
Gravel beach

Glacier
Intermittent streams
Perennial streams
Aqueduct tunnel
Water well—Spring
Falls
Rapids
Intermittent lake
Channel
Small wash
Sounding—Depth curve 10
Marsh (swamp)
Dry lake bed
Land subject to controlled inundation

Woodland
Mangrove
Submerged marsh
Scrub
Orchard
Wooded marsh
Vineyard
Bldg. omission area

VARIATIONS WILL BE FOUND ON OLDER MAPS

Fig. 16.9 Index of the symbols that appear on U.S. Geological Survey topographic maps.

PLANNING MAPS

Most of the maps used in planning are of the conventional types. Land use, land ownership, and zoning, for example, are usually portrayed with choropleth maps, while topography and rainstorm intensities are portrayed with standard isopleth maps. In some planning problems, maps are used that are not of the conventional sorts. Many of these are special purpose maps used in the formulation of land use and site plans. Vector maps, for instance, are designed to show the directional aspect of a process or feature such as traffic flow, pedestrian views, noise, wind, and sunlight, as they relate to a specific point or area (Fig. 16.10). Linkage maps are used to show the spatial relations and interactions between certain features, land use activities, and/or systems within a prescribed area. In both vector maps and linkage maps, quantities may be portrayed symbolically or noted directly on the map.

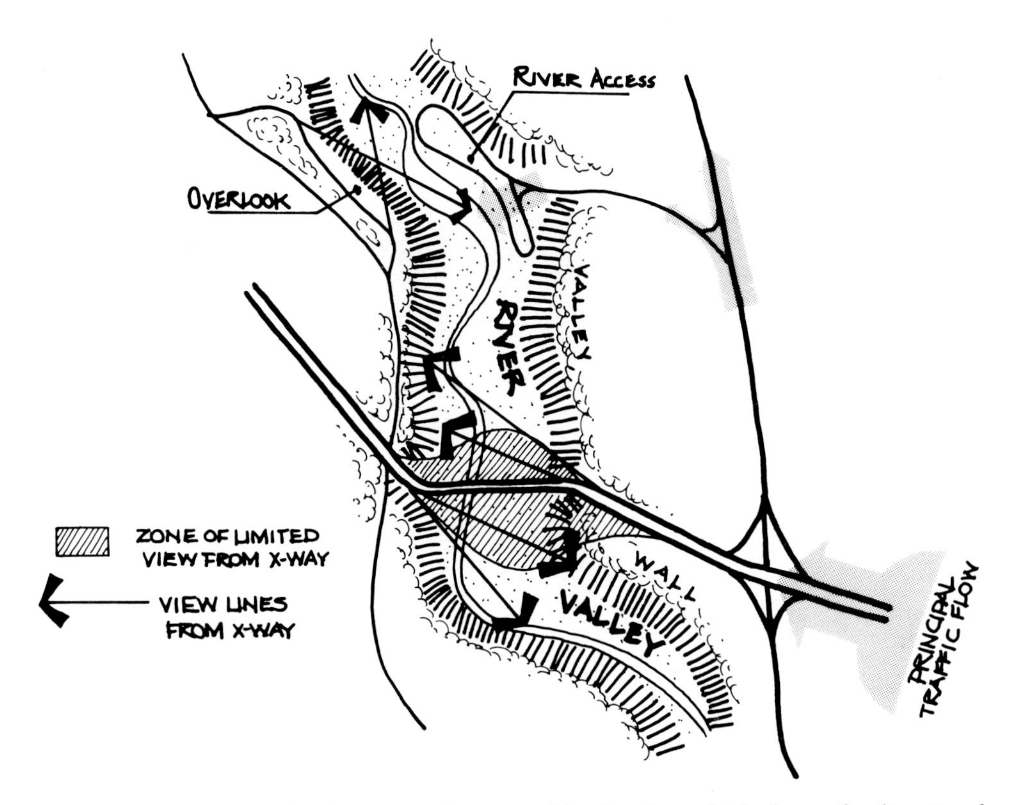

Fig. 16.10 An example of one type of map used in planning which shows the directional components of circulation and views.

GEOLOGIC MAPS

The most common variety of geologic map is basically a complex choropleth map showing the distribution of rock types and rock units of different ages. Additional symbols, such as lines representing major faults and arrows showing the directions of strike and dip, are often superimposed on the choropleth base.

Various types of isopleth maps are also used in geology, including *isopach* maps, which show the surface configuration of a subsurface rock formation, topographic contour maps, geothermal heat flow maps, and maps showing the thickness of deposits such as volcanic ash or glacial drift.

SELECTED REFERENCES FOR FURTHER READING

Gould, P., and White, R. *Mental Maps.* Harmondsworth, Great Britain: Penguin, 1974.

Lawrence, G.R.P. *Cartographic Methods.* 2nd ed. New York: Methuen, 1979.

Monkhouse, F. J., and Wilkinson, H. R. *Maps and Diagrams.* 3rd ed. London: Methuen, 1971.

Muehrcke, P. C. *Map Use—Reading, Analysis, and Interpretation.* Madison, Wisc.: 1978.

Robinson, A.; Sale, R.; and Morrison, J. *Elements of Cartography.* 4th ed. New York: Wiley, 1978.

Zuylen, L. Van. "Production of Photomaps." Cartography Journal, vol. 6, pp. 92–102.

17

INTERPRETING AND DISPLAYING DATA WITH GRAPHS

INTRODUCTION

Much of science is devoted to finding relationships between selected aspects of nature. In physical geography and landscape planning we are especially interested in spatial relations; that is, how the distribution of one thing, say, ground material, varies in relationship to another, such as surface temperature or plant types. The first step in the search for a relationship is to collect data on these features, and the second step is to display the data in some fashion so the observer has an opportunity to see what relationship, if any, exists. Data may be displayed in a number of different ways; two of the most useful are maps and graphs. In addition to helping the investigator identify relationships, maps and graphs are also important as communication devices. In planning problems, they are standard tools in both the technical and decision-making parts of the process.

TRANSECT GRAPH

Four basic types of graphs are used to show relationships in landscape analysis. A *transect graph* is appropriate for data that are distributed along a line cutting through a sample area. For example, let us say that air temperature at ground level and soil material were sampled at the points shown on the map in Fig. 17.1. To construct the graph, the sample points and the surface material at each point would be located along the base of the graph. Next, the vertical axis of the graph would be scaled for temperature; and finally, the temperature reading for each point would be plotted on the graph (Fig. 17.2). If these points, or the means (averages) of these points for each material, are connected with a line, the individual points can be generalized into a continuous profile that helps us to see trends that otherwise might not be apparent. The accuracy of the graph line depends on the number of sample points; the more points, the greater the accuracy.

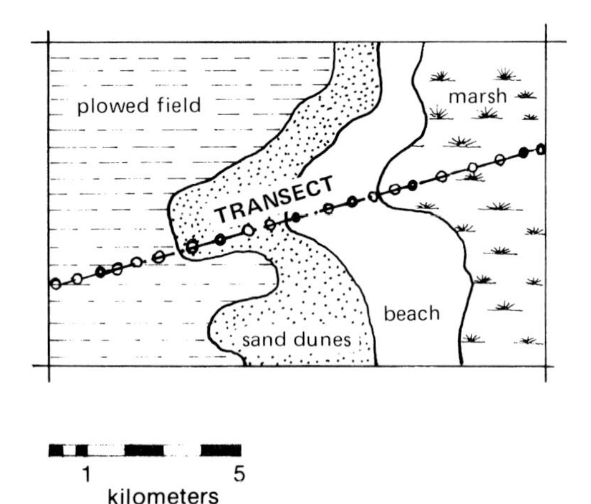

Fig. 17.1 Map showing the locations of sample points along a transect across four types of ground.

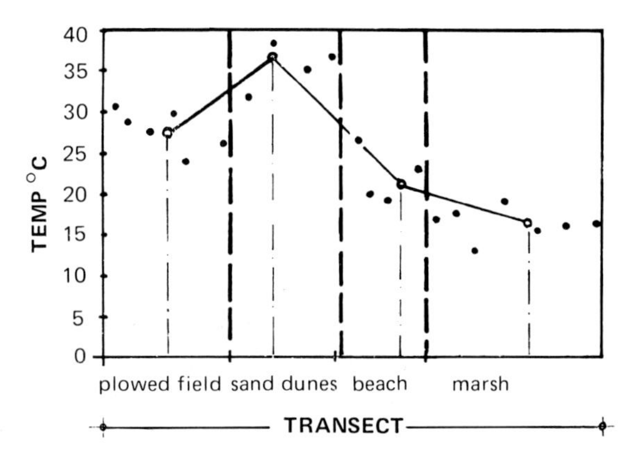

Fig. 17.2 Transect graph comprised of air temperature readings taken at the sample points shown in Fig. 17.1. The graph line represents the mean temperature for each type of ground.

SCATTER DIAGRAM

The transect graph (Fig. 17.2) shows the relationship between a quantitative expression (air temperature) and qualitative expression (surface materials). While it does enable us to identify a relationship or trend, it does not tell us much about the nature of that relationship. Another type of graph, called a *scatter diagram,* shows the relationship, or the lack of it, between two quantitative expressions or variables. More precisely, it shows how one variable, called the *dependent variable,* changes as a function of the other, called the *independent variable.*

In constructing a scattered diagram it is customary to use the horizontal axis of the graph for the independent variable and the vertical axis for the dependent variable. In the problem involving air temperature and surface material, a scatter diagram cannot be constructed unless we can find a meaningful, quantitative expression for soil material. Since we know that moisture content is an important control on soil temperature, let us look for a relationship between soil moisture, expressed as a percentage of the total soil weight, and air temperature near the ground. We assume in this case that the air gains its heat from the ground under it.

The graph in Fig. 17.3 shows one shape this relationship might take. The points are represented by a single line, called a *regression line* (or line of best fit), which slants from left to right. The slope of the line represents the rate of temperature change relative to soil moisture, and can be described by subtracting one temperature on the graph line from another and then dividing the product by the number of percentage points separating the same two temperatures:

$$\Delta t = \frac{T_1 - T_2}{P_1 - P_2}$$

where:

Δt = rate of temperature change
T_1 = higher temperature on graph line
T_2 = lower temperature on graph line
P_1, P_2 = % moisture corresponding to the temperatures T_1 and T_2

Although many types of relationships can be identified with scatter diagrams, two are especially easy to identify: negative and positive relationships. Negative relationships are those in which the dependent variable decreases with increasing values of the independent variable. The graph shown in Fig. 17.3 represents a negative relationship. Positive relationships are essentially the opposite; the independent variable increases with the dependent variable.

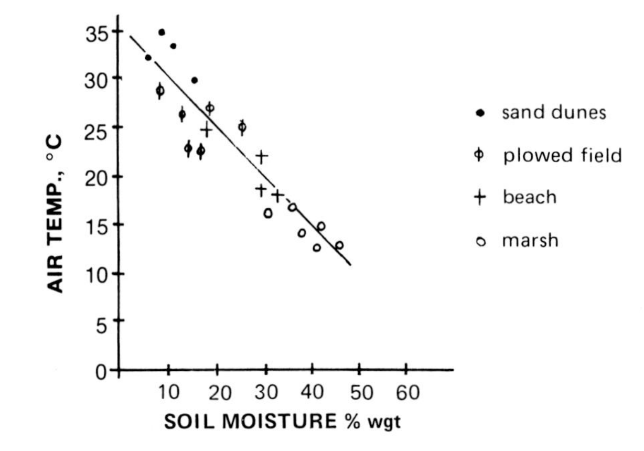

Fig. 17.3 Scatter diagram representing the relationship between air temperature near ground level and soil moisture. The graph line, called a *regression line*, is the best fit that can be drawn among the dots with a single line.

FREQUENCY GRAPH

Another type of graph is the *frequency graph*, which is designed to show what portion (percentage) of a sample falls into various size classes. The height of students in a class (where the class is the sample) is often used to illustrate a frequency graph. In the environmental sciences, the frequency graph is often used to show the distribution (numerical or statistical in this case, not geographical) of something over a period of time; for example, hourly air temperatures at a particular place, the monthly distribution of peak river flows, or hourly wind speeds. In other instances, it may involve the particle sizes in a soil sample or tree sizes in a forest tract. Construction of this graph involves first selecting the appropriately sized categories, which must, by the way, include all readings. Next, the horizontal axis is scaled to accommodate these categories and the vertical axis is scaled for percent (Fig. 17.4).

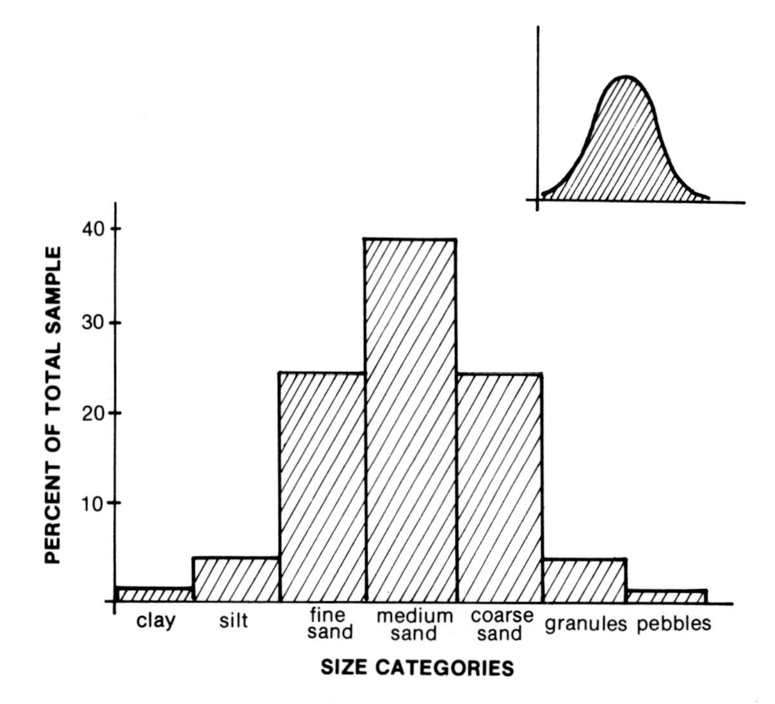

Fig. 17.4 Frequency graph showing the percentage of particles in a soil sample that fall into each size classes.

Most frequency distributions are thick in the middle (for the simple reason that most people are of medium height or soil particles are of medium size) and taper off on both ends. Those distributions, which assume a perfect bell shape, are called *normal distributions.* (See the inset graph in Fig. 17.4.) The middle of the graph, which would be a line through the center of the bell, represents the arithmetic average, or the *mean,* of the distribution. In physical geography we often encounter distributions that are asymmetrical. Such distributions have a disproportionately large number of large or small values and are said to be *skewed.* Peak annual river flows, hourly wind speeds, and ocean wave sizes, for instance, tend to be heavy in the small categories, whereas the large categories trail off far to the right.

TIME SERIES GRAPH

If a graph can be used to show changes in a variable over space, as along a transect, then a graph could also be used to show changes over time. *Time series graphs* can be used to show variations, for example, in air temperature, wind speed, or soil moisture over any time period. The graph is set up the same as a transect graph except the horizontal axis is scaled for time. The graph in Fig. 17.5 is designed to show the change in air and soil temperatures over twenty-four-hour intervals. Like many phenomena, daily air temperature and soil heat tend to follow a more or less cyclical pattern, with a high near mid-day and a low at night. Many other phenomena, however, including tornadoes, earthquakes, and the flows of some rivers, do not vary in cyclical fashion, but tend to vary sporadically over time.

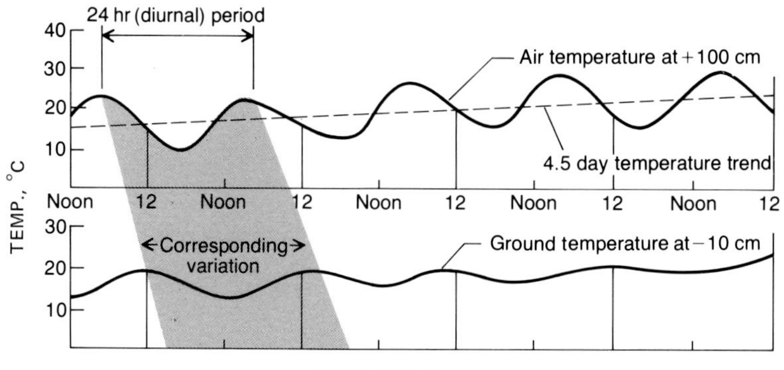

Fig. 17.5 Time series graph showing the change in air temperature over twenty-four-hour intervals and the corresponding change in soil temperature.

SELECTED REFERENCES FOR FURTHER READING

Davis, J. C., and McCullagh (eds.). *Display and Analysis of Spatial Data.* New York: Wiley, 1975.

Earle, James H. "Graphs," in *Drafting Technology.* Reading, Mass.: Addison-Wesley, 1982.

Hunkins, Dalton R., and Thomas L. Pinot. *Mathematics Tools and Models.* Reading, Mass.: Addison-Wesley, 1977.

Ore, O. *Graphs and Their Uses.* Westminster, Md.: Random House, 1963.

APPENDIX A: U.S. NATIONAL AIR QUALITY, WATER QUALITY, AND NOISE STANDARDS

National Ambient Air Quality Standards

Pollutant	*Primary Standard, micrograms per cubic meter*	*Secondary Standard, micrograms per cubic meter*
Particulate Matter		
annual geometric mean	75	60
maximum 24-hour concentration[a]	260	150
Sulfur Oxides		
annual arithmetic mean	80 (0.03 ppm)	—
max. 24-hour concentration[a]	365 (0.14 ppm)	—
max. 3-hour concentration[a]	—	1,300 (0.5 ppm)
Carbon Monoxide		
max. 8-hour concentration[a]	10 (9 ppm)	10
max. 1-hour concentration[a]	40 (35 ppm)	40
Ozone		
max. hourly avg. concentration[a]	235 (0.12 ppm)	235
Nitrogen Dioxide		
annual arithmetic mean	100 (0.05 ppm)	100
Hydrocarbons		
max. 3-hour concentration[a]	160 (0.24 ppm)	160
(6–9 a.m.)	160 (0.24 ppm)	160
Lead		
Max. arithmetic means (average over calendar quarter)	1.5	1.5

[a]Not to be exceeded more than once a year per site.
[b]Not to be exceeded, on an average, more than once per year, based upon readings from the last three years.
Note: ppm indicates parts of pollutant per million parts of air.
Source: Environmental Protection Agency Regulations on National Primary and Secondary Ambient Air Quality Standards.

Raw-Surface-Water Criteria for Public Water Supplies

Substance	*Surface-Water Criteria, mg/liter*	
	Permissive Criteria	*Desirable Criteria*
Coliforms (MPN)	10,000	<100
Fecal coliforms (MPN)	2,000	<20
Inorganic chemicals (mg/l)		
Ammonia-N	0.5	<0.01
Arsenic[a]	0.05	Absent
Barium[a]	1.0	Absent
Boron[a]	1.0	Absent
Cadmium[a]	0.01	Absent
Chloride[a]	250	250
Chromium[a] (hexavalent)	0.05	Absent
Copper[a]	1.0	Virtually absent
Dissolved oxygen	≥4	Near saturation
Iron	0.3	Virtually absent
Lead[a]	0.05	Absent
Manganese[a]	0.05	Absent
Nitrate[a]-N	10	Virtually absent
Selenium[a]	0.01	Absent
Silver[a]	0.05	Absent
Sulfate[a]	250	<50
Total dissolved solids[a]	500	<200
Urany ion[a]	5	Absent
Zinc[a]	5	Virtually absent
Organic chemicals (mg/l)		
ABS		
Carbon chloroform extract[a]	0.15	<0.04
Cyanide[a]	0.20	Absent
Herbicides		
2,4-D + 2,4,5-T + 2,4-TP[a]	0.1	Absent
Oil and gases[a]	Virtually absent	Absent
Pesticides[a]		
Adrian	0.017	Absent
Chlordane	0.003	Absent
DDT	0.042	Absent
Dieldrin	0.017	Absent
Endrin	0.001	Absent
Heptachlor	0.018	Absent
Lindane	0.056	Absent
Methoxychlor	0.035	Absent
Toxaphene	0.005	Absent
Phenols[a] 0.001Absent		

[a]Substances that are not significantly affected by the following treatment process: coagulation (less than about 50 mg/liter of alum, ferric sulfate, or copperas, with alkali addition as necessary but without coagulant aids or activated carbon), sedimentation (6 hours or less), rapid sand filtration (3 gpm/ft^2 or less), and disinfection with chlorine (without consideration to concentration or form of chlorine residual).

Source: "Raw Water Quality Criteria for Public Supplies," National Technical Advisory Committee Report (U.S. Department of the Interior, issued by the Federal Water Pollution Control Administration, 1968).

Noise Criteria and Standards: Noise Levels and Risk

Sound Levels (decibels)	*Source*	*Risk from Exposure*
140	Jet engine (25 m distance)	Harmful to hearing
130	Jet takeoff (100 m away)	
	Threshold of pain	
120	Propeller aircraft	
110	Live rock band	Chance of hearing loss
100	Jackhammer/Pneumatic chipper	
90	Heavy-duty truck	
	Los Angeles, 3rd floor apartment next to freeway	
	Average street traffic	
80	Harlem, 2nd floor apartment	Damage possible with prolonged exposure
70	Private car	
	Boston row house on major avenue	—
	Business office	
	Watts—8 mi. from touchdown at major airport	
60	Conversational speech or old residential area	
50	San Diego—wooded residential area	—
40	California tomato field	
	Soft music from radio	
30	Quiet whisper	
20	Quiet urban dwelling	—
10	Rustle of leaf	
0	Threshold of hearing	

Source: U.S. Environmental Protection Agency

OSHA's Noise Exposure Limits for Workers with Unprotected Ears

Noise (dB_A)	*Permissible Exposure (hours and minutes)*
85	16 hrs
87	12 hrs 6 min
90	8 hrs
93	5 hrs 18 min
96	3 hrs 30 min
99	2 hrs 18 min
102	1 hr 30 min
105	1 hr
108	40 min
111	26 min
114	17 min
115	15 min
118	10 min
121	6.6 min
124	4 min
127	3 min
130	1 min

Exposures above or below the 90 dB limit have been "time weighted" to give what OSHA believes are equivalent risks to a 90 dB eight-hour exposure. From U.S. Federal Registrar.

APPENDIX B: UNITS OF MEASUREMENT AND CONVERSION

CONVERSION FACTORS AND DECIMAL NOTATIONS

Energy, Power, Force, and Pressure

Energy Units and Their Equivalents

- *joule* (abbreviation J): 1 joule = 1 unit of force (a newton) applied over a distance of 1 meter = 0.239 calorie
- *calorie* (abbreviation cal): 1 calorie = heat needed to raise the temperature of 1 gram of water from 14.5°C to 15.5°C = 4.186 joules
- *British Thermal Unit* (abbreviation BTU): 1 BTU = heat needed to raise the temperature of 1 pound of water 1° Fahrenheit from 39.4° to 40.4°F = 252 calories = 1055 joules

Power

- *watt* (abbreviation W): 1 watt = 1 joule per second
- *horsepower* (abbreviation hp): 1 hp = 746 watts

Force and Pressure

- *newton* (abbreviation N): 1 newton = force needed to accelerate a 1-kilogram mass over a distance of 1 meter in 1 second squared
- *bar* (abbreviated b): 1 bar = pressure equivalent to 100,000 newtons on an area of 1 square meter

- *millibar* (abbreviation mb): 1 millibar = one-thousandth $\left(\frac{1}{1000}\right)$ of a bar
- *pascal* (abbreviation Pa): 1 pascal = force exerted by 100,000 newtons on an area of 1 square meter
- *atmosphere* (abbreviation Atmos.): 1 atmosphere = 14.7 pounds of pressure per square inch = 1013.2 millibars

Length, Area, and Volume

1 micrometer (μm) = 0.000001 meter = 0.0001 centimeter

1 millimeter (mm) = 0.03937 inch = 0.1 centimeter

1 centimeter (cm) = 0.39 inch = 0.01 meter

1 inch (in.) = 2.54 centimeters = 0.083 foot

1 foot (ft) = 0.3048 meter = 0.33 yard

1 yard (yd) = 0.9144 meter

1 meter (m) = 3.2808 feet = 1.0936 yards

1 kilometer (km) = 1000 meters = 0.6214 mile (statute) = 3281 feet

1 mile (statute) (mi) = 5280 feet = 1.6093 kilometers

1 mile (nautical) (mi) = 6076 feet = 1.8531 kilometers

Area

1 square centimeter (cm^2) = 0.0001 square meter = 0.15550 square inch

1 square inch ($in.^2$) = 0.0069 square foot = 6.452 square centimeters

1 square foot (ft^2) = 144 square inches = 0.0929 square meter

1 square yard (yd^2) = 9 square feet = 0.8361 square meter

1 square meter (m^2) = 1.1960 square yards = 10.764 square feet

1 acre (ac) = 43,560 square feet = 4046.95 square meters

1 hectare (ha) = 10,000 square meters = 2.471 acres

1 square kilometer (km^2) = 1,000,000 square meters = 0.3861 square mile

1 square mile (mi^2) = 640 acres = 2.590 square kilometers

Volume

1 cubic centimeter (cm^2) = 1000 cubic millimeters = 0.0610 cubic inch

1 cubic inch ($in.^3$) = 0.0069 cubic foot = 16.387 cubic centimeters

1 liter (l) = 1000 cubic centimeters = 1.0567 quarts

1 gallon (gal) = 4 quarts = 3.785 liters

1 cubic ft (ft^3) = 28.31 liters = 7.48 gallons = 0.02832 cubic meter

1 cubic yard (yd^3) = 27 cubic feet = 0.7646 cubic meter

1 cubic meter (m^3) = 35.314 cubic feet = 1.3079 cubic yards

1 acre-foot (ac-ft) = 43,560 cubic feet = 1234 cubic meters

Mass and Velocity

Mass (Weight)

1 gram (g) = 0.03527 ounce* = 15.43 grains

1 ounce (oz) = 28.3495 grams = 437.5 grains

1 pound (lb) = 16 ounces = 0.4536 kilogram

1 kilogram (kg) = 1000 grams = 2.205 pounds

1 ton* (ton) = 2000 pounds = 907 kilograms

1 tonne = 1000 kilograms = 2205 pounds

Velocity

1 meter per second (m/sec) = 2.237 miles per hour

1 km per hour (km/hr) = 27.78 centimeters per second

1 mile per hour (mph) = 0.4470 meter per second

1 knot (kt) = 1.151 miles per hour = 0.5144 meter/second

Quantities, Decimal Equivalents, and Scientific Notation

Quantity	*Decimal Notation*	*Scientific Notation*		*Prefix*
One trillion (U.S.)	1,000,000,000,000	10^{12}	T	tera-
One billion (U.S.)	1,000,000,000	10^{9}	G	giga-
One million	1,000,000	10^{6}	M	mega-
One thousand	1,000	10^{3}	k	kilo-
One hundred	100	10^{2}	h	hecto-
Ten	10	10	da	deka-
One tenth	0.1	10^{-1}	d	deci-
One hundredth	0.01	10^{-2}	c	centi-
One thousandth	0.001	10^{-3}	m	milli-
One millionth	0.000001	10^{-6}	μ	micro-
One billionth (U.S.)	0.000000001	10^{-9}	n	nano-
One trillionth (U.S.)	0.000000000001	10^{-12}	p	pico-

*Avoirdupois, i.e., the customary system of weights and measures in most English-speaking countries.

GLOSSARY

Acid rain Precipitation with pH levels much below average as a result of the formation of sulfuric acid in polluted air.

Active layer The surface layer in a permafrost environment, which is characterized by freezing and thawing on a seasonal basis.

Aggradation Filling in of a stream channel with sediment, usually associated with low discharges and/or heavy sediment loads.

Albedo The percentage of incident radiation reflected by a material. Usage in earth science is usually limited to shortwave radiation and landscape materials.

Alluvial fan A fan-shaped deposit of sediment laid down by a stream at the foot of a slope; very common features in dry regions, where streams deposit their sediment load as they lose discharge downstream.

Angiosperm A flowering, seed-bearing plant; the angiosperms are presently the principal vascular plants on earth.

Angle of repose The maximum angle at which a material can be inclined without failing; in civil engineering the term is used in reference to clayey materials.

Aquifer Any subsurface material that holds a relatively large quantity of groundwater and is able to transmit that water readily.

Atterberg Limits Test A test used to determine a soil's response to the addition of water based on its changes in physical state, as from plastic to liquid.

Backscattering That part of solar radiation directed back into space as a result of diffusion by particles in the atmosphere.

Backshore The zone behind the shore—between the beach berm and the backshore slope.

Backshore slope The bank or bluff landward of the shore that is comprised of *in situ* material.

Backswamps A low, wet area in the floodplain, often located behind a levee.

Bankfull discharge The flow of a river when the water surface has reached bank level.

Baseflow The portion of streamflow contributed by groundwater; it is a steady flow that is slow to change even during rainless periods.

Bay-mouth bar A ribbon of sand deposited across the mouth of a bay.

Berm A low mound that forms along sandy beaches; also used to describe elongated mounds constructed along water features and site borders.

Boreal forest Subarctic conifer forests of North America and Eurasia; floristically homogeneous forests dominated by fir, spruce, and tamarack; in Russia, it is called *taiga.*

Buildable land units Parcels of various size within a designated project area that are suitable for development as defined by a prescribed development program.

Boundary layer The lower layer of the atmosphere; the lower 300 m of the atmosphere where airflow is influenced by the earth's surface.

Boundary sublayer The stratum of calm air immediately over the ground which increases in depth with the height and density of the vegetative cover.

Bowen ratio The ratio of sensible-heat flux to latent-heat flux between a surface and the atmosphere.

Capillarity The capacity of a soil to transfer water by capillary action; capillarity is greatest in medium-textured soils.

Carrying capacity The level of development density or use an environment is able to support without suffering undesirable or irreversible degradation.

Choropleth map A map comprised of areas of any size or shape representing qualitative phenomena (e.g., soils) or quantitative phenomena (e.g., population); map often has a patchwork appearance.

Climate The representative or general conditions of the atmosphere at a place on earth; it is more than the average conditions of the atmosphere, for climate may also include extreme and infrequent conditions.

Closed forest A forest structure with multiple levels of growth from the ground up; a forest in which undergrowth closes out the area between the canopy and the ground.

Coastal dune A sand dune that forms in coastal areas and is fed by sand from the beach.

Coefficient of runoff A number given to a type of ground surface representing the proportion of a rainfall converted to overland flow; it is a dimensionless number between 0 and 1.0 that varies inversely with the infiltration capacity; impervious surfaces have high coefficients of runoff.

Colluvium An unsorted mix of soil and mass-movement debris.

Concentration time The time taken for a drop of rain falling on the perimeter of a drainage basin to go through the basin to the outlet.

Conduction A mechanism of heat transfer involving no external motion or mass transport; instead, energy is transferred through the collision of vibrating molecules.

Conveyance zone The central route of drainage, usually a channel and valley, in a drainage basin.

dBA Decibel scale that has been adjusted for sensitivity of the human ear.

Decibel Unit of measurement for the loudness of sound based on the pressure produced in air by a noise; denoted dB.

Declination of the sun The location (latitude) on earth where the sun on any day is directly overhead; declinations range from 23.27° S latitude to 23.27° N latitude.

Degradation Scouring and downcutting of a stream channel, usually associated with high discharges.

Density *See* **Development density**

Design storm A rainstorm of a given intensity and frequency of recurrence used as the basis for sizing stormwater facilities such as stormsewers.

Detention A strategy used in stormwater management in which runoff is detained on site to be released later at some prescribed rate.

Development density A measure of the intensity of development or land use; defined on the basis of area covered by impervious surface, population density, or building floor area coverage, for example.

Discharge The rate of water flow in a stream channel; measured as the volume of water passing through a cross-section of a stream per unit of time, commonly expressed as cubic feet (or meters) per second.

Diurnal damping depth The maximum depth in the soil which experiences temperature change over a 24-hour (diurnal) period.

Drainage basin The area that contributes runoff to a stream, river, or lake.

Drainage density The number of miles (or km) of stream channels per square mile (or km^2) of land.

Drainage divide The border of a drainage basin or watershed where overland separates between adjacent areas.

Drainage network A system of stream channels usually connected in a hierarchical fashion. *See also* **Principle of stream orders**

Drainfield The network of pipes or tiles through which wastewater is dispersed into the soil.

Ecosystem A group of organisms linked together by a flow energy; also a community of organisms and their environment.

Ecotone The transition zone between two groups, or zones, of vegetation.

Energy balance The concept or model that concerns the relationship among energy input, energy storage, work, and energy output of a system such as the atmosphere or oceans.

Environmental assessment A preliminary study or review of a proposed action (project) and the influence it could have on the environment. Often conducted to determine the need for more detailed environmental impact analysis.

Environmental impact statement A study required by U.S. Federal law for projects (proposed) involving federal funds to determine types and magnitudes of impacts that would be expected in the natural and human environment and the alternative courses of action, including no action.

Environmental inventory Compilation and classification of data and information on the natural and human features in an area proposed for some sort of planning project.

Ephemeral stream A stream without baseflow; one that flows only during or after rainstorms or snowmelt events.

Erodibility The relative susceptibility of a soil to erosion.

Erodibility factor A value used in the universal soil loss equation for different soil types representing relative erodibility; called the K-factor by the U.S. Soil Conservation Service.

Erosion The removal of rock debris by an agency such as moving water, wind, or glaciers; generally, the sculpting or wearing down of the land by erosional agents.

Eutrophication The increase in biomass of a waterbody leading to infilling of the basin and the eventual disappearance of open water; sometimes referred to as the aging process of a waterbody.

Evapotranspiration The loss of water from the soil through evaporation and transpiration.

Facility planning Planning for facilities such as power generating stations or sewage treatment plants; usually carried out by engineers.

Feasibility study A type of technical planning aimed at identifying the most appropriate use of a site.

Fetch The distance of open water in one direction across a water body; it is one of the main controls on wave size.

Floodway fringe The zone designated by U.S. Federal flood policy as the area in a river valley that would be lightly inundated by the 100-year flood.

Floristic system The principal botanical classification scheme in use today; under this scheme the plant kingdom is made up of divisions, each of which is subdivided into smaller and smaller groups arranged according to the apparent evolutionary relationships among plants.

Formation A structural unit of vegetation that may be considered a subdivision of a biochore; a formation may be made up of several communities. In the traditional terminology, it is called a physiognomic unit; in geology, a major unit of rock.

Frequency The term used to express how often a specified event is equaled or exceeded.

Frost wedging A mechanical weathering process in which water freezes in a crack and exerts force on the rock, which may result in the breaking of the rock; a very effective weathering process in alpine and polar environments.

Geomorphic system A physical system comprised of an assemblage of landform linked together by the flow of water, air, or ice.

Geomorphology The field of earth science that studies the origin and distribution of landforms, with special emphasis on the nature of erosional processes; traditionally, a field shared by geography and geology.

Global coordinate system The network of east-west and north-south lines (parallels and meridians) used to measure locations on earth; the system uses degrees, minutes, and seconds as the units of measurement.

Grafting The practice of attaching additional channels to a drainage network; in agricultural areas new channels appear as drainage ditches; in urban areas, as stormsewers.

Greenbelt A tract of trees and associated vegetation in urbanized areas; it may be a park, nature preserve, or part of a transportation corridor.

Groin A wall or barrier built from the beach into the surf zone for the purpose of slowing down longshore transport and holding sand.

Gross sediment transport The total quantity of sediment transported along a shoreline in some time period, usually a year.

Ground frost Frost that penetrates the ground in response to freezing surface temperatures.

Ground sun angle The angle formed between a beam of solar radiation and the surface that it strikes in the landscape.

Groundwater The mass of gravity water that occupies the subsoil and upper bedrock zone; the water occupying the zone of saturation below the soil-water zone.

Gullying Soil erosion characterized by the formation of narrow, steep-sided channels etched by rivulets or small streams of water. Gullying can be one of the most serious forms of soil erosion of cropland.

Habitat The environment with which an organism interacts and from which it gains its resources; habitat is often variable in size, content, and location, changing with the phases in an organism's life cycle.

Hardpan A hardened soil layer characterized by the accumulation of colloids and ions.

Hazard assessment Study and evaluation of the hazard to land use and people from environmental threats such as floods, tornadoes, and earthquakes.

Heat island The area or patch of relatively warm air which develops over urbanized areas.

Heat syndrome Various disorders in the human thermoregulatory system brought on by the body's inability to shed heat or by a chemical imbalance from too much sweating.

Heat transfer The flow of heat within a substance or the exchange of heat between substances by means of conduction, convection, or radiation.

Hillslope processes The geomorphic processes that erode and shape slopes; mainly mass movements such as soil creep and landslides and runoff processes such as rainwash and gullying.

Horizon A layer in the soil that originates from the differentiation of particles and chemicals by moisture movement within the soil column.

Hydraulic radius The ratio of the cross-sectional area of a stream to its wetted perimeter.

Hydrograph A streamflow graph which shows the change in discharge over time, usually hours or days. *See also* **Hydrograph method**

Hydrograph method A means of forecasting streamflow by constructing a hydrograph that shows the representative response of a drainage basin to a rainstorm; the use of a "normalized" hydrograph for flow forecasting in which the size of the individual storm is filtered out. *See also* **Hydrograph**

Hydrologic cycle The planet's water system, described by the movement of water from the oceans to the atmosphere to the continents and back to the sea.

Hydrologic equation The amount of surface runoff (overland flow) from any parcel of ground is proportional to precipitation minus evapotranspiration loss, plus or minus changes in storage water (groundwater and soil water).

Hydrometer method A technique used to measure the clay content in a soil sample that involves dispersing the clay particles in water and drawing off samples at prescribed time intervals.

Hypothermia A physiological disorder associated with cold conditions and characterized by the decline of body temperature, slowed heart beat, lowered blood pressure and other symptoms.

Infiltration capacity The rate at which a ground material takes in water through the surface; measured in inches or centimeters per minute or hour.

Inflooding Flooding caused by overland flow concentrating in a low area.

Infrared film Photographic film capable of recording near infrared radiation (just beyond the visible to a wavelength of 0.9 micrometer), but not capable of recording thermal infrared wavelengths.

Infrared radiation Mainly longwave radiation of wavelengths between 3.0–4.0 and 100 micrometers, but also includes near infrared radiation, which occurs at wavelengths between 0.7 and 3.0–4.0 micrometers.

In situ A term used to indicate that a substance is in place as contrasted with one, such as river sediment, that is in transit.

Interception The process by which vegetation intercepts rainfall or snow before it reaches the ground.

Interflow Infiltration water that moves laterally in the soil and seeps into stream channels; in forested areas this water is a major source of stream discharge.

Isopleth map A map comprised of lines, called isolines, that connect points of equal value.

Land cover The materials such as vegetation and concrete that cover the ground. *See also* **Land use**

Landscape The composite of natural and human features that characterize the surface of the land at the base of the atmosphere; includes spatial, textural, compositional, and dynamic aspects of the land.

Landslide A type of mass movement characterized by the slippage of a body of material over a rupture plane; often a sudden and rapid movement.

Land use The human activities that characterize an area, e.g., agricultural, industrial, and residential.

Latent heat The heat released or absorbed when a substance changes phase as from liquid to gas. For water at 0°C, heat is absorbed or released at a rate of 2.5 million joules per kilogram (597 calories per gram) in the liquid/vapor phase change.

Leachate Fluids that emanate from decomposing waste in a sanitary or chemical landfill.

Leaching The removal of minerals in solution from a soil; the washing out of ions from one level to another in the soil.

Levee A mound of sediment which builds up along a river bank as a result of flood deposition.

Life form The form of individual plants or the form of the individual organs of a plant; in general, the overall structure of the vegetative cover may be thought of as life form as well.

Line scanner A remote sensing device that records signals of reflected radiation in scan lines that sweep perpendicular to the path (flight line) of the aircraft.

Littoral drift The material that is moved by waves and currents in coastal areas.

Littoral transport The movement of sediment along a coastline; it is comprised of two components: longshore transport and onshore-offshore transport.

Loess Silt deposits laid down by wind over extensive areas of the midlatitudes during glacial and postglacial times.

Longshore current A current that moves parallel to the shoreline; velocities generally range between 0.25 and 1 m/sec.

Longshore transport The movement of sediment parallel to the coast.

Magnetic declination The deviation in degrees east or west between magnetic north and true north.

Manning formula A formula used to determine the velocity of streamflow based on the gradient, hydraulic radius, and roughness of the channel; an empirical formula widely used in engineering for sizing channels and pipes.

Mass balance The relative balance in a system, based on the input and output of material such as sediment or water; the state of equilibrium between the input and output of mass in a system.

Mass movement A type of hillslope process characterized by the downslope movement of rock debris under the force of gravity; it includes soil creep, rock fall, landslides, and mudflows; also termed *mass wasting.*

Meander A bend or loop in a stream channel.

Meander belt The width of the train of active meanders in a river valley.

Microclimate The climate of small spaces such as an inner city, residential area, or mountain valley.

Mitigation A measure used to lessen the impact of an action.

Montmorillonite A type of clay that is notable for its capacity to shrink and expand with wetting and drying.

Moraine The material deposited directly by a glacier; also, the material (load) carried in or on a glacier; as landforms moraines usually have hilly or rolling topography.

Mudflow A type of mass movement characterized by the downslope flow of a saturated mass of clayey material.

Multispectral scanning system (MSS) A line-scanning remote sensing system capable of simultaneously recording reflected radiation in several discrete bands (wavelengths).

Nearshore circulation cell The circulation pattern of water and sediment formed by the combined action of rip currents, waves, and longshore currents.

Net sediment transport The balance between sediment moved one way and the other along a shoreline.

Net sediment transport The balance between the quantities of sediment moved in two (opposite) directions along a shoreline.

Nonpoint source Water pollution that emanates from a spatially diffuse source such as the atmosphere or agricultural land.

Nutrients Various types of materials that become dissolved in water and induce plant growth; phosphorus and nitrogen are two of the most effective nutrients in aquatic plants.

Open forest A forest structure with a strong upper one or two stories and limited undergrowth; a forest that is largely open at ground level.

Open space Term applied to undeveloped land, usually land designated for parks, greenbelts, water features, nature preserves and the like.

Open system A system characterized by a through-flow of material and/or energy; a system to which energy or material is added and released over time.

Opportunities and constraints A type of study often carried out in planning projects to determine the principal advantages and drawbacks to a development program proposed for a particular site.

Outflooding Flooding caused by a stream or river overflowing its banks.

Outwash plain A fluvioglacial deposit comprised of sand and gravel with a flat or gently sloping surface; usually found in close association with moraines.

Overland flow Runoff from surfaces on which the intensity of precipitation or snowmelt exceeds the infiltration capacity; also called Horton overland flow, for hydrologist Robert E. Horton.

Oxbow A crescent-shaped lake or pond in a river valley formed in an abandoned segment of channel.

Ozone One of the minor gases of the atmosphere; a pungent, irritating form of oxygen that performs the important function of absorbing ultraviolet radiation.

Parallels The east-west-running lines of the global coordinate system; the equator, the Arctic Circle, and the Antarctic Circle are parallels; all parallels run parallel to one another.

Parent material The particulate material in which a soil forms; the two types of parent material are *residual* and *transported.*

Passive solar collector A solar collector that operates without the aid of powered machinery.

Peak annual flow The largest discharge produced by a stream or river in a given year.

Peak discharge The maximum flow of a stream or river in response to an event such as a rainstorm or over a period of time such as a year.

Peak flow. *See* **Peak discharge**

Pedon The smallest geographic unit of soil defined by soil scientists of the U.S. Department of Agriculture.

Percolation rate The rate at which water moves into soil through the walls of a test pit; used to determine soil suitability for wastewater disposal.

Percolation test A soil-permeability test performed in the field to determine the suitability of a material for wastewater disposal; the test most commonly used by sanitarians and planners to size soil-absorption systems.

Perennial stream A stream that receives inflow of groundwater all year; a stream that has a permanent baseflow.

Performance concept The concept of setting standards on how an environment or land use is expected to perform; includes formulation of goals, standards, and controls.

Periglacial environment An area where frost-related processes are a major force in shaping the landscape.

Permafrost A ground-heat condition in which the soil or subsoil is permanently frozen; long-term frozen ground in periglacial environments.

Permeability The rate at which soil or rock transmits groundwater (or gravity water in the area above the water table); measured in cubic feet (or meters) of water transmitted through a specified cross-sectional area when under a hydraulic gradient of 1 foot per 1 foot (or 1 m per 1 m).

Photopair A set of overlapping aerial photographs that are used in stereoscopic interpretation of aerial photographs.

Photosynthesis The process by which green plants synthesize water and carbon dioxide and, with the energy from absorbed light, convert it into plant materials in the form of sugar and carbohydrates.

Piping The formation of horizontal tunnels in a soil due to sapping, i.e., erosion by seepage water; piping often occurs in areas where gullying is or was active and is limited to soils resistant to cave-in.

Plane coordinate system A grid coordinate system designed by the United States National Ocean Survey in which the basic unit is a square measuring 10,000 feet on a side.

Planned unit development (PUD) A residential planning strategy aimed at increasing the amount of undeveloped land or common space by clustering development in carefully planned units.

Plant production The rate of output of organic material by a plant; the total amount of organic matter added to the landscape over some period of time, usually measured in grams per square meter per day or year.

Plume The stream of exhaust (smoke) emanating from a stack or chimney.

Point source Water pollution that emanates from a single source such as a sewage plant outfall.

Porosity The total volume of pore (void) space in a given volume of rock or soil; expressed as the percentage of void volume to the total volume of the soil or rock sample.

Principle of limiting factors The biological principle that the maximum obtainable rate of photosynthesis is limited by whichever basic resource of plant growth is in least supply.

Principle of stream orders The relationship between stream order and the number of streams per order; the relationship for most drainage nets is an inverse one, characterized by many low-order streams and fewer and fewer streams with increasingly higher orders. *See also* **Stream order**

Principal point The center of an aerial photograph, located at the intersection of lines drawn from the fiducial marks on the photo margin.
Progradation The process of seaward growth of a shoreline.
Pruning In hydrology the cutting back of a drainage net by diverting or burying streams; usually associated with urbanization or agricultural development.

Quadrat sampling A field sampling technique in which small plots, called quadrats, are laid out in the landscape and from which the sample is drawn.

Radiation The process by which radiant (electromagnetic) energy is transmitted through free space; the term used to describe electromagnetic energy, as in infrared radiation or shortwave radiation.
Radiation beam The column of solar radiation flowing into or through the atmosphere.
Rainfall erosion index A set of values representing the computed erosive power of rainfall based on total rainfall and the maximum intensity of the thirty-minute rainfall.
Rainfall intensity The rate of rainfall measured in inches or centimeters of water deposited on the surface per hour or minute.
Rainshadow The dry zone on the leeward side of a mountain range of orographic precipitation.
Rainsplash Soil erosion from the impact of raindrops.
Rainwash Soil erosion by overland flow; erosion by sheets of water running over a surface; usually occurs in association with rainsplash; also called *wash.*
Rating curve A graph that shows the relationship between the discharge and stage of various flow events on a river; once this relationship is established, it may be used to approximate discharge using stage data alone.
Rational method A method for computing the discharge from a small drainage basin in response to a given rainstorm; computation is based on the coefficient of runoff, rainfall intensity, and basin area.
Recharge The replenishment of groundwater with water from the surface.
Recurrence The number of years on the average that separate events of a specific magnitude, e.g., the average number of years separating river discharges of a given magnitude or greater.
Regulatory floodway A zone designated by the U.S. Federal flood policy as the lowest part of the floodplain where the deepest and most frequent floodflows are conducted.
Relief The range of topographic elevation within a prescribed area.
Retention A strategy used in stormwater management in which runoff is retained on site in basins, underground, or released into the soil.
Rip current A relatively narrow jet of water that flows seaward through the breaking waves; it serves as a release for water that builds up near shore.
Riprap Rubble such as broken concrete and rock placed on a surface to stabilize it and reduce erosion.
Runoff In the broadest sense runoff refers to the flow of water from the land as both surface and subsurface discharge; the more restricted and common use, however, refers to runoff as surface discharge in the form of overland flow and channel flow.

Sapping An erosional process that usually accompanies gullying in which soil particles are eroded by water seeping from a bank.
Scatter diagram A graph characterized by a series of plotted points showing the relationship between two quantitative variables.

Scattering The process by which minute particles suspended in the atmosphere diffuse incoming solar radiation.

Sediment sink A coastal environment that favors the massive accumulation of sediment.

Seepage The process by which groundwater or interflow water seeps from the ground.

Septic tank A vat, usually placed underground, used to store wastewater.

Septic system Specifically, a sewage system that relies on a septic tank to store and/or treat wastewater; generally, an on-site (small-scale) sewage-disposal system that depends on the soil to dispose of watewater.

Sensible heat Heat that raises the temperature of a substance and thus can be sensed with a thermometer. In contrast to latent heat, it is sometimes called the heat of dry air.

Sensitive environment Special environments such as wetlands or coastal lands that require protection from development because of their aesthetic and ecological value.

Setback A term used in site planning to indicate the critical distance that a structure or facility should be separated from an edge such as a backshore slope or lake shore.

Shoreland The discontinuous belt of land around a waterbody that is not drained via stream basins.

Side-looking airborne radar The radar system used in remote sensing; so named because the energy pulse is beamed obliquely on the landscape from the side of the aircraft.

Sieve method A technique used to separate the various sizes of coarse particles in a soil sample.

SLAR. *See* **Side-looking airborne radar**

Slope failure A slope that is unable to maintain itself and fails by mass movement such as a landslide, slump, or similar movement.

Slope form The configuration of a slope, e.g., convex, concave, or straight.

Sluiceway A large drainage channel or spillway for glacier meltwater.

Slump A type of mass movement characterized by a back rotational motion along a rupture plane.

Small circle Any circle drawn on the globe that represents less than the full circumference of the earth; thus the plane of a small circle does not pass through the center of the earth. All parallels except the equator are small circles.

Soil absorption system Small scale wastewater disposal system that relies directly on soil to absorb and disperse sewage water from a house or larger building.

Soil creep A type of mass movement characterized by a very slow downslope displacement of soil, generally without fracturing of the soil mass; the mechanisms of soil creep include freeze-thaw activity and wetting and drying cycles.

Soil-forming factors The major factors responsible for the formation of a soil: climate, parent material, vegetation, topography, and drainage.

Soil-heat flux The rate of heat flow into, from, or through the soil.

Soil material Any sediment, rock, or organic debris in which soil formation takes place.

Soil profile The sequence of horizons, or layers, of a soil.

Soil structure The term given to the shape of the aggregates of particles that form in a soil; four main structures are recognized: blocky, platy, granular, and prismatic.

Soil texture The cumulative sizes of particles in a soil sample; defined as the percentage by weight of sand, silt, and clay-sized particles in a soil.

Solar constant The rate at which solar radiation is received on a surface (perpendicular to the radiation) at the edge of the atmosphere. Average strength is 1353 joules/m² · sec, which can also be stated as 1.94 cal/cm² · min.

Solar gain A general term used to indicate the amount of solar radiation absorbed by a surface or setting in the landscape.

Solar heating The process of generating heat from absorbed solar radiation; a widely used term in the solar energy literature.

Solifluction A type of mass movement in periglacial environments, characterized by the slow flowage of soil material and the formation of lobeshaped features; prevalent in tundra and alpine landscapes.

Solstice The dates when the declination of the sun is at 23.27°N latitude (the Tropic of Cancer) and 23.27°S latitude (the Tropic of Capricorn)—June 21–22 and December 21–22, respectively. These dates are known as the winter and summer solstices, but which is which depends on the hemisphere.

State plane coordinate system. *See* **Plane coordinate system**

Stereoscope A viewing device used to gain a three-dimensional image from a photopair.

Stormflow The portion of streamflow that reaches the stream relatively quickly after a rainstorm, adding a surcharge of water to baseflow.

Stratified sampling A sampling technique in which the population or study area is divided into sets or subareas before drawing the sample.

Stream order The relative position, or rank, of a stream in a drainage network. Streams without tributaries, usually the small ones, are first-order; streams with two or more first-order tributaries are second-order, and so on.

Subarctic zone The belt of latitude between 55° and the Arctic and Antarctic circles.

Sub-basin A small drainage basin within the watershed of a lake or impoundment.

Sun angle The angle formed between the beam of incoming solar radiation and a plane at the earth's surface or a plane of the same altitude anywhere in the atmosphere.

Sun pocket A small space designed especially to take advantage of solar radiation and heating.

Surge A large and often destructive wave caused by intensive atmospheric pressure and strong winds.

Suspended load The particles (sediment) carried aloft in a stream of wind by turbulent flow; usually clay- and silt-sized particles.

Taxon Any unit (category) of classification system, usually biological.

Technical planning Data collection, analysis, and related activities used in support of the decision-making process in planning.

Temperate forest A forest of the midlatitude regions that could be described as climatically temperate, e.g., broadleaf deciduous forests of Europe and North America, comprised of beeches, maples, and oaks.

Temperature inversion An atmospheric condition in which the cold air underlies warm air; inversions are highly stable conditions and thus not conducive to atmospheric mixing.

Thermal gradient The change in temperature over distance in a substance; usually expressed in degrees Celsius per centimeter or meter.

Thermal infrared system Line scanner capable of recording thermal infrared energy at wavelengths of 3–5 and 8–14 micrometers.

Threshold The level or magnitude of a process at which sudden or rapid change is initiated.

Tolerance The range of stress or disturbance a plant is able to withstand without damage or death.

Topsoil The uppermost layer of the soil, characterized by a high organic content; the organic layer of the soil.

Township and range A system of land subdivision in the United States which uses a grid to classify land units; standard subdivisions include townships and sections.

Transect sampling A field sampling technique in which the sample is drawn from strips or transects laid out across the study area.

Transmission The lateral flow of groundwater through an aquifer; measured in terms of cubic feet (or meters) transmitted through a given cross-sectional area per hour or day.

Transpiration The flow of water through the tissue of a plant and into the atmosphere via stomatal openings in the foliage.

Transported soil Soil formed in parent material comprised of deposits laid down by water, wind, or glaciers.

Tree line The upper limit of tree growth on a mountain where forest often gives way to alpine meadow.

Tundra Landscape of cold regions, characterized by a light cover of herbaceous plants and underlain by permafrost.

Turbidity A measure of the clearness or transparency of water as a function of suspended sediment.

Turbulent flow Flow characterized by mixing motion in which the primary source of flow resistance is the mixing action between slow-moving and faster-moving molecules in a fluid.

Universal Soil Loss Equation A formula for estimating soil erosion by runoff based on rainfall, plant cover, slope, and soil erodibility.

Urban boundary layer A general term referring to the layer of air over a city that is strongly influenced by urban activities and forms.

Urban canyon City street lined with tall buildings; an urban terrain feature that has a pronounced effect on airflow, radiation and microclimate as a whole.

Urban climate The climate in and around urban areas, it is usually somewhat warmer, foggier, and less well lighted than the climate of the surrounding region.

Urban design An area of professional activity by architects, landscape architects, and urban planners dealing with the forms, materials, and activities of cities.

Urbanization The term used to describe the process of urban development, including suburban residential and commercial development.

Vascular plants Plants in which cells are arranged into a pipelike system of conducting, or vascular, tissue; xylem and phloem are the two main types of vasular tissue.

Wave period The time it takes a wave to travel the distance of one wavelength.

Wave refraction The bending of a wave, which results in an approach angle more perpendicular to the shoreline.

Watertable The upper boundary of the zone of groundwater; in fine-textured materials it is usually a transition zone rather than a boundary line. The configuration of the water table often approximates that of the overlying terrain.

Wetland A term generally applied to an area where the ground is permanently wet or wet most of the year and is occupied by water-loving (or tolerant) vegetation such as cattails, mangrove, or cypress.

Wetted perimeter The distance from one side of a stream to the other, measured along the bottom.

Windshield survey A rapid and general sampling method for vegetation and land use based on observations from a moving automobile.

Zenith For any location on earth, the point that is directly overhead to an observer. The zenith position of the sun is the one directly overhead.

Zenith angle The angle formed between a line perpendicular to the earth's surface (at any location) and the beam of incoming solar radiation (on any date).

INDEX